T0321652

# *E*-RECURSION, FORCING AND C*-ALGEBRAS

**LECTURE NOTES SERIES**
**Institute for Mathematical Sciences, National University of Singapore**

Series Editors: Chitat Chong and Wing Keung To
*Institute for Mathematical Sciences*
*National University of Singapore*

**ISSN: 1793-0758**

---

*Published*

Vol. 17 Interface Problems and Methods in Biological and Physical Flows
*edited by Boo Cheong Khoo, Zhilin Li & Ping Lin*

Vol. 18 Random Matrix Theory and Its Applications:
Multivariate Statistics and Wireless Communications
*edited by Zhidong Bai, Yang Chen & Ying-Chang Liang*

Vol. 19 Braids: Introductory Lectures on Braids, Configurations and Their Applications
*edited by A Jon Berrick, Frederick R Cohen, Elizabeth Hanbury, Yan-Loi Wong & Jie Wu*

Vol. 20 Mathematical Horizons for Quantum Physics
*edited by H Araki, B-G Englert, L-C Kwek & J Suzuki*

Vol. 21 Environmental Hazards:
The Fluid Dynamics and Geophysics of Extreme Events
*edited by H K Moffatt & E Shuckburgh*

Vol. 22 Multiscale Modeling and Analysis for Materials Simulation
*edited by Weizhu Bao & Qiang Du*

Vol. 23 Geometry, Topology and Dynamics of Character Varieties
*edited by W Goldman, C Series & S P Tan*

Vol. 24 Complex Quantum Systems: Analysis of Large Coulomb Systems
*edited by Heinz Siedentop*

Vol. 25 Infinity and Truth
*edited by Chitat Chong, Qi Feng, Theodore A Slaman & W Hugh Woodin*

Vol. 26 Notes on Forcing Axioms
*by Stevo Todorcevic, edited by Chitat Chong, Qi Feng, Theodore A Slaman, W Hugh Woodin & Yue Yang*

Vol. 27 *E*-Recursion, Forcing and C*-Algebras
*edited by Chitat Chong, Qi Feng, Theodore A. Slaman, W. Hugh Woodin & Yue Yang*

*For the complete list of titles in this series, please go to
http://www.worldscientific.com/series/LNIMSNUS

Lecture Notes Series, Institute for Mathematical Sciences,
National University of Singapore

Vol. 27

# E-RECURSION, FORCING AND C*-ALGEBRAS

**Editors**

Chitat Chong
National University of Singapore, Singapore

Qi Feng
Chinese Academy of Sciences, China

Theodore A Slaman
University of California, Berkeley, USA

W Hugh Woodin
Harvard University, USA

Yue Yang
National University of Singapore, Singapore

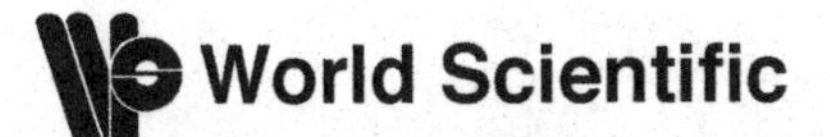


NEW JERSEY • LONDON • SINGAPORE • BEIJING • SHANGHAI • HONG KONG • TAIPEI • CHENNAI

*Published by*

World Scientific Publishing Co. Pte. Ltd.
5 Toh Tuck Link, Singapore 596224
*USA office:* 27 Warren Street, Suite 401-402, Hackensack, NJ 07601
*UK office:* 57 Shelton Street, Covent Garden, London WC2H 9HE

**Library of Congress Cataloging-in-Publication Data**
*E*-recursion, forcing and C*-algebras / editors, Chitat Chong, National University of Singapore, Singapore [and four others].
pages cm. -- (Lecture notes series, Institute for Mathematical Sciences, National University of Singapore ; vol. 27)
Includes bibliographical references and index.
ISBN 978-9814602631 (hardcover : alk. paper) -- ISBN 9814602639 (hardcover : alk. paper) -- ISBN 978-9814603256 (pbk. : alk. paper) -- ISBN 9814603252 (pbk. : alk. paper)
1. Recursion theory. 2. Forcing (Model theory) 3. C*-algebras. I. Chong, C.-T. (Chi-Tat), 1949– editor.
QA9.6.E74 2014
511.3--dc23

2014026066

**British Library Cataloguing-in-Publication Data**
A catalogue record for this book is available from the British Library.



Printed in Singapore

# CONTENTS

## FOREWORD

The Institute for Mathematical Sciences (IMS) at the National University of Singapore was established on 1 July 2000. Its mission is to foster mathematical research, both fundamental and multidisciplinary, particularly research that links mathematics to other efforts of human endeavor, and to nurture the growth of mathematical talent and expertise in research scientists, as well as to serve as a platform for research interaction between scientists in Singapore and the international scientific community.

The Institute organizes thematic programs of longer duration and mathematical activities including workshops and public lectures. The program or workshop themes are selected from among areas at the forefront of current research in the mathematical sciences and their applications.

Each volume of the *IMS Lecture Notes Series* is a compendium of papers based on lectures or tutorials delivered at a program/workshop. It brings to the international research community original results or expository articles on a subject of current interest. These volumes also serve as a record of activities that took place at the IMS.

We hope that through the regular publication of these *Lecture Notes* the Institute will achieve, in part, its objective of reaching out to the community of scholars in the promotion of research in the mathematical sciences.

January 2014

Chitat Chong
Wing Keung To
*Series Editors*

## PREFACE

The series of Asian Initiative for Infinity (AII) Graduate Logic Summer School was held annually from 2010 to 2012. The lecturers were Moti Gitik, Denis Hirschfeldt and Menachem Magidor in 2010, Richard Shore, Theodore A. Slaman, John Steel, and W. Hugh Woodin in 2011, and Ilijas Farah, Ronald Jensen, Gerald E. Sacks and Stevo Todorcevic in 2012. In all, more than 150 graduate students from Asia, Europe and North America attended the summer schools. In addition, two postdoctoral fellows were appointed during each of the three summer schools. These volumes of lecture notes serve as a record of the AII activities that took place during this period.

The AII summer schools was funded by a grant from the John Templeton Foundation and partially supported by the National University of Singapore. Their generosity is gratefully acknowledged.

May 2014

Chitat Chong
National University of Singapore, Singapore

Qi Feng
Chinese Academy of Sciences, China

Theodore A. Slaman
University of California, Berkeley, USA

W. Hugh Woodin
Harvard University, USA

Yue Yang
National University of Singapore, Singapore

*Volume Editors*

# SELECTED APPLICATIONS OF LOGIC TO CLASSIFICATION PROBLEM FOR C*-ALGEBRAS[a]

Ilijas Farah
*Department of Mathematics and Statistics*
*York University, 4700 Keele Street*
*North York, Ontario, Canada M3J 1P3*
*and Matematicki Institut, Kneza Mihaila 34, Belgrade, Serbia*
*ifarah@yorku.ca*
*http://www.math.yorku.ca/~ifarah*

Basics of Elliott's classification program are outlined and juxtaposed with the abstract classification theory from descriptive set theory. Some preliminary estimates on the complexity of the isomorphism relation of separable C*-algebras are given.

## Contents



[a]Partially supported by NSERC.

## 0. Introduction

In recent years we have witnessed a number of applications of set theory to operator algebras. In the present notes I will focus on one very specific (yet particularly exciting) aspect of this development, applications of descriptive set theory (more precisely, theory of abstract classification) to Elliott's classification program for nuclear C*-algebras. These lecture notes contain a plenty of exercises, many (but not all) of them fairly straightforward. They are meant as a bridge towards more advanced literature on classification of C*-algebras, a subject with an abundance of excellent literature. Hints to exercises often refer to the material covered at a later point in the notes and thus provide additional motivation for the introduced notions. The initial parts of the lecture notes, and §2 in particular, are very sketchy and are meant to outline the main ideas and provide a guide to the literature rather than to serve as a textbook.

The basic premise is that every classical classification program in mathematics deals with analytic equivalence relations on Polish spaces. Moreover, classification invariants are usually coded by elements of a Polish space and the computation of invariants is given by a Borel-measurable function.

I shall always start counting at zero. Therefore $i < n$ means that $i$ assumes $n$ distinct values: $0, 1, \ldots, n-1$. Also, an $a \in M_n(\mathbb{C})$ is identified with its matrix entries $a_{ij}$ for $i < n$, $j < n$.

*Suggested references*

For functional analysis, [64]. For the theory of C*-algebras, [62] and [12], and [68] for their K-theory. General theory of operator algebras is excellently surveyed in [4] and standard reference for the Elliott program as of 2002 is [70]. Very detailed notes on topics in operator theory starting with simple (partial isometries) to very advanced can be found at [76]. Classical reference for classical descriptive set theory is [48]. Somewhat dated surveys of some other applications of set theory to C*-algebras are given in [84] and [34]. Papers [33] and [32] contain more details and proofs of the results presented in §7 and §8.2. Survey of the latest developments in Elliott's program will be available shortly in [88].

*Acknowledgments*

These notes are based on two series of lectures given in 2012. At the Luminy Young Set Theory workshop in April 2012 I covered sections dealing with general theory of C*-algebras and Elliott's classification of AF algebras by K-theoretic invariants. At the Asian Initiative for Infinity course in addition I covered this in more detail, and presented the abstract classification viewpoint of Elliott's program. I am indebted to the organizers of both of these meetings for inviting me. I would like to thank to Asger Törnquist for kindly permitting me to use his writeup of Lemma 8.6, taken from the appendix of the original version of [33]. Finally, I would like to thank John Campbell, Boris Kadets, Vladislav Kalashnyk and Jiewon Park for noticing several mistakes in an earlier version of these notes. Last, but not least, I am indebted to Lai Fun Kwong from World Scientific for her outstanding help with editing the final version of these notes.

## 1. Operators on Hilbert Spaces

We begin with a review of the basic properties of operators on a Hilbert space. Throughout we let $H$ denote a complex infinite-dimensional separable Hilbert space, and we let $(e_n)$ be an orthonormal basis for $H$ (see Example 1.1). For $\xi, \eta \in H$, we denote their inner product by $(\xi|\eta)$. We recall that

$$(\eta|\xi) = \overline{(\xi|\eta)}$$

and the norm defined by

$$\|\xi\| = \sqrt{(\xi|\xi)}.$$

The Cauchy–Schwartz inequality says that

$$|(\xi|\eta)| \leq \|\xi\|\|\eta\|.$$

**Example 1.1.** The space

$$\ell^2(\mathbb{N}) = \left\{(\alpha_k)_{k\in\mathbb{N}} : \alpha_k \in \mathbb{C}, \|\alpha\|^2 = \sum |\alpha_k|^2 < \infty\right\}$$

(sometimes denoted simply by $\ell^2$) is a Hilbert space under the inner product $(\alpha|\beta) = \sum \alpha_k \overline{\beta_k}$. If we define $e^n \in \ell^2(\mathbb{N})$ by $e^n_k = \delta_{nk}$ (the Kronecker's $\delta$), then $(e^n)$ is an orthonormal basis for $\ell^2$. For any $\alpha \in \ell^2$, $\alpha = \sum \alpha_n e^n$.

Any Hilbert space has an orthonormal basis, and this can be used to prove that all separable infinite-dimensional Hilbert spaces are isomorphic. Moreover, any two infinite-dimensional Hilbert spaces with the same density character (the minimal cardinality of a dense subset) are isomorphic.

**Example 1.2.** If $(X,\mu)$ is a measure space,

$$L^2(X,\mu) = \left\{f : X \to \mathbb{C} \text{ measurable} : \int |f|^2 d\mu < \infty\right\} / \{f : f = 0 \text{ a.e.}\}$$

is a Hilbert space under the inner product $(f|g) = \int f\overline{g}d\mu$ and with the norm defined by $\|f\|^2 = \int |f|^2 d\mu$.

We will let $a, b, \ldots$ denote linear operators $H \to H$. We recall that

$$\|a\| = \sup\{\|a\xi\| : \xi \in H, \|\xi\| = 1\}.$$

If $\|a\| < \infty$, we say $a$ is *bounded.* An operator is bounded if and only if it is continuous. We denote the algebra of all bounded operators on $H$ by $\mathcal{B}(H)$ (some authors use $L(H)$), and throughout the paper all of our operators will be bounded. We define the *adjoint* $a^*$ of $a$ to be the unique operator satisfying

$$(a\xi|\eta) = (\xi|a^*\eta)$$

for all $\xi, \eta \in H$. Note that since an element of $H$ is determined by its inner products with all other elements of $H$ (e.g., take an orthonormal basis), an operator $a$ is determined by the values of $(a\xi|\eta)$ for all $\xi, \eta$ or even by the values $(ae_m|e_n)$ for $m$ and $n$ in $\mathbb{N}$.

**Lemma 1.3.** *For all $a, b$ in $\mathcal{B}(H)$ we have*

*(1)* $(a^*)^* = a$
*(2)* $(ab)^* = b^*a^*$
*(3)* $\|a\| = \|a^*\|$
*(4)* $\|ab\| \leq \|a\| \cdot \|b\|$
*(5)* $\|a^*a\| = \|a\|^2$

*Proof.* These are all easy calculations. For example, for (5), for $\|\xi\| = 1$,

$$\|a\xi\|^2 = (a\xi|a\xi) = (\xi|a^*a\xi) \leq \|\xi\| \cdot \|a^*a\xi\| \leq \|a^*a\|,$$

the first inequality holding by Cauchy–Schwartz. Taking the sup over all $\xi$, we obtain $\|a\|^2 \leq \|a^*a\|$. Conversely,

$$\|a^*a\| \leq \|a^*\|\|a\| = \|a\|^2$$

by (3) and (4). □

Entries (1)–(3) state that $\mathcal{B}(H)$ is a *Banach *-algebra* (or a *Banach algebra with involution* *) and (5) is sometimes called the *C*-equality.*

### 1.1. *Subspaces and subalgebras of $\mathcal{B}(H)$*

Before focusing on C*-algebras, I shall list some other structures based on $\mathcal{B}(H)$. Each one of these categories is replete with possible applications of logic.

1.1.1. *Operator space* is a closed linear subspace of $\mathcal{B}(H)$. In addition to its Banach space structure, an operator space $X$ is considered with the Banach space structure on $M_n(X)$, $n \times n$ matrices of elements of $X$ identified with operators from $H^n$ to $H^n$ with respect to the operator norm. Morphisms in this category are *completely bounded* maps—linear maps whose canonical extension to $M_n(X)$ is bounded for all $n$. The standard reference on operator spaces is [17].

1.1.2. *Operator system* is an operator space that is in addition closed under the adjoint. Morphisms are *completely positive* maps—linear maps that preserve positivity of operators and matrices. They are indispensable in the theory of C*-algebras, and more on this can be found in [63] and [5].

1.1.3. *Concrete C*-algebras* are in addition closed under multiplication of operators. Morphisms are *-homomorphisms—maps that preserve all algebraic structure (+, $\cdot$ and *). Remarkably, such maps are automatically completely positive and completely contractive.

We import the terminology from §1.5.1 wholesale and talk about operators in a C*-algebra that are normal, self-adjoint, positive, projections, etc. This will also apply to abstract C*-algebras once they are introduced in Definition 2.2.

1.1.4. *Non-self-adjoint subalgebras* are norm-closed and closed under + and $\cdot$, but not necessarily closed under *. See [66] for more on this subject.

1.1.5. *von Neumann algebras* are C*-algebras that are closed in weak operator topology. An excellent introductory source is [46].

Theories of these categories have many connections between each other. For example, the study of tensor products of C*-algebras and their finite-dimensional approximation is deeply steeped in the study of operator systems. In the present lecture notes I shall focus on C*-algebras.

### 1.2. *Exercises*

**1.2.1.** Prove that for all $\xi, \eta \in H$ the following so-called *polarization identity* holds (note that $i = \sqrt{-1}$)

$$(\xi|\eta) = \frac{1}{4}\sum_{k=0}^{3} i^k(\xi + i^k\eta|\xi + i^k\eta).$$

The right-hand side is a linear combination of norms of vectors $\xi + i^k\eta$ for $0 \leq k \leq 3$. Consequentially, the scalar product on $H$ is uniquely determined by its Hilbert space norm.

The *dimension* of a Hilbert space is the least cardinality of an orthonormal basis.

**1.2.2.** All orthonormal bases in a fixed Hilbert space have the same cardinality. Two complex Hilbert spaces are isomorphic if and only if they have the same dimension. Prove the following statements.

**1.2.3.** Prove that a Hilbert space is separable if and only if it has a finite or countable orthonormal basis.

**1.2.4** (Inner automorphisms). Assume $u$ is a linear isometry from $H$ onto itself. Prove that $\operatorname{Ad} u\colon \mathcal{B}(H) \to \mathcal{B}(H)$ defined by

$$(\operatorname{Ad} u)a = uau^*$$

is an automorphism of C*-algebra $\mathcal{B}(H)$.

**1.2.5.** If $u\colon H_1 \to H_2$ is an isomorphism between Hilbert spaces, then

$$(\operatorname{Ad} u)a = uau^*$$

is an isomorphism between $\mathcal{B}(H_1)$ and $\mathcal{B}(H_2)$. The operator $\operatorname{Ad} u(a)$ is just $a$ with its domain and range identified with $H_2$ via $u$.

**1.2.6.** An operator $u$ in a unital C*-algebra is a *unitary* if $uu^* = u^*u = 1$. Prove that $\operatorname{Ad} u$ is an automorphism of $A$.

**1.2.7.** Prove that all automorphisms of $\mathcal{B}(H)$ are inner.

(Hint 1: As a warmup prove that all automorphisms of the Boolean algebra $\mathcal{P}(\mathbb{N})$ are "trivial." You first have to replace the quotation marks with an appropriate definition of "trivial."

Hint 2 (for a better proof): Fix a unit vector $\xi$ and let $\eta$ be a unit vector such that $\Phi(p_\xi) = p_\eta$ (here $p_\xi$ denotes the projection to the subspace spanned by $\xi$). With $v_{\xi,\zeta}$ denoting the rank 1 linear map sending $\xi$ to $\zeta$, let $u(\alpha) = \beta$, where $\beta$ is the unique vector such that $\Phi(v_{\xi,\alpha}) = v_{\eta,\beta}$.

A **-polynomial* is a term in the language $\{+, \cdot, *\}$. If $P(x)$ is a *-polynomial and $a$ is an operator then $P(a)$ is naturally defined (again, logic for metric structures provides the right setting for this; see [29]). For a set of operators $X$ in $\mathcal{B}(H)$ or in a C*-algebra $A$ let $C^*(X)$ denote the C*-algebra generated by $X$. If $X = \{x_1, \dots, x_n\}$ is finite we may write $C^*(x_1, \dots, x_n)$ instead of $C^*(X)$.

**1.2.8.** Prove that $C^*(X)$ is the norm-closure of $\{P(\bar{a}) : P(\bar{x})$ is a *-polynomial in $n$ (non-commuting) variables with complex coefficients and $\bar{a}$ is an $n$-tuple in $X$ for $n \in \mathbb{N}\}$.

**1.2.9.** Prove that $C^*(a)$ is abelian if and only if $a$ is normal.

An *ideal* in a C*-algebra is a two-sided, norm closed, self-adjoint ideal (there is some redundancy in this definition).

**1.2.10.** Prove that an ideal generated by an element $a$ of a C*-algebra is equal to the closure of the linear span of $\{bac : b, c \in A\}$.

### 1.3. *Spectrum and spectral radius*

We start with a useful lemma.

**Lemma 1.4.** *If* $\|a\| < 1$ *then* $1 - a$ *is invertible in* $\mathcal{B}(H)$.

*Proof.* The series $b = \sum_{n=0}^{\infty} a^n$ is convergent and hence in $\mathcal{B}(H)$. By considering partial sums one sees that $(1-a)b = b(1-a) = 1$. □

The *spectrum* of an operator $a$ in $\mathcal{B}(H)$ is

$$\mathrm{sp}(a) = \{\lambda \in \mathbb{C} : a - \lambda I \text{ is not invertible}\}.$$

The spectrum of a bounded linear operator is always a compact subset of $\mathbb{C}$ (Exercise 1.4.4), and it is moreover always nonempty (the latter fact follows from a clever use of Liouville's theorem; see e.g., [64, Theorem 4.1.13]). Also, for a normal operator $a$ we have that $\|a\| = r(a)$ (Exercise 1.4.8), where $r(a)$ is the *spectral radius* of $a$ defined as follows:

$$r(a) = \max\{|\lambda| : \lambda \in \mathrm{sp}(a)\}.$$

### 1.4. *Exercises*

**1.4.1.** Prove that the spectrum of a finite-dimensional matrix is equal to the set of its eigenvalues.

**1.4.2.** Prove that the invertible elements form an open subset of $\mathcal{B}(H)$.
(Hint: If $ab = 1 = ba$, find an $\varepsilon > 0$ such that $\|b - c\| < \varepsilon$ implies $\|ac - 1\| < 1$ and $\|ca - 1\| < 1$. Then apply Lemma 1.4.)

**1.4.3.** Prove that $\mathrm{sp}(a) \subseteq \{\lambda : |\lambda| \leq \|a\|\}$ for all $a$.

**1.4.4.** Prove that $\mathrm{sp}(a)$ is compact for all $a$.
(Hint: Use Exercise 1.4.2 and Exercise 1.4.3.)

**1.4.5.** Prove that $\|a\| \geq r(a)$ for all $a$.

**1.4.6.** Find an example of a nonzero $a$ such that $\mathrm{sp}(a) = \{0\}$, and hence $\|a\| > r(a)$.
(Hint: A well-chosen $2 \times 2$ matrix would do.)

**1.4.7.** Prove that $\lim_n \|a^n\|^{1/n}$ exists and is equal to $r(a)$.
(Hint: See [64, Lemma 4.1.13].)

**1.4.8.** Assume $a$ is normal. Prove $\|a\| = r(a)$.
(Hint: First prove that $\|a^{2^n}\| = \|a\|^{2^n}$ and then use Exercise 1.4.7. Prove this equality in the self-adjoint case first, using the C*-equality.)

**1.4.9.** Show that if $\lambda \notin \mathrm{sp}(a)$ then $\|(a - \lambda \cdot 1)^{-1}\| \leq 1/\operatorname{dist}(\lambda, \mathrm{sp}(a))$.

**1.4.10.** Find a normal operator on a Hilbert space that has no eigenvectors.
(Hint: Example 1.5.)

### 1.5. *Normal operators and the spectral theorem*

In this section we introduce some distinguished classes of operators in $\mathcal{B}(H)$, such as normal and self-adjoint operators (cf. §1.5.1).

**Example 1.5.** Assume $(X, \mu)$ is a probability measure space. If $H_0 = L^2(X, \mu)$ and $f\colon X \to \mathbb{C}$ is bounded and measurable, then

$$H_0 \ni g \overset{m_f}{\longmapsto} fg \in H_0$$

is a bounded linear operator. We have $\|m_f\| = \|f\|_\infty$ and

$$m_f^* = m_{\bar{f}}.$$

Hence $m_f^* m_f = m_f m_f^* = m_{|f|^2}$. We call operators of this form *multiplication operators.*

Recall that an operator $a$ is *normal* if $aa^* = a^*a$. Clearly, all multiplication operators are normal. When $H$ is a complex Hilbert space, normal operators have a nice structure theory. It is summarized in the following theorem, stated a bit prematurely since its proof involves Proposition 2.5.

**Theorem 1.6** (Spectral Theorem). *If $a$ is a normal operator then there is a probability measure space $(X, \mu)$, a measurable function $f$ on $X$, and a Hilbert space isomorphism $\Phi\colon L^2(X, \mu) \to H$ such that $\Phi a \Phi^{-1} = m_f$.*

*Proof.* For a proof see e.g., [2, Theorem 2.4.5]. □

Therefore every normal operator is a multiplication operator for some identification of $H$ with an $L^2$ space. Conversely, every multiplication operator is clearly normal. If $X$ is discrete and $\mu$ is the counting measure, the characteristic functions of the points of $X$ form an orthonormal basis for $L^2(X, \mu)$ and the spectral theorem says that $a$ is diagonalized by this basis. In general, the spectral theorem says that normal operators are "measurably diagonalizable".

An operator $a$ is *self-adjoint* if $a = a^*$. Self-adjoint operators are obviously normal. For any $b \in \mathcal{B}(H)$, the "real" and "imaginary" parts of $b$, defined by $b_0 = (b + b^*)/2$ and $b_1 = (b - b^*)/2i$, are self-adjoint and satisfy $b = b_0 + ib_1$. Thus any operator is a linear combination of two self-adjoint operators. It is easy to check that an operator is normal if and only if its real and imaginary parts commute, so the normal operators are exactly the linear combinations of commuting self-adjoint operators.

1.5.1. *Types of bounded operators* The following distinguished classes of operators will play an important role.

(1) $a$ is *normal* if $aa^* = a^*a$,
(2) $a$ is *self-adjoint* (or *Hermitian*) if $a = a^*$,
(3) $a$ is *positive* if $a = b^*b$ for some $b \in H$,
(4) $a$ is a *projection* if $a^2 = a^* = a$,
(5) $a$ is *positive* (or $a \geq 0$) if $a = b^*b$ for some $b$,
(6) $a$ is *unitary* if $aa^* = a^*a = 1$,
(7) $a$ is an *isometry* if $aa^*$ is a projection and $a^*a = 1$.
(8) $a$ is *partial isometry* if both $aa^*$ and $a^*a$ are projections.

Note that a positive element is automatically self-adjoint. For self-adjoint elements $a$ and $b$ write $a \leq b$ if $b - a$ is positive.

Any complex number $z$ can be written as $z = re^{i\theta}$ for $r \geq 0$ and $|e^{i\theta}| = 1$. Considering $\mathbb{C}$ as the set of operators on a one-dimensional Hilbert space, there is an analogue of this on an arbitrary Hilbert space.

**Theorem 1.7** (Polar Decomposition). *Any $a \in \mathcal{B}(H)$ can be written as $a = bv$ where $b$ is positive and $v$ is a partial isometry. Moreover, $v$ can be chosen so that* $\ker(v) = \ker(a)$. *This additional requirement makes $v$ unique.*

*Proof.* See [64], but apply it to $a^*$ instead. □

### 1.6. *Exercises*

**1.6.1.** The real and imaginary parts of a multiplication operator $m_f$ are $m_{\Re f}$ and $m_{\Im f}$. A multiplication operator $m_f$ is self-adjoint if and only if $f$ is real (a.e.). By the spectral theorem, all self-adjoint operators are of this form up to the unitary equivalence.

**1.6.2.** Two projections $p$ and $q$ commute if and only if $pq$ is a projection.

**1.6.3.** For projections $p$ and $q$ write $p \leq q$ if $pq = p$ (equivalently, $qp = p$). Prove the following.

(1) $p \leq q$ implies that $p$ and $q$ commute.
(2) This is a partial ordering on any set of projections.
(3) $p \leq q$ if and only if the self-adjoint operator $q - p$ is positive.
(4) $p \leq q$ and $\|p - q\| < 1$ implies $p = q$.

**1.6.4.** If $p$ and $q$ are projections such that $pq = q$ and $(1-p)(1-q) = 1-q$ then $p = q$.

**1.6.5.** Every self-adjoint unitary is of the form $u = 1-2p$ for a projection $p$.

## 2. Preliminaries on C*-Algebras

### 2.1. *Positivity, states and the GNS theorem*

Let $X$ be a locally compact Hausdorff space. Recall that $C_0(X)$ denotes the space of continuous complex-valued functions on $X$ such that for every $\varepsilon > 0$ the set $\{x \in X : |f(x)| \geq \varepsilon\}$ is compact. It is considered as a Banach algebra with respect to $+$, $\cdot$ and adjoint defined as pointwise conjugation. If $X$ is compact then we write $C(X)$. By the following remarkable result, these algebras are exactly the abstract abelian C*-algebras.

**Theorem 2.1** (Gelfand–Naimark). *Every abelian C*-algebra is isomorphic to $C_0(X)$ for a unique locally compact Hausdorff space $X$. The algebra is unital if and only if $X$ is compact.* □

Space $X$ is equal to the space of characters on $A$ (see Exercise 2.2.2). A proof of this theorem can be found in e.g., [64] or [2]. In fact, the Gelfand–Naimark theorem is functorial: the category of abelian C*-algebras is contravariantly isomorphic to the category of locally compact Hausdorff spaces (cf. Exercise 2.2.6).

**Definition 2.2.** An *abstract* C*-algebra is a Banach algebra satisfying the conclusion of Lemma 1.3.

Concrete C*-algebras were introduced in §1.1.3. By the definition every concrete C*-algebra is an abstract C*-algebra. The fact that the converse is true can hardly be overestimated.

**Theorem 2.3** (Gelfand–Naimark–Segal). *Every abstract C*-algebra $A$ is isomorphic to a concrete C*-algebra.* □

From now on we will usually refer to abstract C*-algebras as C*-algebras. An occasional concrete C*-algebra will also be treated in the abstract way, independently from its actual representation on $\mathcal{B}(H)$.

To a logician, Theorem 2.3 states that C*-algebras form an axiomatizable class in an appropriately chosen logic. This fact was made precise and taken advantage of in [28] and [29].

### 2.2. *Exercises*

**2.2.1.** Check the easy direction of Theorem 2.3.

In the following three exercises we define the Gelfand transform and give a (very rough!) outline of the proof of Gelfand–Naimark theorem. *Character* of a C*-algebra $A$ is a *-homomorphism $\phi\colon A \to \mathbb{C}$. Let $\hat{A}$ denote the set of characters of $A$, and note that $\hat{A} = \{0\}$ for many C*-algebras $A$ (e.g., all matrix algebras).

**2.2.2.** Prove that every character is continuous and has norm $\leq 1$, and that $\hat{A}$ is a weak*-compact subset of $A^*$.

(Hint: For the first part one needs to check that the kernel is closed. For this apply Lemma 1.4. In the second part, by Alaoglu's theorem (the unit ball of $A^*$ is weak*-compact) one only needs to check that the characters form a closed subset of the unit ball of $A^*$.)

**2.2.3.** If $A$ is an abelian C*-algebra then $\hat{A}$ is equal to the set of pure states of $A$.

(Hint: This is a consequence of the Riesz Representation Theorem. See Exercise 3.17.)

**2.2.4.** If $X \subseteq \mathbb{C}$ is compact, then $C(X) \cong C^*(\iota_X, 1)$, where $\iota_X$ is the identity function on $X$ and 1 is the constantly one function.

(Hint: Stone–Weierstrass.)

**2.2.5.** If $A$ is a C*-algebra define the map $\Gamma\colon A \to C(\hat{A})$ by

$$\Gamma(a)(\phi) = \phi(a).$$

Show that $\Gamma$ is a *-homomomorphism. If $A$ is moreover abelian, show that $\Gamma$ is an isometric isomorphism.

(Hint: For the second part, combine Exercise 2.2.3 and Exercise 1.4.8.)

**2.2.6.** Assume $X$ and $Y$ are compact Hausdorff spaces and $\Phi\colon C(X) \to C(Y)$ is a *-homomorphism.

(1) Prove that there exists a unique continuous $f: Y \to X$ such that $\Phi(a) = a \circ f$ for all $a \in C(X)$.
(2) Prove that $\Phi$ is a surjection if and only if $f$ is an injection.
(3) Prove that $\Phi$ is an injection if and only if $f$ is a surjection.
(4) Prove that for every $f: Y \to X$ there exists a unique $\Phi: C(X) \to C(Y)$ such that (1)–(3) above hold.

Let $A$ be a C*-algebra. A continuous linear functional $\phi : A \to \mathbb{C}$ is *positive* if $\phi(a) \geq 0$ for all positive $a \in A$. It is a *state* if it is positive and of norm 1. We denote the space of all states on $A$ by $\mathbb{S}(A)$.

**2.2.7.** If $\xi \in H$ is a unit vector, define a functional $\omega_\xi$ on $\mathcal{B}(H)$ by

$$\omega_\xi(a) = (a\xi|\xi).$$

Then $\omega_\xi(a) \geq 0$ for a positive $a$ and $\omega_\xi(I) = 1$; hence it is a state. We call a state of this form a *vector state.*

**2.2.8.** If $A$ is unital then $\mathbb{S}(A) = \{\phi \in A^* : \|\phi\| = 1 = \phi(1)\}$.

**2.2.9.** Prove that if $A$ is a unital subalgebra of $B$ then all states of $A$ extend to states of $B$.

(Hint: Exercise 2.2.8.)

In the following $\mathfrak{X}^*$ denotes the Banach space dual of Banach space $\mathfrak{X}$.

**2.2.10.** Assume $\Phi: A \to B$ is a unital *-homomorphism. Define $\Phi^*: B^* \to A^*$ via $\Phi^*(\psi) = \psi: \Phi$.

(1) Prove that $\Phi^*$ maps $\mathbb{S}(B)$ into $\mathbb{S}(A)$.
(2) Prove that $\Phi^*$ maps $T(B)$ into $T(A)$ (the definition of a trace and $T(A)$ is given in §3.6).
(3) Prove that $\Phi^*$ is injective if and only if $\Phi$ is surjective.
(4) Prove that $\Phi^*$ is surjective if and only if $\Phi$ is injective.

## 2.3. *Continuous functional calculus*

We are about to introduce one of the key tools in the theory of C*-algebras building on Gelfand–Naimark theorem in Proposition 2.5 below. Following §1.3, a spectrum $\mathrm{sp}_A(a)$ of an element $a$ of an arbitrary unital C*-algebra $A$ can be defined as

$$\mathrm{sp}_A(a) = \{\lambda \in \mathbb{C} | a - \lambda \cdot 1 \text{ is not invertible in } A\}.$$

The reason one this notation is not in standard usage is contained in Lemma 2.6 below. Let us first prove its special case.

**Lemma 2.4.** *Let $a$ be a normal element in a unital $C^*$-algebra $A$ and let $B = C^*(a, I)$ (the algebra generated by $a$ and the identity). Then $\mathrm{sp}_A(a) = \mathrm{sp}_B(a)$.*

*Proof.* Since $B$ is a subalgebra of $A$, we have that $\mathrm{sp}_B(a) \supseteq \mathrm{sp}_A(a)$. In order to show the converse inclusion, we need to show that an operator $b \in B$ that is not invertible in $B$ is not invertible in $A$.

Assume the contrary, and fix a $b \in B$ which is not invertible in $B$ but has an inverse $d$ in $A$. By Theorem 2.1, $B$ is isomorphic to $C(X)$ for a compact Hausdorff space $X$, and we can identify $b$ with a function $f$ on $X$. Since $f$ is not invertible, we have $f(x) = 0$ for some $x \in X$. Fix $\varepsilon > 0$ pick an open neighbourhood $U$ of $x$ such that $|f(y)| < \varepsilon$ for all $y \in U$. Now let $g \in C(X)$ be a continuous function such that $g(x) = 1$, $0 \leq g(y) \leq 1$ for all $y$ and $g(z) = 0$ for $z \notin U$. Let $c \in B$ correspond to $g$.

Then $\|cbd\| = \|c\| = 1$. On the other hand, $\|cb\| = \max_{y \in X} |g(y)f(y)\| \leq \varepsilon$. Therefore, $\|d\| \geq 1/\varepsilon$. Since $\varepsilon$ was arbitrarily small and did not depend on $d$, this is a contradiction. □

**Proposition 2.5.** *If $a \in \mathcal{B}(H)$ is normal then $C(\mathrm{sp}(a)) \cong C^*(a, I)$. The isomorphism sends function $f \in C(\mathrm{sp}(a))$ to $f(a)$, where $f(a)$ is defined naturally in case when $f$ is a *-polynomial.*

*Proof.* By Lemma 2.4 it suffices to prove that $C^*(a, I)$ is isomorphic to $C(\mathrm{sp}_0(a))$, where $\mathrm{sp}_0(a)$ denotes the spectrum of $a$ as defined in $C^*(a, I)$. Let $X$ be a compact Hausdorff space such that $C^*(a, I) \cong C(X)$, as guaranteed by Gelfand–Naimark theorem. For any $\lambda \in \mathrm{sp}(a)$, $a - \lambda \cdot 1$ is not invertible so there exists $\phi_\lambda \in X$ such that $\phi_\lambda(a - \lambda \cdot 1) = 0$, or $\phi_\lambda(a) = \lambda$. Conversely, if there is $\phi \in X$ such that $\phi(a) = \lambda$, then $\phi(a - \lambda \cdot 1) = 0$ so $\lambda \in \mathrm{sp}(a)$. Since any nonzero homomorphism to $\mathbb{C}$ is unital, an element $\phi \in X$ is determined entirely by $\phi(a)$. Since $X$ has the weak* topology, $\phi \mapsto \phi(a)$ is thus a continuous bijection from $X$ to $\mathrm{sp}(a)$, which is a homeomorphism since $X$ is compact. □

**Lemma 2.6.** *Suppose $A$ is a unital subalgebra of $B$ and $a \in A$ is normal. Then $\mathrm{sp}_A(a) = \mathrm{sp}_B(a)$, where $\mathrm{sp}_A(a)$ and $\mathrm{sp}_B(a)$ denote the spectra of $a$ as an element of $A$ and $B$, respectively.*

*Proof.* Since an element invertible in the smaller algebra is clearly invertible in the larger algebra, we have that $\mathrm{sp}_B(a) \subseteq \mathrm{sp}_A(a)$ and we only need to check that $\mathrm{sp}_A(a) \subseteq \mathrm{sp}_B(a)$. Pick $\lambda \in \mathrm{sp}_A(a)$. We need to prove that $a - \lambda \cdot 1$ is not invertible in $B$. Assume the contrary and let $b$ be the inverse of $a - \lambda \cdot 1$. Fix $\varepsilon > 0$ and let $U \subseteq \mathrm{sp}_A(a)$ be the open ball around $\lambda$ od radius $\varepsilon$. Let $g \in C(\mathrm{sp}_A(a))$ be a function supported by $U$ such that $\|g\| = 1$. Then $g = b(a - \lambda \cdot 1)g$, hence $\|b(a - \lambda \cdot 1)g\| = 1$. On the other hand, $(a - \lambda \cdot 1)g = f \in C(\mathrm{sp}_A(a))$ so that $f$ vanishes outside of $U$ and $\|f(x)\| < \varepsilon$ for $x \in U$, hence $\|(a - \lambda \cdot 1)g\| < \varepsilon$. Thus $\|b\| > 1/\varepsilon$ for every $\varepsilon > 0$, a contradiction. □

Note that the isomorphism defined in Proposition 2.5 is canonical and maps $a$ to the identity function on $\mathrm{sp}(a)$. It follows that for any polynomial $p$, the isomorphism maps $p(a)$ to the function $z \mapsto p(z)$. More generally, for any continuous function $f : \mathrm{sp}(a) \to \mathbb{C}$, we can then define $f(a) \in C^*(a, I)$ as the preimage of $f$ under the isomorphism. For example, we can define $|a|$ and if $a$ is self-adjoint then it can be written as a difference of two positive operators as

$$a = \frac{|a| + a}{2} - \frac{|a| - a}{2}.$$

If $a \geq 0$, then we can also define $\sqrt{a}$. Lemma 2.7 is an another application of the "continuous functional calculus" of Corollary 2.5. A remark about terminology is in order. It is customary among C*-algebraists to call 1-Lipshitz maps *contractions*. Recall that a map $\Phi$ is *Lipshitz* if $d(\Phi(x), \Phi(y)) \leq d(x, y)$ for all $x$ and $y$. Although this terminology makes various fixed-point theorems false[b], I shall use it in order to be compatible with the standard terminology.

**Lemma 2.7.** *Any *-homomorphism $\Phi : A \to B$ between C*-algebras is a contraction (in particular, it is continuous). Therefore, any (algebraic) isomorphism between C*-algebras is an isometry.*

*Proof.* By passing to the unitizations, we may assume $A$ and $B$ are unital and $\Phi$ is unital as well (i.e., $\Phi(I_A) = 1_B$).

Note that for any $a \in A$, $\mathrm{sp}(\Phi(a)) \subseteq \mathrm{sp}(a)$ (by the definition of the spectrum). Thus for $a$ normal using Exercise 1.4.8 we have

$$\begin{aligned}
\|a\| &= \sup\{|\lambda| : \lambda \in \sigma(a)\} \\
&\geq \sup\{|\lambda| : \lambda \in \sigma(\Phi(a))\} \\
&= \|\Phi(a)\|.
\end{aligned}$$

[b]Outside of the theory of operator algebras, contractions are usually required to strictly decrease the distance.

For general $a$, $aa^*$ is normal so by the C*-equality we have

$$\|a\| = \sqrt{\|aa^*\|} \geq \sqrt{\|\Phi(aa^*)\|} = \|\Phi(a)\|,$$

concluding the proof. □

For subsets $K$ and $L$ of a metric space $(X, d)$ and $\varepsilon > 0$ we write $K \subseteq_\varepsilon L$ if and only if $\inf_{y \in L} d(x, y) \leq \varepsilon$ for all $x \in K$. Note that $X$ and $d$ are implicit in this notation.

**Lemma 2.8.** *Assume $a$ and $b$ are normal and $\|a - b\| \leq \varepsilon$. Then* $\mathrm{sp}(a) \subseteq_\varepsilon \mathrm{sp}(b)$ *and* $\mathrm{sp}(b) \subseteq_\varepsilon \mathrm{sp}(a)$.

*Proof.* It suffices to prove that for an arbitrary $\lambda \in \mathbb{C}$ such that $\mathrm{dist}(\lambda, \mathrm{sp}(a)) > \varepsilon$ we have $\lambda \notin \mathrm{sp}(b)$.

Fix such $\lambda$ and let $c = (a - \lambda \cdot 1)^{-1}$. By Exercise 1.4.9 we have that $\|c\| < 1/\varepsilon$. Then $c(b - \lambda \cdot 1) = c(a - \lambda \cdot 1) - c(a - b) = 1 - c(a - b)$. The right-hand side is invertible by Lemma 1.4, and therefore $b - \lambda \cdot 1$ is invertible as well. □

The following slightly amusing remark can be safely ignored. Consider $K(\mathbb{C})$, the space of compact subsets of $\mathbb{C}$, as a Polish space with respect to the Hausdorff distance

$$d_H(K, L) = \max\{\inf\{\varepsilon : K \subseteq_\varepsilon L \text{ and } L \subseteq_\varepsilon K\}.$$

Lemma 2.8 states that the map $a \mapsto \mathrm{sp}(a)$ from normal operators in $A$ into $K(\mathbb{C})$ is a contraction (cf. the paragraph before Lemma 2.7).

## 2.4. *Exercises*

**2.4.1.** Prove the following are equivalent for all $a \in A$.

(1) $a = b^*b$ for some $b \in A$.
(2) $a$ is normal and $\mathrm{sp}(a) \subseteq [0, \infty)$.

**2.4.2.** Assume $a$ is a normal operator. Characterize $a$ being self-adjoint, projection, positive, unitary, in terms of the spectrum of $a$.

**2.4.3.** A multiplication operator $m_f$ is invertible if and only if there is some $\varepsilon > 0$ such that $|f| > \varepsilon$ (a.e.). Thus since $m_f - \lambda I = m_{f-\lambda}$, $\mathrm{sp}(m_f)$ is the essential range of $f$ (the set of $\lambda \in \mathbb{C}$ such that for every neighborhood $U$ of $\lambda$, $f^{-1}(U)$ has positive measure).

**2.4.4.** If $f$ is a continuous function on $\mathrm{sp}(a)$, prove that $\mathrm{sp}(f(a)) = \{f(\lambda) : \lambda \in \mathrm{sp}(a)\}$.

**2.4.5.** Assume $a$ is normal. Show that $C^*(a) \cong C_0(\mathrm{sp}(a) \setminus \{0\})$.

**2.4.6.** Show that Theorem 1.6 is a corollary of Proposition 2.5.

## 2.5. *Constructions of C*-algebras*

In many of the constructions of C*-algebras given below the fact that the algebraic structure determines the norm considerably simplifies the discussion.

2.5.1. *Unitization* Every C*-algebra $A$ can be embedded in a unital C*-algebra $\tilde{A}$ in a minimal way as follows. On $A \times \mathbb{C}$ define the operations as follows: $(a, \lambda)(b, \xi) = (ab + \lambda b + \xi a, \lambda\xi)$, $(a, \lambda)^* = (a^*, \bar{\lambda})$ and $\|(a, \lambda)\| = \sup_{\|b\| \leq 1} \|ab + \lambda b\|$ and check that this is still a C*-algebra.

A straightforward calculation shows that $(0, 1)$ is the unit of $\tilde{A}$ and that $A \ni a \mapsto (a, 0) \in \tilde{A}$ is an isomorphic embedding of $A$ into $\tilde{A}$.

2.5.2. *Direct sums* Given C*-algebras $A$ and $B$ we define their direct sum $A \oplus B$ to be the set of all pairs $(a, b)$ with the pointwise defined operations and norm defined by $\|(a, b)\| = \max\{\|a\|, \|b\|\}$. This is easily seen to be an abstract C*-algebra. If $A$ and $B$ are given with concrete representations on spaces $H$ and $K$, respectively, then it is equally easy to represent $A \oplus B$ on $H \oplus K$ as block-matrices

$$(a, b) = \begin{pmatrix} a & 0 \\ 0 & b \end{pmatrix}.$$

As is customary we write $a + b$ instead of $(a, b)$. One similarly defines a direct sum of any finite number of C*-algebras. Given an infinite family of C*-algebras $A_i$, for $i \in I$, the direct sum is defined as

$$\bigoplus_{i \in I} A_i = \{(a_i : i \in I) : a_i \in A_i \text{ for all } i, \text{ and } \{i : \|a_i\| > 1/n\} \text{ is finite for all } n\}.$$

Operations are defined pointwise and the norm is the supremum norm. It is again easy to see that this is a C*-algebra.

2.5.3. *Direct products* Finite direct products coincide with finite sums. Given an infinite family of C*-algebras $A_i$, for $i \in I$, one defines

$$\prod_{i \in I} A_i = \{(a_i : i \in I) : \sup_i \|a_i\| < \infty\}.$$

With operations and norm defined as in §2.5.2 this is a C*-algebra that has $\bigoplus_{i \in I} A_i$ as an ideal. One can prove that $\prod_{i<I} A_i$ is nonseparable unless all but finitely many $A_i$ are isomorphic to $\mathbb{C}$.

2.5.4. *Direct limits (also called inductive limits)* Let $A_i$, for $i \in \Lambda$ be family of C*-algebras indexed by a directed set. Also assume that for $i < j'$ we have a *-homomorphism $F_{ij}: A_i \to A_j$ and that these *-homomorphisms commute. Then $A_i$ together with $F_{ij}$ form a *directed system* of C*-algebras.

Then the *direct limit* of this system, $A = \lim_i A_i$, is defined as follows. Consider the set of all $(a_i : i \geq j)$ such that $j \in \Lambda$, $a_j \in A_j$ and $a_i = F_{ji}(a_j)$ for all $i \geq j$. This limit comes equipped with canonical *-homomorphisms $F_i: A_i \to A$ for all $i$ which commute with all $F_{ij}$.

One should keep in mind that this is an abuse of notation, since the direct limit depends on connecting maps $F_{ij}$ as well as the algebras $A_i$ (see the examples in §4.4).

2.5.5. *Matrix algebra over A* Given a C*-algebra $A$ and $n \in \mathbb{N}$ we define C*-algebra $M_n(A)$ as follows. Its elements are $n \times n$ matrices over $A$. The algebraic operations are defined to be the usual matrix operations. In order to define norm fix a faithful representation of $A$ on a Hilbert space $H$. Now interpret each $a \in M_n(A)$ naturally as an operator on the direct sum of $n$ copies of $H$. We equip $M_n(A)$ with the corresponding operator norm.

By the automatic continuity (Lemma 2.7) the norm on $M_n(A)$ is canonical. However, it is notoriously nontrivial to compute. For example, a deceivingly simple Anderson's paving conjecture is equivalent to the positive solution to the central Kadison–Singer problem on extensions of pure states.

2.5.6. *Stabilization* Story goes that in the olden days, whenever encountered with a non-unital C*-algebra one would immediately unitize it. Nowadays, whenever encountered with a unital C*-algebra one stabilizes it and hence turns it into a non-unital C*-algebra. The motivation for this behaviour will become apparent in §5.

Given $A$, define a direct limit as follows. Let $A_n$ be $M_n(A)$ and let $F_{n,n+1}: A_n \to A_{n+1}$ be given by adding the $n+1$-st zero row and zero column to $a$, or in block-matrix notation $F_n(a) = \begin{pmatrix} a & 0 \\ 0 & 0 \end{pmatrix}$.

Then maps $F_{n,n+1}$ for $n \in \mathbb{N}$ define a commuting system of non-unital *-homomorphisms. The direct limit is a non-unital algebra called the *stabilization* of $A$. This algebra is just a special case of minimal C*-algebraic tensor product defined in §2.5.7 below. More precisely, $M_n(A)$ is (isomorphic to) $A \otimes M_n(\mathbb{C})$ and the stabilization of $A$ is (isomorphic to) $A \otimes \mathcal{K}$, where $\mathcal{K}$ denotes the algebra of compact operators on an infinite-dimensional, separable, complex Hilbert space.

For future use in §5 we record a bit of notation. By $M_\infty(A)$ we denote $\bigcup_n M_n(A)$, with the connecting maps as defined above.

A C*-algebra is $A$ *stable* if it is isomorphic to $A \otimes \mathcal{K}$. Since $\mathcal{K}$ itself is stable, $A$ is stable if and only if it is isomorphic to $B \otimes \mathcal{K}$ for some $B$.

2.5.7. *Minimal tensor product* An *algebraic tensor product* of C*-algebras $A$ and $B$ is defined as a quotient of the linear span of elementary tensors $a \otimes b$. It is customary to denote this algebraic tensor product by $A \odot B$. On this complex *-algebra one wants to define a norm satisfying axioms listed in Lemma 1.3 and take the completion. By the GNS theorem (Theorem 2.3) such completion is a C*-algebra. It turns out that in some cases there is no unique C*-norm on $A \odot B$; for example, this is the case with $\mathcal{B}(H) \odot \mathcal{B}(H)$. This is even more remarkable in light of the fact that the tensor product of Hilbert spaces $H$ and $K$ is uniquely defined: If $(e_\xi)$ is an orthonormal basis of $H$ and $(f_\eta)$ is an orthonormal basis of $K$ then $e_\xi \otimes f_\eta$ is an orthonormal basis of $H \otimes K$.

I shall cut the corners and only describe construction of the so-called *minimal* C*-algebraic tensor product, without even explaining why is it minimal. As a matter of fact, I shall not even prove that it is uniquely defined (this requires showing a true, albeit not obvious, fact that $A \otimes B$ does not depend on the choice of representations of $A$ and $B$).

Assume $A$ and $B$ are unital C*-algebras. By the GNS theorem (Theorem 2.3) we can fix *-isomorphisms $\Phi\colon A \to \mathcal{B}(H)$ and $\Psi\colon B \to \mathcal{B}(K)$. Without a loss of generality, we may assume these *-homomorphisms are unital. We can canonically identify $\mathcal{B}(H)$ with a subalgebra of $\mathcal{B}(H \otimes K)$, by sending each $a$ to the operator such that $a(e \otimes f) = a(e) \otimes f$ for all $e \in H$ and $f \in K$. Similarly we identify $\mathcal{B}(K)$ with a subalgebra of $\mathcal{B}(H \otimes K)$. This defines representations of $A$ and $B$ on $\mathcal{B}(H \otimes K)$, and we identify $A$ and $B$ with their respective images. Then $ab = ba$ for all $a \in A$ and $b \in B$ and we define $A \otimes B$ to be the C*-algebra generated by $A$ and $B$. This product is sometimes denoted $A \otimes_{\min} B$.

One can similarly define a tensor product of a family (finite or infinite) of C*-algebras, $\bigotimes_{i \in \mathcal{I}} A_i$. If $\mathcal{I}$ is infinite then one needs to assume that all but finitely many of $A_i$ are unital and let $\bigodot_{i \in \mathcal{I}} A_i$ be the span of the set of all elementary tensors $\bigotimes_{i \in \mathcal{I}} a_i$ where $a_i \in A_i$ and $a_i = I_A$ for all but finitely many $i$.

The assumption that $A$ is unital was needed in order to have an isomorphic copy of $B$, $1 \otimes B$, inside $A \otimes B$. The unitality of $B$ is used in the analogous way.

**Definition 2.9.** C*-algebra $A$ is *nuclear* if for every C*-algebra $B$ there is a unique C*-algebra norm on $A \odot B$.

It is not difficult to check that all finite-dimensional C*-algebras are nuclear and that the class of nuclear algebras is closed under taking tensor products and direct limits. Also, all abelian C*-algebras are nuclear. Therefore, all algebras (except $\mathcal{B}(H)$) considered in these notes are nuclear and all tensor product norms used here will be uniquely determined.

The theory of tensor products of C*-algebras is full of surprises and two of the most important classes of C*-algebras, nuclear and exact algebras, are defined by their behaviour with respect to the tensor products. This exciting subject is beyond the scope of the present paper and the reader may want to consult [5] for more details. See also Exercises 2.6.5 and 2.6.6.

2.5.8. *Continuous fields of C*-algebras* Given a compact space $X$ and a C*-algebra $A$, let $C(X, A)$ denote the algebra of all continuous functions $f: X \to A$. The operations are given pointwise and the norm is the supremum norm, $\|f\| = \sup_{x \in X} \|f(x)\|$.

One can vary this definition by restricting the range of functions $f$ to obtain more general C*-algebras.

2.5.9. *Corners* This is a special case of a hereditary subalgebra. Given a C*-algebra $A$ and a projection $p \in A$, we can consider the subalgebra $pAp = \{pap : a \in A\}$. This is a unital C*-algebra, although it is typically not a unital subalgebra of $A$, even if $A$ has a unit.

2.5.10. ... *and so on* Some important constructions of C*-algebras, such as maximal tensor products, group C*-algebras (both full and reduced), multiplier algebras and coronas will not be used in these notes.

## 2.6. *Exercises*

**2.6.1.** Let $X$ be a locally compact, non-compact, Hausdorff space. By Gelfand–Naimark theorem, the unitization of $C_0(X)$ is isomorphic to $C(Y)$ for some compact Hausdorff space $Y$. What is the relation between $X$ and $Y$?

**2.6.2.** Prove that a direct product of infinitely many C*-algebras is nonseparable unless all but finitely many of them are isomorphic to $\mathbb{C}$.

**2.6.3.** Assume $A = \lim_i A_i$ is unital. Prove that there is $i_0 \in \lambda$ such that for all $i_0 < i$ algebra $A_i$ is unital and for all $i_0 < i < j$ the map $F_{ij}$ is unital.

**2.6.4.** Check that $M_n(M_k(\mathbb{C}))$ is isomorphic to $M_{nk}(\mathbb{C})$.

**2.6.5.** Prove that $M_n(A)$ is algebraically isomorphic to $M_n(\mathbb{C}) \otimes A$ and that it carries a uniquely defined C*-norm.

(Hint: To prove uniqueness of the norm use Lemma 2.7.)

**2.6.6.** Show that the CAR algebra can be identified with the unital direct limit of algebras $A_n = \bigotimes_n M_2(\mathbb{C})$, which in turn can be identified with the infinite tensor product $\bigotimes_{\mathbb{N}} M_2(\mathbb{C})$. (Just like in the case of direct limits, C*-algebraic tensor product is the norm-completion of the algebraic tensor product.)

**2.6.7** (Matrix units I)**.** Prove that $M_n(\mathbb{C})$ is the unique C*-algebra generated by elements $(e_{ij})_{i<n,j<n}$ with the following properties.

(1) Each $e_{ii}$, for $i < n$, is a projection.
(2) $\sum_{i<n} e_{ii} = 1$.
(3) $e_{ij}e_{kl} = \delta_{jk}e_{il}$, where $\delta_{jk} = 1$ if $j = k$ and $\delta_{jk} = 0$ if $j \neq 0$ (Kronecker's delta).
(4) $e_{ij}^* = e_{ji}$.

**2.6.8** (Matrix units II)**.** Prove that $M_n(\mathbb{C})$ is generated (as a C*-algebra) by matrix units $e_{1j}$, for $j < n$, as in Exercise 2.6.7.

**2.6.9.** Prove that an element $a$ of $C(X, A)$ is normal (self-adjoint, unitary, projection,...) if and only if $a(x)$ is normal (self-adjoint, unitary, projection,...) for all $x \in X$.

The above exercise makes connection between projections in $C(X, M_n(\mathbb{C}))$ and vector bundles over $X$ (see §6.0.1).

**2.6.10.** Prove that $C(X) \otimes C(Y) \cong C(X \times Y)$.

**2.6.11.** Prove that $M_m(\mathbb{C}) \otimes M_n(\mathbb{C}) \cong M_m(M_n(\mathbb{C})) \cong M_{mn}(\mathbb{C})$.

(Hint: This is Exercise 2.6.4.)

**2.6.12.** Assume there is a unital *-homomorphism $\Phi\colon M_k(\mathbb{C}) \to A$ and let $p \in A$ be an image of a minimal projection in $A$. Prove that $A \cong M_k(pAp)$.

(Hint: Exercise 2.6.5.)

**2.6.13.** With the notation used in §2.5.7, prove that $\mathcal{B}(H) \otimes \mathcal{B}(H)$ is a proper subset of $\mathcal{B}(H \otimes H)$ if $H$ is infinite-dimensional.

(Hint: Let $e_n$, for $n \in \mathbb{N}$, be an orthonormal sequence in $H$. Let $p$ be the orthogonal projection to the closed linear span of $\{e_n \otimes e_n : n \in \mathbb{N}\}$.

This operator is in $\mathcal{B}(H \otimes H)$ but it cannot be approximated by finite linear combinations of elementary tensors.)

**2.6.14.** Prove that the stabilization of $A$ (§2.5.6) is isomorphic to $A \otimes \mathcal{K}$.

**2.6.15.** Prove that $\mathcal{K} \cong \mathcal{K} \otimes \mathcal{K}$.

## 3. Local Theory of C*-Algebras

### 3.1. *Polar decomposition*

I now collect several somewhat technical (yet illuminating) results on the local structure of C*-algebras. By polar decomposition theorem (Theorem 1.7) every operator $a$ in $\mathcal{B}(H)$ can be represented as a product of a positive element and a partial isometry, $a = |a|v$. The positive part of $a$ is given by formula $(aa^*)^{1/2}$ and therefore belongs to $C^*(a)$. In Exercise 3.3.2 we shall see that the partial isometry $v$ need not belong to $C^*(a)$. We now discuss to what extent this representation, analogous to $z = \rho e^{i\theta}$ for complex numbers, works in arbitrary C*-algebras.

**Lemma 3.1.** *If $b$ is invertible, then $b = cu$ for a unitary $u$ and a positive $c$. Also, $c = (bb^*)^{1/2}$.*

*Proof.* Since $bb^*$ is positive, $c = (bb^*)^{1/2}$ is well-defined by continuous functional calculus. Also, since $b$ is invertible, so are $b^*$, $bb^*$ and $c$. Then $u = c^{-1}b$ satisfies

$$uu^* = (bb^*)^{-1/2}bb^*(bb^*)^{-1/2} = 1$$

and

$$u^*u = b^*(bb^*)^{-1/2}(bb^*)^{-1/2}b^* = 1,$$

and is therefore a unitary. □

**Lemma 3.2.** *If 0 is not an accumulation point of $\mathrm{sp}((aa^*)^{1/2})$ then there is a partial isometry $v$ in $C^*(a)$ such that $a = |a|v$.*

*Proof.* For a moment consider $C^*(a)$ as a concrete C*-algebra on some Hilbert space and let $a = |a|v$ be the polar decomposition of $a$. We have that $b = (aa^*)^{1/2}$ is in $C^*(a)$ and it remains to prove that $v \in C^*(a)$.

Let $f\colon \mathrm{sp}(b) \to \mathbb{C}$ be defined by $f(0) = 0$ and $f(t) = 1/t$ if $t \neq 0$. By our assumption $f$ is continuous on $\mathrm{sp}(b)$ and therefore by the continuous functional calculus we have $f(b) \in C^*(a)$. Since $f(b)$ and $b$ commute, $f(b)b$ is a self-adjoint element whose spectrum is included in $\{0, 1\}$ and it is therefore a projection. Denote this projection by $p$ and note that $p \in C^*(a)$. Since $\ker(p) = \ker((aa^*)^{1/2}) = \ker(a^*)$ we have $pv = v$. Finally, $pv = f(b)|a|v = f(b)a$ and therefore $pv \in C^*(a)$. □

### 3.2. *Stability*

If $b$ is self-adjoint and $\|a - b\| < \varepsilon$ then $c = (a + a^*)/2$ is self-adjoint, belongs to $C^*(a)$, and satisfies $\|c-b\| < \varepsilon$. This is an instance of the stability phenomenon: if an operator $b$ belongs to a distinguished class of operators and $\|a-b\|$ is small, then there is $c \in C^*(a)$ in the same distinguished class as $b$ such that $\|c - b\|$ is small.

**Lemma 3.3.** *Assume $p$ is a projection and $\|a-p\| < \varepsilon$ with $\varepsilon < 1/2$. Then there is a projection $q \in C^*(a)$ such that $\|p - q\| < 2\varepsilon$.*

*Proof.* We may assume $p \neq 0$ (otherwise take $q = 0$). By replacing $a$ with $(a + a^*)/2$ we may assume $a$ is self-adjoint. By Lemma 2.8 we have that $\mathrm{sp}(a) \subseteq (-\varepsilon, \varepsilon) \cup (1 - \varepsilon, 1 + \varepsilon)$. Since $\varepsilon < 1/2$ the function $f$ on $\mathrm{sp}(a)$ that sends $(-\varepsilon, \varepsilon)$ to 0 and $(1 - \varepsilon, 1 + \varepsilon)$ to 1 is well-defined and continuous. By continuous functional calculus we have $q = f(a) \in C^*(a)$ such that $\mathrm{sp}(q) = \{0, 1\}$. Therefore $q$ is a projection. A straightforward computation (Exercise 3.3.1) shows that $\|q - a\| < \varepsilon$ and therefore $q$ is as required. □

A straightforward modification of the proof of Lemma 3.3 gives the following.

**Lemma 3.4.** *Let $F$ be a finite subset of $\mathbb{C}$. Then for every $\varepsilon > 0$ there exists $\delta = \delta(F, \varepsilon) > 0$ with the following property. If $b$ is normal and such that $\mathrm{sp}(b) = F$ and $\|a - b\| < \delta$, then there exists a normal $c \in C^*(a)$ such that $\|c - b\| < \varepsilon$ and $\mathrm{sp}(c) = F$.* □

Although partial isometries are not necessarily normal, a result similar to the above still applies.

**Lemma 3.5.** *For every $\varepsilon > 0$ there exists $\delta > 0$ with the following property. Assume $v$ is a partial isometry and $\|a - v\| < \delta$. Then there is a partial isometry $w \in C^*(a)$ such that $\|v - w\| < \varepsilon$. Moreover, if $p$ and $q$ are projections such that $\|vv^* - p\| < \delta$ and $\|v^*v - q\| < \delta$ then we can choose partial isometry $w \in C^*(a, p, q)$ so that $ww^* = p$ and $w^*w = q$.*

*Proof.* Fix $\delta < \min(e^2/2, 1/12)$. If $\|a - v\| < \delta$ then $\|aa^* - vv^*\| \leq \|a(a^* - v^*)\| + \|(a - v)v^*\| \leq \delta\|a\| + \delta \leq 3\delta$ since $\|a\| \leq 1 + \delta$. By Lemma 2.8 this implies $\mathrm{sp}(aa^*) \subseteq_{3\delta} \{0, 1\}$ and therefore Exercise 2.4.4 implies $\mathrm{sp}((aa^*)^{1/2}) \subseteq_{\sqrt{3\delta}} \{0, 1\}$. Thus $\mathrm{sp}((aa^*)^{1/2})$ is included in two short intervals centered at 0 and 1. Let $f \in C((\mathrm{sp}(aa^*)^{1/2})$ be the function that maps the first interval to 0 the second to 1. Then $f((aa^*)^{1/2})$ is a self-adjoint element whose spectrum is $\{0,1\}$, and it is therefore a projection. It is also within $\sqrt{3\delta}$ of $(aa^*)^{1/2}$ and it therefore satisfies the assumptions of Lemma 3.2. Hence there is a partial isometry $w \in C^*(a)$ such that $a = (aa^*)^{1/2}w$.

For the moreover part, pick $\delta$ small enough so that $pvq$ satisfies the assumptions of the first part of the lemma. Applying it to $pvw$ we obtain $w \in C^*(a,p,q)$ such that $pwq = w$, $\|ww^* - p\| < 1$ and $\|w^*w - q\| < 1$. By Exercise 1.6.3, we have $ww^* = p$ and $w^*w = q$. □

The last few lemmas are concerned with stability of classes of operators in C*-algebras. For more information on this exciting subject the reader can consult the excellent [56].

The following lemma will be important in the analysis of the structure of UHF algebras. We write $A_1 = \{a \in A : \|a\| \leq 1\}$.

**Lemma 3.6.** *For every $n \in \mathbb{N}$ and $\varepsilon > 0$ there exists $\delta > 0$ with the following property. If $A$ and $B$ are unital subalgebras of $C$ such that $A$ is isomorphic to $M_n(\mathbb{C})$ and $A_1 \subseteq_\delta B$, then there exists a unitary $u \in C$ such that $uAu^* \subseteq B$ and $\|1 - u\| \leq \varepsilon$.*

*Proof.* I shall try to avoid the computation of $\delta$ (it is given in [12] and in [39]). Assume $\delta_1 > 0$ is very small. Consider $a \in A$ defined by $a = \mathrm{diag}(1,2,3,\ldots,n)$. Then $a = \sum_{j<n} jp_j$, where $p_j$ for $j < n$ is a projection in $A$ and $\sum_{j<n} p_j = 1$. By Lemma 3.4 there exists $b \in B$ such that $\|b - a\| < \delta_2$ and $b$ is self-adjoint with $\mathrm{sp}(b) = \{1,2,\ldots,n\}$. Then $b = \sum_{j<n} jq_j$, where $q_j$ for $j < n$ is a projection and $\sum_{j<n} q_j = 1$. We have $\|p_j - q_j\| < \delta_2$ for all $j$. As in Exercise 2.6.8, let $e_{1j}$ for $j < n$ be partial isometries[c] generating $A$. Then $e_{1j}^* e_{1j} = p_j$ for $j < n$ and $e_{1j} = p_1 e_{1j} p_j$ for $j < n$.

By choosing $\delta$ small enough so that Lemma 3.5 can be applied to each $e_{1j}$, we find partial isometries $e_{1j}$ for $1 \leq j \leq n$ in $B$ as required. Then define $e_{ij} = e_{1i}^* e_{1j}$ for all $i$ and $j$, note that $p_i = e_{ii}$ and that $e_{ij}$ are matrix units as required. □

The following lemmas will be important in the analysis of Murray-von Neumann equivalence of projections (see §3.4).

[c]Projections are partial isometries, too.

**Lemma 3.7.** *If $a$ is self-adjoint and there exist $x$ and a self-adjoint $b$ such that $xa = bx$ then $x^*x$ commutes with $a$.*

*If $x$ is moreover invertible then $b$ is unitarily equivalent to $a$, i.e., there exists unitary $u \in A$ such that $b = uau^*$.*

*Proof.* By the self-adjointness of $a$ and $b$ we have $ax^* = x^*b$. Therefore

$$x^*xa = x^*bx = ax^*x$$

as required.

Now assume $x$ is invertible. By Lemma 3.1 we have $x = u|x|$ for some $u \in A$. By the first part, $x^*x$ commutes with $a$. Since $|x| = (x^*x)^{1/2}$ belongs to the C*-algebra generated by $x^*x$, it also commutes with $a$. We therefore have $b = xax^{-1} = u|x|a|x|^{-1}u^* = uau^*$. □

We state an immediate consequence of the above lemma for future reference.

**Lemma 3.8.** *If $a$ and $b$ are self-adjoint and $b = xax^{-1}$ for an invertible $x$ then $b = uau^*$ for a unitary $u$.* □

### 3.3. *Exercises*

**3.3.1.** Assume $a$ is normal and $f$ and $g$ are continuous functions on $\mathrm{sp}(a)$. Then $\|f(a) - g(a)\| = \|f - g\|_\infty$.

**3.3.2.** Find an operator $a$ such that $C^*(a)$ does not contain partial isometry $v$ such that $a = |a|v$.

(Hint: Choose $a$ to be compact but of infinite rank.)

**3.3.3.** Show that invertible elements form an open set in a unital C*-algebra. (Hint: Use the proof of Lemma 1.4 to show that $b$ invertible and $\|b - c\| < \|b\|$ implies $c$ is invertible.)

**3.3.4.** Prove that $a$ is invertible if and only if $a^*$ is invertible if and only if $aa^*$ invertible if and only if $|a|$ is invertible.

**3.3.5.** Prove Lemma 3.4 and express $\delta$ in terms of $\varepsilon$ and $F$.

### 3.4. *Murray–von Neumann equivalence of projections*

Murray–von Neumann equivalence of projections is a noncommutative analogue of equinumerosity relation of sets and also a continuous variant of the dimension of a closed subspace of the Hilbert space (see Example 3.9 (1)). It was introduced by Murray and von Neumann in their seminal series of papers 'Rings of operators' (an old name for von Neumann algebras) where it played a fundamental role in type classification (see e.g., [46]). While C*-algebras are not nearly as well-behaved as von Neumann algebras Murray–von Neumann equivalence of projections is a useful tool in classification problem for some well-behaved classes of C*-algebras.

Assume $A$ is a C*-algebra. Two projections $p$ and $q$ in $A$ are *Murray-von Neumann equivalent* if there exists $v \in A$ such that

$$vv^* = p \text{ and } v^*v = q. \qquad \text{(MvN)}$$

In this case we write $p \sim q$ and keep in mind that the relation depends on the ambient algebra $A$.

Note that a witness $v$ of Murray-von Neumann equivalence is necessarily a partial isometry.

In some C*-algebras $p \sim q$ is strictly weaker than the requirement that $p$ and $q$ are conjugate by a unitary (see Example 3.9 (1)).

**Example 3.9.** (1) If $A = \mathcal{B}(H)$ then $p \sim q$ if and only if the range of $p$ and the range of $q$ have the same dimension, where the dimension of a closed subspace of the Hilbert space is the minimal cardinality of an orthonormal basis. This is an immediate consequence of the fact that two complex Hilbert spaces with the same dimension are linearly isometric.

(2) A special case of (1) is $A = M_n(\mathbb{C})$, where two projections are Murray-von Neumann equivalent if and only if they have the same rank. This extends to the algebra $\mathcal{K}$ of compact operators.

(3) If $A$ is abelian then $p \sim q$ if and only if $p = q$.

The following example requires some minimal knowledge of vector bundles; see e.g., [41].

**Example 3.10.** Projections of $C(X, M_n(\mathbb{C}))$ are maps $f\colon X \to M_n(\mathbb{C})$ such that $f(x)$ is a projection for all $x \in X$. By identifying a projection in $M_n(\mathbb{C})$ with a subspace of $\mathbb{C}^n$ one sees that projections of $C(X, \mathcal{K})$ are vector bundles over $X$. Murray-von Neuman equivalence of these projections is the usual equivalence of vector bundles.

**Lemma 3.11.** *Assume $p$ and $q$ are projections in $A$ such that $\|p-q\| < 1/2$. Then $p \sim q$. If $A$ is unital then there is moreover a unitary $u$ such that $u^*pu = q$.*

*Proof.* We first prove the case when $A$ is unital. Let $a = pq + (1-p)(1-q)$. Since $1-p$ and $1-q$ are at a distance $\|p-q\| < 1/2$, the distance from $a$ to $1 = p^2 + (1-p)^2$ is $< 1$. By Lemma 1.4 $a$ is invertible. By Lemma 3.1 we have $a = |a|u$ for a unitary $u$.

One easily checks that $p_1 = aqa^{-1}$ satisfies $p_1^2 = p_1 = p_1^*$, and is therefore a projection. Similarly, $p_2 = a(1-p)a^{-1}$ is a projection and $p_1 + p_2 = 1$. By inspecting the definition of $a$ one sees that $pp_1 = p_1$ and $(1-p)p_2 = p_2$, By Exercise 1.6.4 we conclude that $p = aqa^{-1}$. By Lemma 3.8 we conclude $p = uqu^*$. Then $v = uq$ is a partial isometry such that $vv^* = p$ and $v^*v = q$.

Now assume $A$ is not unital. By the above, in the unitization of $A$ there exists a unitary $u$ such that $uqu^* = p$. The partial isometry $v = uq$ as above belongs to $A$ and witnesses $p \sim q$. □

Lemma 3.11 can be improved; see Exercise 3.5.7.

### 3.5. *Exercises*

**3.5.1.** Prove that $p \sim q$ is equivalent to the existence of $v$ such that $v^*pv = q$ and $vqv^* = p$. Also prove that such $v$ is necessarily a partial isometry.

**3.5.2.** If $F\colon A \to B$ is a *-homomorphism and $p$ and $q$ are projections in $A$, show that $p \sim q$ implies $F(p) \sim F(q)$. Give an example showing that the converse may fail.

**3.5.3.** Let $\Phi\colon A \to B$ be a unital *-homomorphism between C*-algebras and let $a$ and $b$ be such that $b = \Phi(a)$.

(1) Prove that if $a$ is normal (self-adjoint, positive, unitary, projection, partial isometry) then $b$ is normal (self-adjoint, positive, unitary, projection, partial isometry).
(2) Assume $b$ is self-adjoint (positive) Prove that we can choose $a'$ such that $\Phi(a') = b$ and $a'$ is self-adjoint (positive).
(3) Provide examples showing that $b$ can be normal (unitary, projection, partial isometry, respectively) while no $a'$ satisfying $\Phi(a') = b$ is normal (unitary, projection, partial isometry, respectively).
(Hint: For projections and partial isometries consider abelian algebras and use Exercise 2.2.6.)

**3.5.4.** Find a C*-algebra $A$ and two Murray-von Neumann equivalent projections that are not conjugate.
(Hint: Try $\mathcal{B}(H)$.)

Let $\mathcal{P}(A)$ denote the set of all projections of a C*-algebra $A$.

**3.5.5.** Two projections are *homotopic* (in a C*-algebra $A$) if they belong to the same path-connected component of $\mathcal{P}(A)$. Prove that being homotopic implies being conjugate by a unitary.

(Hint: Lemma 3.11.)

**3.5.6.** Prove that $\|p - q\| < 1$ implies $p$ and $q$ are homotopic.

(Hint: The path $tp + (1-t)q$, for $0 \leq t \leq 1$, consists of nonzero positive elements. Use the continuous functional calculus to morph it into a path consisting of projections.)

**3.5.7.** Prove that $\|p - q\| < 1$ implies $p$ and $q$ are conjugate and in particular $p \sim q$.

(Hint: Combine Exercise 3.5.5 and Exercise 3.5.6.)

**3.5.8.** Prove that there exists $\varepsilon > 0$ such that for all $A$ and projections $p$ and $q$ in $A$ we have $p \sim q$ if and only if there exists $a \in A$ such that $\|aa^* - p\| < \varepsilon$ and $\|a^*a - q\| < \varepsilon$.

(Hint: Lemma 3.5 and Lemma 3.11.)

**3.5.9.** Prove that the following two properties of a C*-algebra $A$ are equivalent.

(1) the set of invertible self-adjoint elements in the unitization of $A$ is dense in the set of all self-adjoint elements in the unitization of $A$.
(2) Linear combinations of projections are dense in $A$.

(Hint: Every element of $A$ is a linear combination of two self-adjoint operators. Use continuous functional calculus.)

C*-algebras $A$ satisfying either of the statements from Exercise 3.5.9 have *real rank zero.*

Recall that on projections we define a relation $p \leq q$ if and only if $pq = p$ (Exercise 1.6.3). A nonzero projection $p$ in a C*-algebra is *minimal* if the only projections $q \leq p$ are 0 and $p$.

**3.5.10.** Prove that a projection $p$ in a real rank zero algebra $A$ is minimal if and only if $pAp$ is isomorphic to $\mathbb{C}$. Then prove that all minimal projections in a simple real rank zero algebra are Murray-von Neumann equivalent.

(Hint: If $p$ and $q$ are minimal projections in a real rank zero algebra prove that the vector space $pAq = \{paq : a \in A\}$ is one-dimensional.)

## 3.6. *Traces*

A *trace* of a C*-algebra $A$ is a state $\tau$ such that $\tau(ab) = \tau(ba)$ for all $a$ and $b$. We record an immediate consequence of the definition of $\sim$.

**Lemma 3.12.** *If $\tau$ is a trace on $A$ and $p \sim q$ then $\tau(p) = \tau(q)$.* □

Let

$$T(A) = \{\tau \in A^* : \tau \text{ is a trace}\}.$$

If $\tau$ and $\sigma$ are traces and $0 < t < 1$ then $t\tau + (1-t)\sigma$ is a trace. Therefore $T(A)$ is a convex subset of the unit sphere of $A^*$. Also, since being a trace is a closed condition, by Birkhoff–Alaoglu theorem $T(A)$ is compact in the weak*-topology. Being a compact and convex set, by the Krein–Milman theorem $T(A)$ is the closure of the convex hull of its extreme points.

On $M_n(\mathbb{C})$ define the normalized trace via (below $a$ stands for the matrix $(a_{ij})_{i\leq n, j\leq n}$)

$$\operatorname{tr}(a) = \frac{1}{n}\sum_{j=1}^{n} a_{jj}.$$

**Lemma 3.13.** *If $p$ and $q$ are projections in $M_n(\mathbb{C})$ then $p \sim q$ if and only if* $\operatorname{tr}(p) = \operatorname{tr}(q)$, *and* $\operatorname{tr}(p) = k/n$ *where $0 \leq k \leq n$ is the dimension of the range of $p$.*

*Proof.* The direct implication is Lemma 3.12. The converse implication is an exercise in linear algebra. □

**Lemma 3.14.** *Functional* tr *is a unique trace on $M_n(\mathbb{C})$.*

*Proof.* This is of course a standard linear algebra fact. Fix $a$ and $b$ and note that the $i$-th diagonal entry of $ab$ is equal to $\sum_{j<n} a_{ij}b_{ji}$, and therefore $\operatorname{tr}(ab) = \frac{1}{n}\sum_{i<n}\sum_{j<n} a_{ij}b_{ji}$. Analogous argument shows that $\operatorname{tr}(ba)$ has the same value. A similar computation shows that $\operatorname{tr}(aa^*) = \frac{1}{n}\sum_{i,j} a_{ij}\bar{a}_{ij} \geq 0$, and tr is therefore positive. Finally, $\operatorname{tr}(1) = 1$ is clear.

In order to check the uniqueness assume $\sigma$ is a trace of $M_n(\mathbb{C})$. By Lemma 3.12 and Example 3.9 (2) all rank one projections have the same trace. Therefore $\sigma(p) = 1/n$ for all rank one projections $p$. This implies that for diagonal matrices $a$ $\sigma(a) = \operatorname{tr}(a)$.

It only remains to note that every off-diagonal matrix unit necessarily has trace 0. □

**Lemma 3.15.** *There is a unital *-homomorphism from $M_n(\mathbb{C})$ into $M_k(\mathbb{C})$ if and only if $n$ divides $k$. All unital *-homomorphisms from $M_n(\mathbb{C})$ into $M_k(\mathbb{C})$ are conjugate.*

*Proof.* This can be proved in many ways and our proof is not the shortest. If $\Phi\colon M_n(\mathbb{C}) \to M_k(\mathbb{C})$ is a unital *-homomorphism then $\tau(a) = \operatorname{tr}(\Phi(a))$ defines a trace on $M_n(\mathbb{C})$. By Lemma 3.14 trace $\tau$ coincides with tr. Since a *-homomorphism sends projections to projections, by Lemma 3.13 we conclude that $1/n = m/k$ for some $m$, concluding the proof.

In order to prove the second part, assume $\Phi$ and $\Psi$ are unital *-homomorphisms of $M_n(\mathbb{C})$ into $M_k(\mathbb{C})$. Fix a minimal projection $p \in M_n(\mathbb{C})$, for example $p = \mathrm{diag}(1, 0, 0, \dots, 0)$. Then $\Phi(p)$ and $\Phi(q)$ are projections in $M_k(\mathbb{C})$ each with trace $1/n$ Therefore $A = \Phi(p)M_k(\mathbb{C})\Phi(p)$ and $B = \Psi(p)M_k(\mathbb{C})\Psi(p)$ are both isomorphic to $M_{k/n}(\mathbb{C})$. By Exercise 2.6.12 we have that $M_k(\mathbb{C}) \cong M_n(A) \cong M_n(B)$. Therefore an isomorphism $\alpha\colon A \to B$ extends to an automorphism $\alpha'$ of $M_k(\mathbb{C})$ such that $\Psi = \alpha' \circ \Phi$. By the easy finite-dimensional case of Exercise 1.2.7 we have that $\alpha$ is inner and therefore for some unitary $u \in M_k(\mathbb{C})$ we have $\Psi = \mathrm{Ad}\, u \circ \Phi$, as required. □

Note that if $F\colon B \to C$ is a unital *-homomorphism and $\tau$ is a trace of $C$ then $\tau \circ F$ is a trace of $B$. The map $T(C) \ni \tau \mapsto \tau \circ F \in T(B)$ is continuous and affine.

**Lemma 3.16.** *Assume $A = \lim_n A_n$ is unital. If each $A_n$ has a unique trace then $A$ has a unique trace. More generally, $T(A) = \varprojlim T(A_n)$.*

*Proof.* We prove only the first assertion. Since $A$ is unital, all but finitely many *-homomorphisms from $A_n$ to $A$ are unital. Let $\tau_n$ be the unique trace of $A_n$. Then $\tau_{n+1} \restriction A_n = \tau_n$. Therefore $\tau' = \lim_n \tau_n$ is a well-defined trace on a dense subset of $A$. Since trace is norm-continuous, $\tau'$ has a unique extension to a trace of $A$.

Assume $\sigma$ is a trace of $A$. Then $\sigma \restriction A_n = \tau_n$ for all $n$ and therefore $\sigma$ and $\tau$ agree on a dense subset of $A$. Since $\sigma$ is a continuous functional, $\sigma = \tau$.

In order to prove the second assertion use Exercise 2.2.10 in addition to the above and observe that the functor is contravariant. □

**Example 3.17.** (1) Assume $A$ is unital and abelian. By the Gelfand–Naimark theorem $A = C(X)$ for a compact Hausdorff space $X$. By the Riesz Representation theorem, every continuous functional $\phi$ of $A$ is of the form $\phi(f) = \int f\, d\mu$ for a finite Radon measure $\mu$ on $X$. If $\phi$ is a state then $\mu_\phi$ is a probability measure. Since $A$ is abelian the condition $\phi(ab) = \phi(ba)$ is automatic and therefore all states are traces.

Therefore $T(A)$ is affinely homeomorphic to $P(X)$, the space of Radon probability measures on $X$.

(2) By Lemma 3.14 $T(M_n(\mathbb{C}))$ is a singleton for every $n$.

(3) Furthermore, Lemma 3.16 implies that every UHF algebra carries a unique trace.

(4) Let $A = B \oplus C$. Then clearly $T(A) = \{\lambda\tau + (1-\lambda)\sigma : \tau \in T(B), \sigma \in T(C), 0 < \lambda < 1\}$. Therefore if $A$ is a direct sum of $n$ matrix algebras by (2) we have that $T(A)$ is affinely homeomorphic to the $n$-simplex, $\Delta_n$.

(5) By (4) every trace $\tau$ of $\bigoplus_{i<n} M_{n(i)}(\mathbb{C})$ is determined by vector $(\lambda_i^\tau : i < n)$ in $[0,1]^n$ such that $\sum_{i<n} \lambda_i^\tau = 1$. Therefore if $A = \bigoplus_{i<n} M_{n(i)}(\mathbb{C})$, $B = \bigoplus_{i<k} M_{k(i)}(\mathbb{C})$, $F\colon A \to B$ is a unital *-homomorphism and $\tau$ is a trace on $B$. then $\sigma(a) = \tau(F(a))$ is a trace on $A$. If $p_i$ is the identity of $M_{n(i)}(\mathbb{C})$ and $q_j$ is the identity of $M_{k(j)}(\mathbb{C})$ then $\sigma(p_i)$ is uniquely determined by $\tau(q_j)$ and the Bratteli diagram of $F$.

### 3.7. *Exercises*

**3.7.1.** Assume $n \leq k$. Classify all (not necessarily unital) *-homomorphisms of $M_n(\mathbb{C})$ into $M_k(\mathbb{C})$, up to conjugacy.

(Hint: Consider the image of the identity and apply Lemma 3.15, which gives the unital *-homomorphism case.)

**3.7.2.** Prove that $T(A)$ is a weak*-compact convex subset of $A^*$ (the Banach space dual of $A$).

**3.7.3.** Let $X$ be a compact metric space and assume $A$ has a unique trace. Prove that $T(C(X, A))$ is affinely homeomorphic to $\mathcal{P}(X)$, the space of Borel probability measures on $X$.

(Hint: See Example 3.17 (1).)

## 4. UHF Algebras and AF Algebras

### 4.1. *UHF algebras*

C*-algebras that are infinite tensor products of full matrix algebras $M_n(\mathbb{C})$ are said to be *uniformly hyperfinite* (shortly UHF). For separable algebras this is equivalent to being a unital direct limit of full matrix algebras. In separable case we have therefore have algebras of the form $\lim_j M_{n(j)}(\mathbb{C})$ for a sequence $n(j)$ such that $n(j)$ divides $n(j+1)$ for all $j$ (Lemma 3.15). Also by the same lemma, the choice of *-homomorphisms in the unital case is inconsequential for the isomorphism type of the direct limit.

A *generalized integer* (regrettably also known as *supernatural number*) is a formal product of the form $\prod_j p_j^{k(j)}$ where $p_j$, for $j \in \mathbb{N}$, is an increasing enumeration of the primes and $k(j) \in \mathbb{N} \cup \{\infty\}$ (with $0 \in \mathbb{N}$).

To a separable UHF algebra $A = \lim_j M_{n(j)}(\mathbb{C})$ we associate a generalized integer $\mathbf{k}$ such that $k(j)$ is the largest $k$ such that $p_j^k$ divides $n(l)$

for some $l$, or $\infty$ if the set of such $k$ is unbounded. One can show that the generalized integer uniquely determines a separable UHF algebra up to the isomorphism.

Let me again emphasize that UHF algebras are by definition unital. Nonunital algebras that are direct limits of full matrix algebras are called matroid algebras, approximately matricial (AM) algebras, or stabilized UHF algebras (note that the latter terminology is somewhat misleading, since they are not necessarily stable; see Exercise 4.3.6).

**Example 4.1.** (1) The CAR (Canonical Anticommutation Relation) algebra is the UHF algebra which is a direct limit of $M_{2^n}(\mathbb{C})$ for $n \in \mathbb{N}$. It is often denoted by $M_{2^\infty}$.

(2) One can similarly define $M_{3^\infty}$ as the direct limit of $M_{3^n}(\mathbb{C})$ for $n \in \mathbb{N}$.

(3) The *universal UHF algebra* is the UHF algebra corresponding to the generalized integer $\prod_j p_j^\infty$.

Let $D_n$ denote the subalgebra consisting of all diagonal matrices in $M_n(\mathbb{C})$. Then $D_n$ is a maximal abelian subalgebra of $M_n(\mathbb{C})$ isomorphic to $\mathbb{C}^n$. If $A = \lim_j M_{n(j)}(\mathbb{C})$ is a UHF algebra then algebras $D_{n(j)}$ form a directed system and their limit $D$ is the *diagonal masa* in $A$ (cf. Exercise 4.3.2). If $A$ is unital and infinite-dimensional then by Exercise 4.3.3 its diagonal masa is isomorphic to $C(2^{\mathbb{N}})$. Therefore the CAR algebra can be considered as a noncommutative version of the Cantor space. (It is customary to identify compact Hausdorff space $X$ and the C*-algebra $C(X)$, since compact Hausdorff spaces and unital abelian C*-algebras form equivalent categories.)

**Lemma 4.2.** *Every UHF algebra $A$ has a unique trace $\tau$. The values of $\tau$ on projections of $A$ are all numbers of the form $k/n$, where $k \in \mathbb{N}$ and $n$ is a natural number that divides $n_A$.*

*Proof.* Each $M_n(\mathbb{C})$ has a unique trace (Lemma 3.14) and the conclusion follows by Lemma 3.16. □

We note that Glimm's result applies to an apparently larger class of algebras (see Theorem 4.4).

**Theorem 4.3** (Glimm). *Separable UHF algebras $A$ and $B$ are isomorphic if and only if they have the same generalized integer.*

*Proof.* If $\mathbf{k}_A \neq \mathbf{k}_B$ then by Lemma 4.2 $A$ and $B$ are not isomorphic.

Assume $\mathbf{k}_A = \mathbf{k}_B$. Write $A = \lim_j M_{n(j)}(\mathbb{C})$ and $B = \lim_j M_{m(j)}(\mathbb{C})$. By going to subsequences of $n(j)$ and of $m(j)$ we may assume that for all $j$ we have that $n(j)$ divides $m(j)$ and $m(j)$ divides $n(j+1)$. By Lemma 3.15 we can fix a *-homomorphism $\phi_j\colon M_{n(j)}(\mathbb{C}) \to M_{m(j)}(\mathbb{C})$ and a *-homomorphism $\psi_j\colon M_{m(j)}(\mathbb{C}) \to M_{n(j+1)}(\mathbb{C})$ for every $j$. By the second part of the same lemma we may choose these maps so that all triangles in Figure 1 commute.

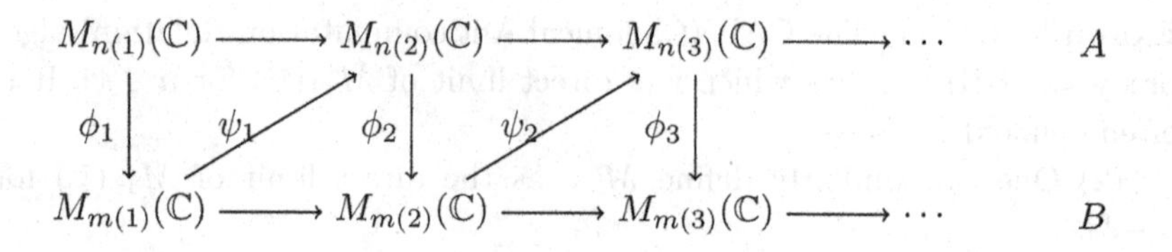


Fig. 1. Proof of Glimm's theorem.

Therefore $\Phi_0 = \bigcup_j \phi_j$ is a well-defined *-homomorphism from a dense subalgebra of $A$ into $B$. It is an isometry by Lemma 2.7 and it therefore extends to a *-homomorphism $\Phi\colon A \to B$. By the same argument we have a *-homomorphism $\Psi\colon B \to A$ extending $\bigcup_j \psi_j$. We claim that $\Psi \circ \Phi$ is the identity on $A$. It suffices to check this for the dense subalgebra $\bigcup_j M_{n(j)}(\mathbb{C})$. Indeed, for any $j$ it is the identity on $M_{n(j)}(\mathbb{C})$ by the commutativity of the above diagram. Similarly $\Phi \circ \Psi$ is the identity on $B$, and therefore $\Phi\colon A \to B$ is a *-isomorphism. □

### 4.2. *Another look at the UHF algebras*

A C*-algebra $A$ is *locally matricial* (or LM) if for every $\varepsilon > 0$ and every finite $F \subseteq A$ there exist $n$ and a *-homomorphism $\Phi\colon M_n(\mathbb{C}) \to A$ such that $F \subseteq_\varepsilon \Phi(M_n(\mathbb{C}))$. (Recall that $K \subseteq_\varepsilon L$ means that $\inf_{y\in L} \|x - y\| \le \varepsilon$ for all $x \in K$.) In other words, for every finite subset of $A$ there exists a full matrix subalgebra $B$ of $A$ such that each element of $F$ is within $\varepsilon$ of $B$.

Infinite tensor products of unital C*-algebras were defined in §2.5.7.

**Theorem 4.4** (Glimm). *For a separable unital C*-algebra the following are equivalent.*

*(1) A is a tensor product of full matrix algebras.*
*(2) A is a direct limit of full matrix algebras (i.e., it is UHF), and*
*(3) A is LM.*

*An algebra as in (1) is necessarily unital, and (2) and (3) remain equivalent in the case when A is not necessarily unital.*

*Proof.* We prove only the equivalence of (2) and (3) (but see the hint to Exercise 4.3.1). By the definition of direct limit (2) implies (3), even without the separability requirement. The implication from (3) to (2) is a consequence of Lemma 3.6. □

As a corollary to Theorem 4.4 and Theorem 4.3, to each unital separable LM algebra one can associate a generalized integer and that this generalized integer is a complete isomorphism invariant for unital separable LM algebras.

The following theorem taken from [30] shows that the situation in non-separable case is quite different. Recall that a *density character* of a C*-algebra is the minimal cardinality of a dense subset.

**Theorem 4.5.** *(1) There exist a unital C*-algebra of density character $\aleph_1$ that is a direct limit of full matrix algebras but not a tensor product of full matrix algebras.*
*(2) Every LM algebra of density character $\leq \aleph_1$ is a direct limit of full matrix algebras.*
*(3) There exists a unital LM algebra of density character $\aleph_2$ that is not a direct limit of full matrix algebras.* □

All algebras constructed in [30] are indistinguishable from the CAR algebra by their Elliott invariant, Cuntz semigroup, or any other known C*-algebraic invariant (see [31]).

### 4.3. *Exercises*

**4.3.1.** Prove the equivalence of (1) and (2) in Theorem 4.4: a unital separable C*-algebra is a tensor product of full matrix algebras if and only if it is a unital direct limit of full matrix algebras.

(Hint: Exercise 2.6.12.)

**4.3.2.** Prove that the diagonal masa (see the paragraph before Lemma 4.2) is a masa (i.e., a maximal abelian C*-subalgebra).

**4.3.3.** Prove that the diagonal masa of a separable infinite-dimensional UHF algebra is isomorphic to $C(2^{\mathbb{N}})$, where $2^{\mathbb{N}}$ denotes the Cantor space.

(Hint: It is a direct limit of finite-dimensional abelian C*-algebras. Prove that it is isomorphic to $C(X)$ for $X$ a compact metrizable zero-dimensional space without isolated points.)

**4.3.4.** Let $A$ be a UHF algebra. Show that its generalized integer $\mathbf{k}_A$ is uniquely defined as the number whose finite divisors are those $n$ such that $M_n(\mathbb{C})$ has a unital *-homomorphism into $A$.

**4.3.5.** Let $A$ and $B$ be unital separable UHF algebras. Prove that $A$ is elementarily equivalent to $B$ (in the logic of metric structures, [29]) if and only $A$ is isomorphic to $B$.

(Hint: [7].)

**4.3.6.** A C*-algebra is *stable* if it is isomorphic to its stabilization (see §2.5.6). Construct a direct limit of full matrix algebras that is neither unital nor stable.

(Hint: First prove that a stable algebra cannot have a finite trace and then construct a nonunital direct limit of full matrix algebras with a finite trace.)

**4.3.7** (Dixmier). Classify separable non-unital direct limits of full matrix algebras.

(Hint: First classify pairs $(A, p)$ where $A$ is a non-unital direct limit of full matrix algebras and $p \in A$ is a projection. Being familiar with classification of rank one torsion-free abelian groups may help, but beware of Exercise 4.3.6.)

**4.3.8.** Characterize when two separable UHF algebras have isomorphic corners (see §2.5.9).

**4.3.9.** Let $A$ and $B$ be separable UHF algebras. Prove that $A$ is isomorphic to a unital subalgebra of $B$ if and only if $\mathbf{k}_A$ divides $\mathbf{k}_B$.

### 4.4. *Bratteli diagrams*

The following lemma is a consequence of the Artin–Wedderburn theorem but we sketch a direct proof in Exercise 4.7.6 below.

**Lemma 4.6.** *Every finite-dimensional C*-algebra is *-isomorphic to a direct sum of finitely many full matrix algebras that each of these full matrix algebras is a minimal (nontrivial) ideal of the algebra.* □

We shall introduce a tool for describing *-homomorphisms $\Phi\colon A \to B$ between finite-dimensional C*-algebras. By Lemma 4.6 every such algebra is a direct sum of its minimal ideals each of which is isomorphic to a full matrix algebras. Recall that there exists a unital *-homomorphism from $M_n(\mathbb{C})$ into $M_k(\mathbb{C})$ if and only if $n$ divides $k$ (Lemma 3.15). If $n \leq k$ and $\Phi\colon M_n(\mathbb{C}) \to M_k(\mathbb{C})$ is a *-homomorphism (not necessarily unital) then its *multiplicity* is the rank of $\Phi$-image of any minimal projection in $M_n(\mathbb{C})$ (since $\Phi$ preserves Murray–von Neumann equivalence, the rank is well-defined). In other words, if $\Phi$ sends $a$ to $\operatorname{diag}(a, a, \dots, a, 0, \dots, 0)$ then its multiplicity is the number of the occurrences of $a$ on the right-hand side. (If you have not tackled Exercise 3.7.1 yet, now is a good time.)

Bratteli diagram of $\Phi\colon A \to B$ is a bipartite graph whose vertices on the left correspond to the minimal ideals of $A$ and vertices on the right correspond to the minimal ideals of $B$. These vertices may be labelled by numbers indicating the dimension of the corresponding algebra. Two vertices are connected by $k$ edges if and only if the multiplicity of the map between them is $k$.

An example is in order. Consider a unital *-homomorphism between $M_2(\mathbb{C}) \oplus M_3(\mathbb{C})$ and $M_6(\mathbb{C}) \oplus M_5(\mathbb{C}) \oplus M_6(\mathbb{C})$ defined by

$$(a, b) \mapsto (\operatorname{diag}(a, a, a), \operatorname{diag}(a, b), \operatorname{diag}(b, b)).$$

The Bratteli diagram describing this map is given in Figure 2.

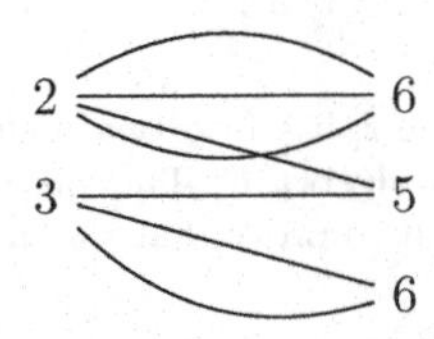


Fig. 2. A Bratteli diagram.

By Lemma 4.7, *-homomorphism described by a Bratteli diagram is unique up to unitary conjugacy.

When describing a unital AF algebra $A = \lim_n A_n$ by a Bratteli diagram we put together diagrams of each $\Phi_n\colon A_n \to A_{n+1}$. For convenience we also let $A_1 = \mathbb{C}$ and assume that all $\Phi_n$ are unital. Under these conventions the labels of vertices can be omitted since the dimension of any of the full matrix algebras can be determined by adding the multiplicities of vertices from the earlier levels.

Some examples of C*-algebras defined via Bratteli diagrams are given in Figures 3–7.

**Lemma 4.7.** *Every unital *-homomorphism between finite direct sums of matrix algebras corresponds to a Bratteli diagram. Moreover, any two unital *-homomorphisms with the same Bratteli diagram are conjugate.*

Fig. 3. This diagram describes the unital directed system $M_2(\mathbb{C}) \to M_4(\mathbb{C}) \to M_8(\mathbb{C})\dots$ and its limit, the CAR algebra $M_{2^\infty}$.

Fig. 4. This diagram represents $M_{3^\infty}$.

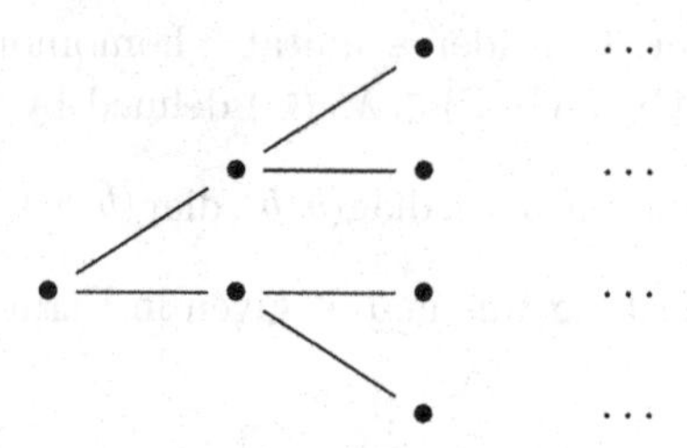

Fig. 5. In this diagram every node splits into two nodes. Note that all nodes in the diagram correspond to the abelian algebra $\mathbb{C}$. Therefore the $n$-th level corresponds to the algebra $\mathbb{C}^{2^n}$ and it is not difficult to prove that the direct limit is the algebra $C(2^{\mathbb{N}})$.

*Proof.* This is an almost immediate consequence of Lemma 3.15. First note that if $F\colon M_n(\mathbb{C}) \to M_k(\mathbb{C})$ is a non-unital *-homomorphism then $p = F(1)$ is a projection. Also, $pM_k(\mathbb{C})p$ is isomorphic to $M_m(\mathbb{C})$ where $m$ is the rank of $p$. By Lemma 3.15 $m$ is a multiple of $n$.

In order to prove the second part of the lemma, fix a unital *-homomorphism $F\colon \bigoplus_{i<N} M_{n(i)}(\mathbb{C}) \to \bigoplus_{i<K} M_{k(i)}(\mathbb{C})$. Let $p_i$ be the identity of $M_{n(i)}(\mathbb{C})$ and let $q_j$ be the identity of $M_{k(j)}(\mathbb{C})$. Then $F(\sum_i p_i) = \sum_i F(p_i) = 1$ and $r_{ij} = q_j F(p_i)$ is a projection since $q_j$ is a central projection. Then $q_j = \sum_i r_{ij}$ and the rank of $r_{ij}$ is by the above a multiple of the rank of $p_i$ for all $i$ and $j$. This data gives a Bratteli diagram. □

### 4.5. *Exercises*

**4.5.1.** If $A$ and $B$ are separable AF algebras given by the same Bratteli diagram then they are isomorphic. The converse may fail.

(Hint: Apply Lemma 4.7 along the diagram.)

**4.5.2.** Characterize Bratteli diagrams of simple AF algebras.

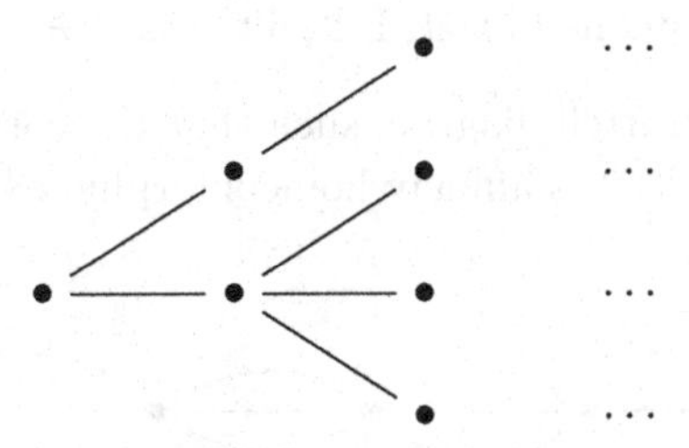

Fig. 6. In this diagram the $n$-th level has $2^n$ nodes, each one corresponding to $\mathbb{C}$. Hence the algebra is a direct limit of $\mathbb{C}^{2^n}$, just like the algebra $C(2^{\mathbb{N}})$ from the previous example. However, in the present case only one of the nodes on the $n$-th level splits, and it splits into $2^n$ other nodes. One can prove that the direct limit is the algebra $C(\omega + 1)$, where $\omega + 1$ is a converging sequence together with its limit.

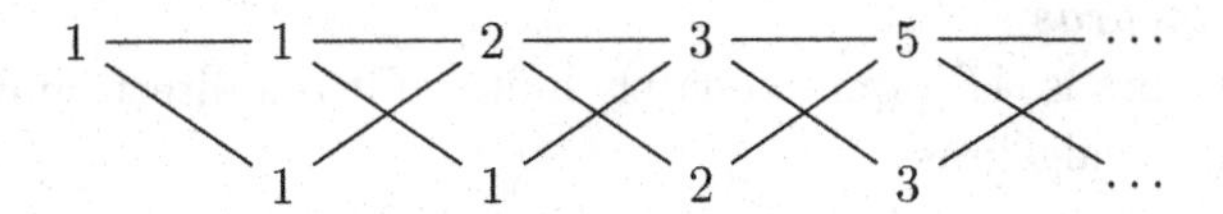


Fig. 7. In this example nodes are marked by numbers for readability. The $n$-th level of this diagram corresponds to algebra $M_{F(n)}(\mathbb{C}) \oplus M_{F(n+1)}(\mathbb{C})$, where $F(n)$ is the $n$-th Fibonacci number. It can be shown that this algebra, called *Fibonacci algebra*, is a simple, unital AF algebra with a unique trace that is not a UHF algebra.

(Hint: Given a Bratteli diagram of $A$, identify its subsets whose direct limits are ideals of $A$.)

**4.5.3.** Which algebra corresponds to the Bratteli diagram given in Figure 8?

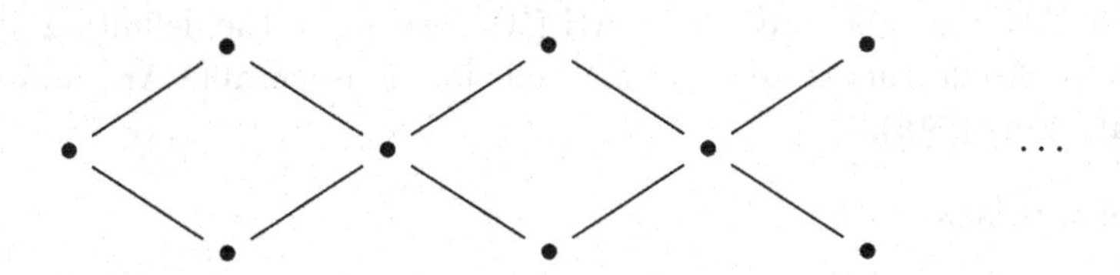

Fig. 8. Diagram for Exercise 4.5.3.

**4.5.4** (Bratteli)**.** By Exercise 4.5.1, to each Bratteli diagram $\mathbb{D}$ one can associate the unique AF algebra $A(\mathbb{D})$. Describe the equivalence relation on Bratteli diagrams defined by $\mathbb{D}_1 \mathrel{E} \mathbb{D}_2$ iff $A(\mathbb{D}_1) \cong A(\mathbb{D}_2)$.

**4.5.5.** Construct a Bratteli diagram such that the corresponding AF algebra $A$ is simple and $T(A)$ is affinely homeomorphic to $[0,1]$.

Hint: See Figure 9.

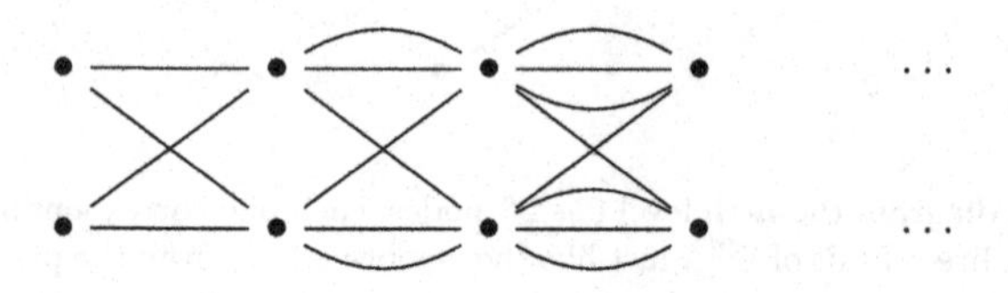

Fig. 9. Hint for Exercise 4.5.5.

## 4.6. *AF algebras*

A C*-algebra is *AF* (approximately finite) if it is a direct limit of finite-dimensional C*-algebras.

If $A$ is a subalgebra of $B$, $X$ is a subset of $B$, and $\varepsilon > 0$, we write

$$X \subseteq_\varepsilon A$$

if every element of $X$ is within $\varepsilon$ of an element of $A$.

**Definition 4.8.** A C*-algebra $B$ is an *LF-algebra* (locally finite) if for every finite $F \subseteq B$ there are a finite-dimensional subalgebra $C$ of $A$ such that $X \subseteq_\varepsilon C$.

Clearly every AF algebra is an LF algebra. Analogously to the case of UHF algebras (Theorem 4.4), the converse is true for separable AF algebras (Exercise 4.7.10). However, this fails for separable AH algebras: direct limits of AH algebras need not be AH ([11]; see §6 for the definition of AH algebras). Analogous statements also fail for nonseparable AF, and even AM, algebras ([30]).

## 4.7. *Exercises*

**4.7.1.** Prove that for $n \in \mathbb{N}$ every $n$-dimensional abelian C*-algebra is isomorphic to $\mathbb{C}^n$ (with the max norm).

A projection $p$ in $A$ is *minimal* if it is nonzero and the only projections $\leq p$ are $p$ and 0.

**4.7.2.** Prove that UHF algebras have no minimal projections.

**4.7.3.** Assume $D$ is a masa (maximal abelian C*-subalgebra) in $A$. Prove that every minimal projection of $D$ is a minimal projection of $A$.

**4.7.4.** Prove that if $A$ is infinite dimensional then every masa in $A$ is infinite-dimensional.

(Hint: Use Exercise 4.7.3. It works even if $A$ has no nontrivial projections.)

In the following two exercises we sketch a direct proof of the instance of Artin–Wederburn theorem for C*-algebras (Lemma 4.6).

**4.7.5.** Assume $A$ is a C*-algebra that is both simple and finite-dimensional. Prove that $A \cong M_n(\mathbb{C})$ for some $n$.

(Hint: First let $D$ be a maximal abelian subalgebra of $A$ and apply Exercise 4.7.1. Then apply Exercise 3.5.10.)

**4.7.6.** Assume $A$ is a finite-dimensional C*-algebra. Prove that it is a direct sum of full matrix algebras.

(Hint: Apply Exercise 4.7.5 to each minimal ideal of $A$.)

Recall that $p \sim q$ implies $\tau(p) = \tau(q)$ for all traces $\tau$. While the converse in general fails, it holds in some well-behaved classes of C*-algebras. The case of UHF algebras was already used earlier.

**4.7.7.** Assume $p$ and $q$ are projections in an AF algebra $A$. Prove that if $\tau(p) = \tau(q)$ for all traces $\tau$ then $p \sim q$.

**4.7.8.** Prove that a Bratteli diagram corresponds to an abelian AF algebra if and only if it is a tree. In this case, the corresponding AF algebra is isomorphic to $C(X)$ where $X$ is the set of all branches through this tree.

**4.7.9.** Prove that a unital abelian algebra is AF if and only if it is of the form $C(X)$ for a zero-dimensional space $X$.

**4.7.10.** Prove that every separable LF algebra is AF.

(Hint: This is similar to the proof in the case of UHF algebras, i.e., that LM implies AM. See Theorem 4.4. All computations are given in detail in [12].)

**4.7.11.** Check that if $A$ is a direct sum of $n$ full matrix algebras then $T(A)$ is affinely homeomorphic to the $n-1$-dimensional simplex.

Given a unital *-homomorphism $\Phi\colon A \to B$ between finite-dimensional C*-algebras, use the above to describe which traces in $T(A)$ are $\Phi^*$-images of traces in $B$ (see Exercise 2.2.10).

**4.7.12.** Prove that the Fibonacci algebra has a unique trace.
(Hint: Exercise 4.7.11.)

## 5. The Functor $K_0$

We are ready to define the classifying invariant for AF-algebras, $K_0(A)$. We shall define it only in the unital case since the general case involves some additional technicalities (see e.g., [68]). We first define the *Murray-von Neumann semigroup* of a C*-algebra $A$, denoted by $V(A)$.

The underlying set of $V(A)$ is $\mathcal{P}(M_\infty(A))/\sim$. (See §2.5.6 for its definition.) One could consider $\mathcal{K} \otimes A$ instead (see §2.5.6). Since $\mathcal{K} \otimes A$ is the completion of $M_\infty(A)$, Lemma 3.3 implies that every projection in $\mathcal{K} \otimes A$ is equivalent to a projection in $M_\infty(A)$.

Note that a projection $p \in M_n(A)$ is Murray-von Neumann equivalent to $\operatorname{diag}(0, p) \in M_{2n}(A)$ (here 0 is the zero matrix in $M_n(A)$). This device of 'moving projections away down the diagonal' is used to define addition in $V(A)$. For projections $p$ and $q$ in $M_\infty(A)$ we can find $n$ such that both $p$ and $q$ belong to $M_n(A)$ and define $[p] \oplus [q]$ in $V(A)$ to be the equivalence class of the projection $\begin{pmatrix} p & 0 \\ 0 & q \end{pmatrix}$ in $M_{2n}(A)$. By conjugating with $\begin{pmatrix} 0 & 1 \\ 1 & 0 \end{pmatrix}$ we see that this projection is equivalent to $\begin{pmatrix} q & 0 \\ 0 & p \end{pmatrix}$. The associativity of $\oplus$ is also easy to check and therefore $\oplus$ defines an operation on $V(A)$ that turns it into an abelian semigroup.

Note that $p + q = r$ implies $[p] \oplus [q] = [r]$, but certainly not vice versa; in particular $p + q$ need not be a projection.

Recall that a *Grothendieck group* of a semigroup $(V, +)$ is defined as follows. On $V^2$ define equivalence relation $\approx$ via $(f, g) \approx (f', g')$ if $f + g' = f' + g$. The addition of equivalence classes is defined coordinatewise, by $[(f_1, g_1)] + [(f_2, g_2)] = [(f_1 + f_2, g_1 + g_2)]$. This construction results in an abelian group.

We define $K_0(A)$ to be the Grothendieck group of $(V(A), \oplus)$. Also, by letting $K_0^+(A)$ be the image of $V(A)$ we provide an ordered group structure. Finally, in the unital case $K_0(A)$ is the ordered abelian group with order unit,

$$(K_0(A), K_0^+(A), [1_A]),$$

where $[1_A]$ denotes the equivalence class of the identity.

A word of caution is due at this point. The ordering on $K_0(A)$ can behave in a very unusual way. In some cases it contains elements $p$ and $q$

such that $np > (n+1)q$ for some $n$, while $p \not\succ q$. This behaviour can be exhibited by analyzing nontrivial vector bundles (cf. Example 3.10) and it is a key ingredient in the construction of many pathological C*-algebras (see §6.0.1).

We shall use an abbreviation and an abuse of notation and write $K_0(A)$ instead of $(K_0(A), K_0(A)^+, [1_A])$ whenever there is no danger of confusion.

### 5.1. *Computation of $K_0$ in some simple cases*

5.1.1. *$K_0$ of $M_n(\mathbb{C})$* Since $M_n(\mathbb{C}) \otimes \mathcal{K}$ is isomorphic to $\mathcal{K}$, we have that $V(M_n(\mathbb{C}))$ is isomorphic to $(\mathbb{N}, +)$ (with $0 \in \mathbb{N}$). Note that $V(M_n(\mathbb{C}))$ and $V(M_k(\mathbb{C}))$ can be distinguished if one keeps track of the $\sim$-equivalence class of the identity of the algebra, $[1_A]$.

After rescaling, $K_0(M_n(\mathbb{C}))$ becomes $(\mathbb{Z}[1/n], \mathbb{Z}^+[1/n], 1)$.

5.1.2. *$K_0$ of $\mathcal{B}(H)$* We are assuming $H$ is infinite-dimensional. Then we have $[p] + [1] = [p]$ for all $p$, and therefore $K_0(\mathcal{B}(H)) = \{0\}$.

5.1.3. *$K_0$ of the Calkin algebra* Calkin algebra, denoted $\mathcal{C}(H)$, is the quotient $\mathcal{B}(H)/\mathcal{K}(H)$. Gelfand–Naimark–Segal theorem implies that it is a C*-algebra. All nonzero projections in the Calkin algebra are Murray–von Neumann equivalent. This extends to $\mathcal{K} \otimes \mathcal{C}(H)$ and therefore $K_0(\mathcal{C}(H)) = \{0\}$.

5.1.4. *$K_0$ of the CAR algebra* By §5.1.1, for every UHF algebra $A$ we have that $K_0(A)$ is a direct limit of copies of $\mathbb{Z}$, with the positive part being exactly the positive integers. For the CAR algebra, the unit of the $n$-th copy of $\mathbb{Z}$ is $2^n$. Therefore $K_0(M_{2^\infty})$ is isomorphic to the group of dyadic rationals, $\{k/2^n : k \in \mathbb{Z}, n \in \mathbb{N}\}$, with $[1_A] = 1$ and the positive part being exactly the positive dyadic rationals.

5.1.5. *$K_0$ of other UHF algebras* Let $A$ be a UHF algebra corresponding to the generalized integer $\mathbf{k}$. The above argument shows that

$$K_0(A) = \{m/k : m \in \mathbb{Z} \text{ and } k \text{ divides } \mathbf{k}\}$$

with $[1_A] = 1$ and the natural positive part.

Note that for projections $p \in A$ the equivalence class $[p]$ exactly corresponds to the normalized trace $\mathrm{tr}(p)$.

5.1.6. *$K_0$ of a *-homomorphism* If $\Phi: A \to B$ is a *-homomorphism then by Exercise 3.5.2 it extends to a semigroup homomorphism from $V(A)$ to $V(B)$. If $A, B$ and $\Phi$ are unital then $\Phi$ can be canonically extended to a *-homomorphism from $M_\infty(A)$ to $M_\infty(B)$ and we therefore have a homomorphism

$$K_0(\Phi): (K_0(A), K_0(A)^+, [1_A]) \to (K_0(B), K_0(B)^+, [1_B]).$$

If $\Phi$ is an isomorphism then so is $K_0(\Phi)$, but the converse may fail. The following lemma shows that $K_0$ is continuous with respect to inductive limits.

**Lemma 5.1.** *If $A_i$, for $i \in I$ and $F_{ij}$, for $i < j$ is a unital directed system and $A = \lim_i A_i$ then $K_0(A_i)$, for $i \in I$ and $K_0(F_{ij})$ for $i < j$ in $I$ is a directed system and $K_0(A) = \lim_i K_0(A_i)$.*

*Proof.* The first claim is an immediate consequence of the above discussion. By Lemma 3.3 every projection in $A \otimes \mathcal{K}$ is Murray-von Neumann equivalent to an image of a projection in some $A_i \otimes \mathcal{K}$ and therefore $V(A) = \lim_i V(A_i)$ and $K_0(A) = \lim_i K_0(A_i)$. □

## 5.2. *Exercises*

**5.2.1.** Describe $K_0(A \oplus B)$ in terms of $A$ and $B$.

**5.2.2.** Use Exercise 5.2.1 and Lemma 5.1 to show that $K_0$ of an AF algebra is a direct limit of groups of the form $\mathbb{Z}^{n(i)}$, for $n(i) \in \mathbb{N}$, with their natural ordering.

## 5.3. *Cancellation property*

An abelian semigroup $(S, +)$ has the *cancellation property* if $x+y = z+y$ implies $x = z$. This is equivalent to stating that in the Grothendieck group of $(S, +)$ no two distinct elements of $S$ belong to the same equivalence class. A C*-algebra $A$ has the cancellation property if its Murray-von Neumann semigroup has cancellation property.

**Lemma 5.2.** *A direct limit of algebras with cancellation property has cancellation property.*

*Proof.* Assume $A = \lim_n A_n$ does not have cancellation property. Not having cancellation property is witnessed by the following objects in $\mathcal{K} \otimes A$: three projections, $p$, $q$ and $r$, one partial isometry, $v$, such that $vv^* = p+r$ and $v^*v = q+r$ and the absence of a partial isometry $w$ such that $ww^* = p$ and $w^*w = q$. By Lemma 3.3, we may assume $p, q$ and $r$ all belong to

$M_n(A_n)$ for a large enough $n$. By Lemma 3.5, we may assume that $v$ also belongs to $M_n(A_n)$. Therefore $A_n$ does not have cancellation property. □

**Lemma 5.3.** *Every AF algebra has cancellation property.*

*Proof.* By Lemma 5.2 it suffices to show that every finite-dimensional C*-algebra $A$ has cancellation property. This can be proved in a variety of ways. For example, $V(A)$ is isomorphic to the free abelian semigroup with $k$ generators where $k$ is the number of full matrix direct summands of $A$, and this semigroup does not have cancellation property. □

In the category of $K_0$-groups homomorphisms are group homomorphisms that preserve both positivity and the unit.

Note that if $p, q$ and $r$ are projections in $M_\infty(A)$ then $p+q=r$ implies $[p]+[q]=[r]$. The converse is false since $p+q$ need not be a projection. The following weak converse indicates why $K_0$ is defined in $\mathcal{P}(M_\infty(A))$ instead of $\mathcal{P}(A)$.

**Lemma 5.4.** *If $p$ and $q$ are projections in $M_\infty(A)$ then there exists $q' \in M_\infty(A)$ such that $q \sim q'$ and $p+q$ is a projection.*

*Proof.* Let $n$ be such that $p$ and $q$ both belong to $M_n(A)$ and identify them with $\begin{pmatrix} p & 0 \\ 0 & 0 \end{pmatrix}$ and $\begin{pmatrix} q & 0 \\ 0 & 0 \end{pmatrix}$, respectively, in $M_{2n}(A)$. Let $v$ be $\begin{pmatrix} 0 & 1 \\ 1 & 0 \end{pmatrix}$ (a block-matrix in $M_{2n}(A)$). Then $q' = (\operatorname{Ad} v)q = \begin{pmatrix} 0 & 0 \\ 0 & q \end{pmatrix}$ is as required. □

**Lemma 5.5.** *Assume $A$ is finite-dimensional and $B$ has cancellation property. Then every homomorphism of preordered abelian groups with order unit $\phi\colon (K_0(A), K_0(A)^+, [1_A]) \to (K_0(B), K_0(B)^+, [1_B])$ is of the form $K_0(\Phi)$ for a unital *-homomorphism $\Phi\colon A \to B$.*

*Proof.* Choose $n$ such that there are minimal projections $p_i$, for $i < n$, in $A$ satisfying $\sum_{i<n} p_i = 1$. Since $\phi$ is positive, Lemma 5.4 implies that for each $i$ we can choose a projection $q_i \in M_\infty(B)$ such that $\phi([p_i]) = [q_i]$. Since $\phi$ is a homomorphism we have that $\sum_{i<n}[q_i] = [1_B]$. (Here $1_B$ denotes the unit of $B$.) This means that for some projection $r$ we have $\operatorname{diag}(q_0, q_1, \dots, q_{n-1}, r) \sim \operatorname{diag}(1_B, r)$, but by the cancellation property of $B$ we have $\operatorname{diag}(q_0, q_1, \dots, q_{n-1}) \sim 1_B$. Denote the right hand side by $q$ and let $v$ be a partial isometry in $M_\infty(B)$ such that $vqv^* = 1_B$. Then $r_i = vq_iv^*$ is a projection in $M_\infty(B)$ such that $\sum_{i<n} r_i = 1_B$. Therefore $r_i$, for $i < n$, are orthogonal projections in $B$.

Now we can shed the scaffolding provided by $M_\infty(B)$ and work in $B$. Also, $r_i \sim r_j$ if and only if $p_i \sim p_j$. Let $\approx$ be an equivalence relation on $n$ defined by $i \approx j$ if and only if $p_i \sim p_j$ in $A$. For each such pair choose a partial isometry $w_{ij} \in A$ such that

(1) $w_{ij}w_{ij}^* = p_i$ and $w_{ij}^* w_{ij} = p_j$, and
(2) $w_{ij}w_{kl} = w_{il}\delta_{jk}$

for all $i, jk, l$.

We can recursively choose partial isometries $v_{ij}$ for all $i \approx j$ that, together with $r_i$ for $i < n$, generate a finite-dimensional unital subalgebra of $B$ isomorphic to $A$ and satisfy equalities corresponding to the above. Now define $\Phi\colon A \to B$ by $\Phi(p_i) = r_i$, $\Phi(w_{ij}) = v_{ij}$ for $i \approx j$ and extend it linearly. Then $\Phi\colon A \to B$ is a unital *-homomorphism and $K_0(\Phi) = \phi$. □

### 5.4. *Classification of AF algebras*

A *dimension group* is a direct limit of ordered groups with order unit of the form $(\mathbb{Z}^n, (\mathbb{Z}^+)^n, [e])$ where $e \in (\mathbb{Z}^=)^n$. It is an easy consequence of Lemma 5.1 that $K_0$ of a separable unital AF algebra is a dimension group. The converse is also not difficult to prove: a countable ordered group is a dimension group if and only if it is equal to $K_0$ of a separable unital C*-algebra (actually one still has the equivalence if both separability and countability are dropped). A first-order characterization of dimension groups was given by Effros–Handelman–Shen (see [16] or [12]).

**Lemma 5.6.** *Assume* $A = \lim_n A_n$ *and* $B = \lim_n B_n$. *Also assume* $\phi_n\colon A_n \to B_n$ *and* $\psi_n\colon B_n \to A_{n+1}$ *are *-homomorphisms such that the diagram in Figure 10 commutes.*

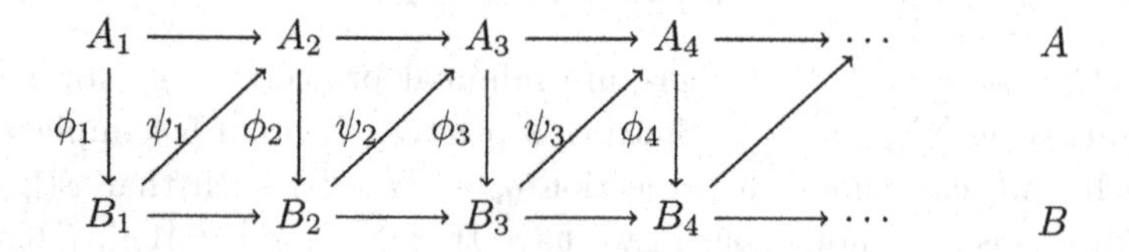


Fig. 10.

*Then* $A$ *is isomorphic to* $B$.

*Proof.* Let $F_n\colon A_n \to A$ denote the canonical *-homomorphism for each $n \in$

$\mathbb{N}$. Since the diagram commutes,

$$\phi(a) = \phi_n(F_n a) \text{ if } a \in A_n$$

is a well-defined *-homomorphism from a dense subset of $A$ into $B$. Since each $\phi_n$ is a contraction, so is $\phi$ and therefore $\phi$ extends to a *-homomorphism from $A$ into $B$. One analogously defines $\psi\colon B \to A$ as a direct limit of $\psi_n$ for $n \in \mathbb{N}$. Since $\psi \circ \phi$ is $\mathrm{id}_A$ and $\phi \circ \psi$ is $\mathrm{id}_B$, we conclude that $\phi$ and $\psi$ are *-isomorphisms. □

Two remarks regarding Lemma 5.6 are in order. First, maps $\phi_n$ and $\psi_n$ are not required to be isomorphisms or even injections. All we need is commutation of the diagram. Second, this lemma is about direct limits of arbitrary structures.

Proof of the following lemma is very similar to the proof of Lemma 5.6.

**Lemma 5.7.** *Assume $A = \lim_n A_n$ and $B = \lim_n B_n$. Also assume $\phi_n\colon A_n \to B_n$ are *-homomorphisms such that the diagram in Figure 11 commutes.*

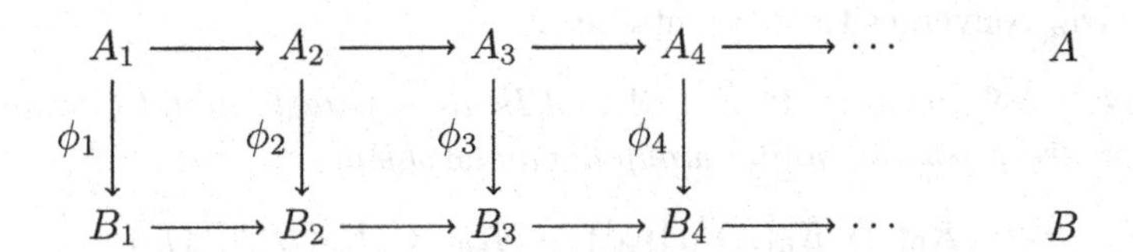


Fig. 11.

*Then there is a *-homomorphism from $A$ to $B$ that extends $\bigcup_n \phi_n$.* □

The assumption that the algebra be unital is not needed in Theorem 5.8 or in Theorem 5.9. However, the definition of $K_0$ of a nonunital algebra is not given in these notes.

**Theorem 5.8** (Elliott, [18]). *Two separable unital AF algebras are isomorphic if and only if their ordered $K_0$ groups are isomorphic.*

*Proof.* Write $A = \lim_n A_n$ and $B = \lim_n B_n$, where all $A_n$ and all $B_n$ are finite dimensional and all connecting maps are unital. Then $K_0(A) = \lim_n K_0(A_n)$ and $K_0(B) = \lim_n K_0(B_n)$. Let $\phi\colon K_0(A) \to K_0(B)$ be an ordered unit group isomorphism. Each $K_0(A_n)$ is finitely generated, and therefore we can go to a subsequence of $A_n$ and $B_n$ so that, after re-enumerating, we have that $\phi$ sends $K_0(A_n)$ into $K_0(B_n)$ and $\phi^{-1}$ sends $K_0(B_n)$ into $K_0(B_{n+1})$ for all $n$.

The plan is to apply Lemma 5.6 with appropriately chosen *-homomorphisms $\Phi_n$ and $\Psi_n$, for $n \in \mathbb{N}$. By applying Lemma 5.5 for each $n$ in both directions we can recursively choose a *-homomorphism $\Phi_n \colon A_n \to B_n$ and a *-homomorphism $\Psi'_n \colon B_n \to A_n$ which lift the corresponding maps between $K_0$ groups. Assume $\Phi_n$ and $\Psi'_n$ were chosen. We shall modify $\Psi'_n$ to make the triangle between $A_n, B_n$ and $A_{n+1}$ in the diagram from Lemma 5.6 commute. We have two *-homomorphisms from $A_n$ to $A_{n+1}$, namely $F_{n,n+1} \colon A_n \to A_{n+1}$ given by the directed system and $\Psi'_n \circ \Phi_n$. Lemma 4.7 applies to this pair and gives a unitary $u_n$ in $A_{n+1}$ such that $\operatorname{Ad} u_n \circ \Psi'_n \circ \Phi_n = F_{n,n+1}$. Then $\Psi_n = \operatorname{Ad} u \circ \Psi'_n$ is as required. Once all $\Phi_n$ and $\Psi_n$ are chosen to make the whole diagram commute, Lemma 5.6 implies $A \cong B$. □

The above proof shows a bit more. Let us say that two *-homomorphisms $\Phi_j \colon A \to B$, for $j = 0, 1$ are *approximately unitarily equivalent* if there is a sequence of unitaries $u_n$, for $n \in \mathbb{N}$, in $B$ such that $\Phi_0 \circ \operatorname{Ad} u_n$ converges to $\Phi_1$ pointwise.

**Theorem 5.9** (Elliott, [18]). *If $A$ and $B$ are separable unital C*-algebras then for every positive unital group homomorphism*

$$\Phi \colon (K_0(A), K_0(A)^+, [1_A]) \to (K_0(B), K_0(B)^+, [1_B])$$

*there exists a unital *-homomorphism $\Phi \colon A \to B$ such that $\phi = K_0(\Phi)$. Moreover, $K_0(\Phi) = K_0(\Psi)$ if and only if $\Phi$ and $\Psi$ are approximately unitarily equivalent. Moreover, if $\phi$ is an isomorphism then so is $\Phi$.*

*Proof.* The proof of the first statement is similar to the above and uses Lemma 5.7. The approximate unitary equivalence of *-homomorphisms whose $K_0$ coincide is a consequence of Lemma 4.7 applied along the finite stages of the diagram. The last sentence is Theorem 5.8. □

### 5.5. *Exercises*

**5.5.1.** Prove that $K_0(A)$ is countable if $A$ is separable.
(Hint: Lemma 3.11.)

**5.5.2.** Prove that if $A$ has cancellation property and $p, q$ are projections in $A$ such that $p \sim q$, then $1 - p \sim 1 - q$. Find a C*-algebra in which this is not true, and which therefore does not have the cancellation property.

(Hint: For the second part use Example 3.9 (1).)

**5.5.3.** Prove that we could have defined $V(A)$ and $K_0(A)$ in $\mathcal{P}(A\otimes\mathcal{K})$ (see §2.5.6) instead of $\mathcal{P}(M_\infty(A))$.

(Hint: Lemma 3.11.)

**5.5.4.** Compute $K_0$ of the UHF algebra corresponding to the generalized integer $\prod_j p_j^{k(j)}$.

**5.5.5.** Prove that $K_0$ of the Fibonacci algebra is (with $s = (1+\sqrt{5})/2$) $(\mathbb{Z}^2, \{(m,n) : sm + n \geq 0\}, (1,0))$.

**5.5.6.** Prove the assertions made in §5.1.6.

**5.5.7.** A *state* on a preordered abelian group with order unit $(G, G^+, [1])$ is a homomorphism $\chi\colon G \to \mathbb{R}$ such that $\chi[G^+] \subseteq \mathbb{R}^+$ and $\chi(1) = 1$. Prove that a trace $\tau \in T(A)$ defines a state of $(K_0(A), K_0(A)^+, [1_A])$.

**5.5.8** (Approximate intertwining, aka Elliott intertwining)**.** Assume $A = \lim_n A_n$ and $B = \lim_n B_n$. Furthermore assume $F_n \subseteq A_n$, for $n \in \mathbb{N}$, is an increasing sequence of finite subsets of $A$ with dense union, and $G_n \subseteq B_n$, for $n \in \mathbb{N}$ is an increasing sequence of finite subsets of $B$ with dense union.

Finally, assume $\phi_n\colon A_n \to B_n$ and $\psi_n\colon B_n \to A_{n+1}$ are *-homomorphisms such that the $n$-th triangle of the diagram in Figure 12 commutes up to $2^{-n}$ on both $F_n$ and $G_n$.

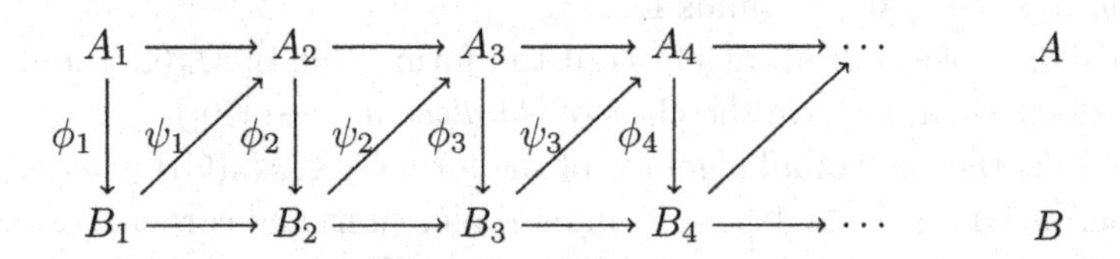


Fig. 12. An approximate intertwining argument.

Prove that $A \cong B$.

**5.5.9.** Prove that $K_0(C([0,1], A)$ is isomorphic to $K_0(A)$ for all $A$.

## 6. Elliott's Program

Around 1990, George Elliott conjectured that separable, unital, nuclear, simple C*-algebras can be classified by K-theoretic invariants. Nuclear C*-algebras can be defined in several equivalent but apparently rather different ways (e.g., Definition 2.9), but for our purposes the actual definition is irrelevant. See e.g., [4] or [70].

The Elliott invariant of an algebra $A$ is the sixtuple

$$\mathrm{Ell}(A) = (K_0(A), K_0(A)^+, [1_A]_0, K_1(A), T(A), r_A\colon T(A) \to S(K_0(A)).$$

We have already defined $K_0$ group and the tracial space $T(A)$. Since $p \sim q$ implies $\tau(p) = \tau(vv^*) = \tau(v^*v) = \tau(q)$, every trace $\tau$ on $A$ defines a state (unital, positive, additive map into $\mathbb{R}$) on $K_0(A)$, and $r_A$ is the coupling map that associates states to traces (see Exercise 5.5.7). $K_1(A)$ is an another countable (in case when $A$ is separable) abelian group (see [70]).

Fix a family of C*-algebras $\mathbb{B}$. We would like to consider it as the family of *building blocks* of a larger family of C*-algebras. This can be formalized in (at least) two ways.

**Definition 6.1.** A C*-algebra is an *A*$\mathbb{B}$*-algebra* if it is a direct limit of a directed system of algebras in $\mathbb{B}$.

**Example 6.2.** (1) If $\mathbb{B}$ consists of all full matrix algebras $M_n(\mathbb{C})$ for $n \in \mathbb{N}$, then A$\mathbb{B}$-algebras are UHF algebras.

(2) if $\mathbb{B}$ consists of all finite-dimensional C*-algebras, then A$\mathbb{B}$-algebras are the *approximately finite*, or *AF*, algebras.

(3) Recall that $\mathbb{T} = \{z \in \mathbb{C} : |z| = 1\}$. Let $\mathbb{B}$ be the class of all direct sums of algebras of the form $C(\mathbb{T}, M_n(\mathbb{C}))$. Then we arrive at the class of *AT algebras* (see [70]; T stands for $\mathbb{T}$).

(4) If $\mathbb{B}$ consists of all algebras of the form $C([0,1], M_n(\mathbb{C}))$ and their direct sums, then we have the class of *AI algebras* (see [70]).

(5) Take the class of all algebras of the form $C(X, M_n(\mathbb{C}))$ where $X$ is a compact metric space and close it under direct sums and corners (§2.5.9) to obtain $\mathbb{B}$. Then A$\mathbb{B}$ is the class of *AH algebras* (H stands for 'homogeneous') (see [70]).

After several spectacular successes ([65], [54], [22], [55], [21], [70]), counterexamples to Elliott's conjecture were found by Rørdam and Toms ([71] and [80]). While these examples rule out functorial classification as conjectured by Elliott, they do not give information about the descriptive complexity of the isomorphism relation of separable nuclear simple unital C*-algebras.

In recent years Elliott's program has been revitalized by influx of new ideas, including invariants such as the Cuntz semigroup (see [6], [9]), regularity properties such as $\mathcal{Z}$-stability ([87]), and a technical tour de force

([23]), among other developments. See [20] for a survey and the upcoming [88] for the current state of the art.

6.0.1. *Failure of cancellation* The following example was first used by Villadsen ([83]) and it appears in one form or another in all known counterexamples to Elliott's program. Recall that $S^2$ is the unit sphere in three-dimensional (real) Euclidean space and consider the C*-algebra $A = C(S^2, M_2(\mathbb{C}))$ (see §2.5.8). By Exercise 2.6.9, projections in $A$ are continuous maps from $S^2$ into $\mathcal{P}(M_2(\mathbb{C}))$, the space of projections in $M_2(\mathbb{C})$. Apart from the 'trivial' projections corresponding to constant maps, this algebra has nontrivial projections. Recall that $\mathbb{C}P^1$ is the *complex projective space*, the space of all lines in $\mathbb{C}^2$. It is homeomorphic to the space $\mathcal{P}(M_2(\mathbb{N}))$ of all 1-dimensional projections in $M_2(\mathbb{C})$. It is also homeomorphic to $S^2$. The *Bott projection* $p$ in $A$ corresponds to a natural homeomorphism of $S^2$ onto $\mathcal{P}(M_2(\mathbb{C}))$ (i.e., to the *Hopff vector bundle*). It is not Murray-von Neumann equivalent to the trivial rank one projection, but there is a partial isometry $v \in A$ such that $p + vpv^* = 1$.

Villadsen used Bott projection and a clever direct limit construction to construct an AH algebra $A$ such that $K_0(A)$ contains a non-positive element $x$ such that $nx$ is positive for some $n$ ([83]).

## 6.1. *Exercises*

Exercises given below illustrate basic constructions of AH algebras. See [82] for analysis of the classification problem for such algebras. In the following exercises $X$ and $Y$ are compact metric spaces, and the property of being metric is mostly unnecessary.

**6.1.1.** Prove that $T(C(X, M_n(\mathbb{C}))$ is affinely homeomorphic to $\mathcal{P}(X)$, the space of Borel probability measures on $X$ (cf. Exercise 3.7.3), where to a measure $\nu$ one associates trace

$$\tau_\nu(f) = \int \operatorname{tr}(f(x))d\nu(x).$$

**6.1.2.** Let $A_n = C(X, M_{2^n}(\mathbb{C}))$. If $x_{n,j}$, for $j < n$ are points in $X$ define a unital *-homomorphism $\Phi_n \colon A_n \to A_{n+1}$ by the block matrix

$$\Phi_n(f) = \operatorname{diag}(f, f(x_{n,0}), f(x_{n,1}), \dots, f(x_{n,n-1})).$$

(i) Prove that the direct limit of this system $A$ is simple if for every $m$ the set $\bigcup_{n \geq m}\{x_{n,j} j < n\}$ is dense in $X$.

(ii) With $\Phi^*$ as in Exercise 2.2.10 prove that for a trace $\tau \in T(A_{n+1})$ $\tau' = \Phi^*(\tau)$ satisfies

$$\tau'(f) = \frac{1}{2}\tau(f) + \frac{1}{2n}\sum_{j<n} \operatorname{tr}(f(x_{n,j})).$$

Assume $m = nk$ and $\phi_j : Y \to X$, for $j < k$, are continuous maps. Then a unital *-homomorphism $\Phi\colon C(X, M_n(\mathbb{C})) \to C(Y, M_m(\mathbb{C}))$ is defined by

$$\Phi(f) = \operatorname{diag}(f \circ \phi_0, f \circ \phi_1, \dots, f \circ \phi_{k-1}).$$

A *-homomorphism of this form is a *standard *-homomorphism with characteristic functions $\phi_j$, for $j < n$.*

**6.1.3.** Given a standard *-homomorphism $\Phi\colon A \to B$ as above, describe the map $\Phi^*\colon T(B) \to T(A)$.

**6.1.4.** Assume $A_n = C([0,1], M_{k(n)}(\mathbb{C}))$ and $\Phi_n\colon A_n \to A_{n+1}$ is a standard unital *-homomorphism and let $A = \lim_n A_n$. Prove that $K_0(A)$ is isomorphic (as a preordered abelian group with order unit) with $K_0$ of the UHF algebra $\lim_n M_{k(n)}(\mathbb{C})$.

(Hint: Recall that the projections in $C([0,1], M_n(\mathbb{C}))$ correspond to continuous functions from $[0,1]$ into $\mathcal{P}(M_n(\mathbb{C}))$, and therefore have well-defined rank. In order to prove that $p \sim q$ if and only if their ranks coincide, one uses Lemma 3.5 together with the compactness and homotopic triviality of $[0,1]$.)

**6.1.5.** Let $A_n = C([0,1], M_{2^n}(\mathbb{C}))$ and let $\Phi_n\colon A_n \to A_{n+1}$ be the standard *-homomorphism with characteristic functions $\phi_0(x) = x/2$ and $\phi_1(x) = (1+x)/2$. Let $A = \lim_n A_n$.

(i) Prove that $A$ is simple.

(ii) Compute $K_0(A)$.

(iii) Prove that $A$ has a unique trace.

By the last exercise, Exercise 5.5.7, and the fact that $K_1$ of both $A$ and the CAR algebra is trivial, one has that $\mathrm{Ell}(A) \cong \mathrm{Ell}(M_{2^\infty})$. It is then a consequence of Elliott's classification of simple AI algebras ([21]) that $A$ is isomorphic to the CAR algebra.

**6.1.6** (Dimension drop algebras). Fix $m$ and $n$ in $\mathbb{N}$ and identify $M_{mn}(\mathbb{C})$ with $M_m(\mathbb{C}) \otimes M_n(\mathbb{C})$ (Exercise 2.6.11). Consider the following subalgebra of $C([0,1], M_{mn}(\mathbb{C}))$.

$$\mathcal{Z}_{m,n} = \{f : f(0) \in M_m(\mathbb{C}) \otimes 1_n \text{ and } f(1) \in 1_m \otimes M_n(\mathbb{C})\}$$

Prove that the only projections in $\mathcal{Z}_{m,n}$ are 0 and 1 (i.e., "$\mathcal{Z}_{m,n}$ is projectionless") if and only if $m$ and $n$ are relatively prime.

**6.1.7.** Prove that $T(\mathcal{Z}_{m,n})$ is affinely homeomorphic to $\mathcal{P}([0,1])$, the space of probability measures on $[0,1]$.
(Hint: See Exercise 6.1.1.)

**6.1.8.** Construct a direct limit of projectionless dimension drop algebras that is simple and has a unique trace.
(Hint: For the trace part keep in mind Exercise 2.2.10(2).)

The solution to Exercise 6.1.8 is uniquely defined up to the isomorphism. It is the notorious *Jiang–Su algebra* $\mathcal{Z}$ ([45]). Its Elliott invariant is equal to the Elliott invariant of $\mathbb{C}$, and moreover $\mathrm{Ell}(\mathcal{Z} \otimes A) = \mathrm{Ell}(A)$ for all well-behaved (as well as many misbehaved) C*-algebras $A$. A C*-algebra $A$ is *$\mathcal{Z}$-stable* if $A \otimes \mathcal{Z} \cong A$, and by the above one can only hope to classify $\mathcal{Z}$-stable algebras by their Elliott invariants. By remarkable results of Wilhelm Winter, this is true in a number of cases ([82], [86], [87]).

## 7. Abstract Classification

Main references for the remaining sections are [33] and [32]. Recall that a topological space $X$ is *Polish* if it is separable and completely metrizable. A subset of $X$ is *analytic* if it is a continuous image of a Borel set in some Polish space. An equivalence relation $E$ on $X$ is analytic if it is an analytic subset of $X^2$.

The theory of abstract classification can be traced back to the work of Mackey on classification of representations of locally compact metrizable groups, and in particular to the following.

**Definition 7.1** (Mackey). An equivalence relation $E$ on $X$ is *smooth* if there is a Borel-measurable $f\colon X \to \mathbb{R}$ such that

$$x\, E\, y \text{ iff } f(x) = f(y).$$

**Example 7.2.** Similarity of $n \times n$ complex Hermitian matrices is smooth. This is because one can associate to $M$ the list of its eigenvalues (in the increasing order, with multiplicities).

Smooth equivalence relations are effectively classifiable.

**Example 7.3.** Classification of rank 1 torsion-free abelian groups.

These are exactly the subgroups of $\mathbb{Q}$. To every such group $\Gamma$ we can associate a generalized integer $\prod_j p_j^{k(j)}$ (see §4.1) as follows. Choose $a \in \Gamma$ and consider $\{n : (\exists b \in \Gamma) nb = a\}$. This set is the set of all divisors of a generalized integer $\mathbf{k}(\Gamma, a)$.

Note that $\mathbf{k}$ depends on the choice of $a$, but only up to a finite factor. A straightforward recursive construction shows that $\Gamma_1$ and $\Gamma_2$ are isomorphic if and only if for some (equivalently, all) $a_1 \in \Gamma_1$ and $a_2 \in \Gamma_2$ the corresponding generalized integers coincide up to a finite factor, i.e., there exist nonzero $m_1$ and $m_1$ in $\mathbb{N}$ such that $m_1\mathbf{k}(\Gamma_1, a_1) = m_2\mathbf{k}(\Gamma_2, a_2)$.[d]

One could prove that the equivalence relation in Example 7.3 is not smooth. However, it is generally accepted as a simple and natural classification result. In order to compare the complexity of equivalence relations on Polish spaces (and therefore of different classification problems), [36] and [40] have independently introduced the following definition

**Definition 7.4.** Assume $E, F$ are equivalence relations on Polish spaces $X, Y$, respectively. Then $E$ is *Borel reducible* to $F$, or $E \leq_B F$, if there is a Borel-measurable map $f \colon X \to Y$ such that

$$x \, E \, y \Leftrightarrow f(x) \, F \, f(y).$$

Note that $E \leq_B = \mathbb{R}$ if and only if $E$ is smooth. The relation $E \leq_B F$ can be interpreted in the following ways.

(1) Borel cardinality of $X/E$ is $\leq$ than the Borel cardinality of $Y/F$.
(2) Classification problem for $E$ is simpler than the classification problem for $F$.
(3) $F$-equivalence classes are complete invariants for $E$-equivalence classes.

**Thesis 7.5.** *Almost all classical classification problems deal with analytic equivalence relations on Polish spaces.*

**Thesis 7.6.** *In almost all cases, the space of invariants has a Polish topology and the computation of invariants is given by a Borel-measurable function.*

In order to verify a particular instance of these two theses, one needs to

(i) Define a Polish space $X$ whose elements correspond to objects that need to be classified,

[d] We consider generalized integers as formal infinite products.

(ii) Define a Polish space $Y$ whose elements correspond to the intended invariants,
(iii) Check that the map that associates invariant $I(x)$ to each object $x \in X$ is Borel.

Once all three steps are verified, the particular classification problem is subject to the well-developed abstract theory of classification (e.g., [44] and [15]), and one can proceed to estimate the complexity of this classification problem.

## 7.1. *Effros Borel space*

A *standard Borel space* is a pair $(X, \Sigma)$ where $X$ is a set and $\Sigma$ is a $\sigma$-algebra of subsets of $X$ with the property that $X$ carries a Polish topology $\tau$ such that $\Sigma$ is the corresponding $\sigma$-algebra of Borel sets. The actual topology on $X$ is irrelevant for our purposes and by a classical result of Kuratowski all uncountable Polish spaces are Borel-isomorphic. Therefore, considering a standard Borel space instead of a Polish space as the ambient space for our classification problems makes our setting more canonical. However, if the topology on $X$ is particularly natural we shall use it.

7.1.1. *Spaces of countable structures* Fix any countable signature $\sigma$. The set $\mathcal{M}_\sigma$ of all countable models of $\sigma$ can naturally be identified with a subset of $\mathcal{P}(\mathbb{N})$, and the latter carries the compact metric topology. For details see [3]. A case of particular interest for us is when $\sigma$ is the signature of ordered groups with order unit. This is worked out in [32].

The isomorphism relation on this space is analytic. To see this, note that the space of all triples $(A, B, f)$ where $A$ and $B$ are in $\mathcal{M}_\sigma$ and $f\colon \mathbb{N} \to \mathbb{N}$ is an isomorphism between $A$ and $B$ is closed. The set $\{(A, B) \in \mathcal{M}_\sigma : A \cong B\}$ is the projection of this set of triples and therefore analytic.

For example, the space $\mathbf{G}$ of preordered countable groups with order unit looks as follows. The space $\mathbf{G}$ consists of all triples $(F, G, H)$ which code $(K, K^+, [1])$ as follows. We may assume that the underlying set of $K$ is $\mathbb{N}$, and $F \subseteq \mathbb{N}^3$ is the set $\{(m, n, k) : m +_K n = k\}$. Also $G = K$ and $H = 1_K$.

7.1.2. *Compact metric spaces* Every compact metric space is homeomorphic with a subspace of the Hilbert cube $[0,1]^{\mathbb{N}}$. Thus the space $K([0,1]^{\mathbb{N}})$ of all closed subsets of $[0,1[^{\mathbb{N}}$ is the space of all compact metric spaces. The Hausdorff metric on this space,

$$d_H(K, L) = \inf_{\varepsilon \geq 0} (K \subseteq_\varepsilon L \text{ and } L \subseteq_\varepsilon L)$$

turns it into a compact metric space.

The homeomorphism relation on this space is analytic. To see this, identify a homeomorphism $f$ between $K$ and $L$ with its graph. This graph is a closed subset of the square and the set of all triples $(K, L, f)$ such that $K$ and $L$ are compact subsets of $[0,1]^{\mathbb{N}}$ and $f\colon K \to L$ is (with the above identification) a closed subset of the compact metric space $K([0,1]^{\mathbb{N}})^2 \times K(([0,1]^{\mathbb{N}})^2)$. Like in the previous example, its continuous image is analytic.

7.1.3. *Separable Banach spaces* In order to define a Borel space of all separable Banach spaces we shall need some lemmas. All Banach spaces are real, although the complex case neither involves any additional complexity nor brings additional simplicity (as in the case of C*-algebras).

**Lemma 7.7.** *Every separable Banach space is isometric with a subspace of $C(\{0,1\}^{\mathbb{N}})$ of all continuous functions from the Cantor space into $\mathbb{R}$.*

*Proof.* This is related to Exercise 2.2.10. A Banach space $X$ is isometric to a subspace of a Banach space $Y$ if and only if there is an affine continuous surjection $f$ from the unit ball of $Y^*$ onto the dual ball of $X^*$. By the Birkhoff–Alaoglu theorem the unit ball of a separable Banach space is compact in the weak*-topology. We can therefore fix a continuous surjection of $\{0,1\}^{\mathbb{N}}$ onto the unit ball of $X^*$ and extend it affinely to the unit ball of $C(\{0,1\}^{\mathbb{N}})$. □

Let $Z = C(\{0,1\}^{\mathbb{N}})$ and let $F(Z)$ denote the space of all closed subsets of $Z$. If $Y$ is not locally compact, the Hausdorff metric used in §7.1.2 will not be separable, In order to define a Borel structure on $F(Z)$ we use the following result.

**Theorem 7.8** (Effros). *If $Y$ is a Polish space then the space $F(Y)$ of closed subsets of $Y$ with respect to the $\sigma$-algebra generated by the sets $\{K \in F(Y) : K \cap U \neq \emptyset\}$ for $U \subseteq Y$ open is a standard Borel space.*

*Proof.* The idea is to fix a metric compactification $Y$ of $X$ and take advantage of the fact that a subspace of a Polish space is Polish if and only if it is $G_\delta$, in both directions. For details see [48, §12.C]. □

Since it is not difficult to check that for a Banach space $Z$ the set $\{Y \in F(Z) : Y \text{ is a linear subspace}\}$ is Effros-Borel, we have a standard Borel space of all separable Banach spaces.

There are (at least) three natural equivalence relations on this space:

(i) (linear) isometry,
(ii) isomorphism (i.e., the existence of a linear homeomorphism) and
(iii) bi-embeddability (the existence of linear isometric embedding of $X$ into $Y$ and a linear isometric embedding of $Y$ into $X$).

It is again not difficult to check that all three relations are analytic. Complexities of these equivalence relations were computed in [57] and [35].

7.1.4. *von Neumann algebras with a separable predual* (A reader not familiar with these may want to skip this paragraph.) Every such von Neumann algebra is isomorphic to a weakly closed subalgebra of $\mathcal{B}(H)$ for a separable complex Hilbert space $X$. Since $\mathcal{B}(H)$ is weakly separable, one can consider this space with respect to the Effros Borel structure. However, unlike in the case of Banach spaces, this space carries a natural Polish topology called *Effros–Maréchal* topology. See [74] and [85] for more.

7.1.5. *Separable C*-algebras* In both cases of Banach spaces and von Neumann algebras there exists a universal separable object ($C(\{0,1\}^{\mathbb{N}})$ and $\mathcal{B}(H)$, respectively hence the Effros Borel structure provides the setting for analysis of classification problems. This is not the case with C*-algebras. By a result of Junge and Pisier ([47]) there is no universal separable C*-algebra. However, there are at least two different Borel spaces of separable C*-algebras.

The following space was defined by Kechris ([49]). It takes advantage of a slight refinement of the GNS theorem (Theorem 2.3), to the effect that every separable C*-algebra is isomorphic to a subalgebra of $\mathcal{B}(H)$ for the separable Hilbert space $H$. The space $\mathcal{B}(H)$ becomes a standard Borel space when equipped with the Borel structure generated by the weakly open subsets. Let

$$\Gamma = \mathcal{B}(H)^{\mathbb{N}},$$

and equip this with the product Borel structure. For each $\gamma \in \Gamma(H)$ we let $C^*(\gamma)$ be the C*-algebra generated by $\gamma$. If we identify each $\gamma \in \Gamma(H)$ with $C^*(\gamma)$, then $\Gamma(H)$ parameterizes all separable C*-algebras acting on $H$. This gives us a standard Borel parameterization of the category of all separable C*-algebras. The relation $\gamma_1 \sim \gamma_2$ is an analytic equivalence relation (see [32] or Exercise 7.4.2).

The above can be considered as the space of concrete separable C*-algebras. One can also define the space of abstract separable C*-algebras and prove that there is a Borel isomorphism between these spaces that respects the corresponding isomorphism relations (see [32]).

### 7.2. *Computation of the Elliott invariant is Borel*

By §7.1.5 and §7.1.1 we have standard Borel spaces $\Gamma$ of separable C*-algebras and $\mathbf{G}$ countable ordered groups with order unit, respectively. The following lemma (and corresponding facts for other C*-algebraic invariants) taken from [32] largely justifies taking the descriptive set theoretic view of Elliott's program.

Let $\Gamma_1 = \{\gamma \in \Gamma : \gamma(0) = 1\}$—the (clearly Borel) space of unital separable C*-algebras.

**Lemma 7.9.** *There is a Borel map* $\mathbb{K}\colon \Gamma_1 \to \mathbf{G}$ *such that* $\mathbb{K}(\gamma)$ *is isomorphic to* $(K_0(C^*(\gamma)), K_0(C^*(\gamma)^+, [1_A])$.

*Sketch of the proof.* The details can be found in [32].

First we need a Borel map $\mathbb{P}$ that sends $\Gamma$ to $\Gamma$ so that $\mathbb{P}(\gamma)$ enumerates a countable dense set of projections in $\mathcal{P}(M_\infty(C^*(\gamma)))$. For simplicity, we shall instead only construct $\mathbb{P}_1$ such that $\mathbb{P}_1(\gamma)$ enumerates a countable dense set of $\mathcal{P}(C^*(\gamma))$. Let $\mathfrak{p}_n$, for $n \in \mathbb{N}$, enumerate all *-polynomials over $\mathbb{Q} + i\mathbb{Q}$ in variables $x_j$, for $j \in \mathbb{N}$, with the property that $\mathfrak{p}_n(x) = \mathfrak{p}_n(x)^*$. Then $\{\mathfrak{p}_j(\gamma) : j \in \mathbb{N}\}$ enumerates a countable dense subset of the set of self-adjoint operators in $C^*(\gamma)$.

Let $f\colon \mathbb{R} \to \mathbb{R}$ be any function whose iterates, $f^n$ for $n \in \mathbb{N}$, uniformly converge to some function $g$ such that $g(x) = 0$ if $x \leq 1/4$ and $g(x) = 1$ if $x \geq 3/4$. If $\mathrm{sp}(\mathfrak{p}_j(\gamma)) \cap [1/4, 3/4] = \emptyset$ then $f^n(\mathfrak{p}_j(\gamma))$ converges to a projection in the norm topology.

A verification that $\mathbb{P}_1(\gamma) = (q_j : j \in \mathbb{N})$ with $q_j = \lim_n f^n(\mathfrak{p}_j(\gamma))$ if this sequence is norm-convergent and $q_j = 0$ otherwise is a Borel map is straightforward. Clearly, $\mathbb{P}(\gamma)$ is an enumeration of a dense set of projections in $C^*(\gamma)$.

By Lemma 3.11, the range of $\mathbb{P}(\gamma)$ intersects all Murray–von Neumann equivalence classes in $\mathcal{P}(M_\infty(C^*(\gamma))$. We need to check that the map $\gamma \mapsto \sim_\gamma$ that associates a binary relation on $\mathbb{N}$ to $\gamma$ such that $m \sim_\gamma n$ if and only if projections $\mathbb{P}(\gamma)(m)$ and $\mathbb{P}(\gamma)(n)$ are Murray-von Neumann equivalent in $M_\infty(C^*(\gamma))$. But this is a consequence of the fact that we have an effective enumeration of a dense subset of $C^*(\gamma)$ akin to $\mathfrak{p}_j(\gamma)$, for $j \in \mathbb{N}$, and Exercise 3.5.8.

Similar arguments show that the operation $\oplus$ and the Groethendieck construction can be effectively defined on $\mathbb{P}(\gamma)$. This gives a Borel map that sends $\gamma$ to an element of the Borel space $\mathbf{G}$ of preordered countable groups with order unit that codes $(K_0(C^*(\gamma)), K_0(C^*(\gamma))^+, [1_{C^*(\gamma)}])$. $\square$

The following was proved in [32].

**Theorem 7.10.** *There is a standard Borel space of Elliott invariants* Ell *and there is a Borel map* $\Phi\colon \Gamma \to \mathrm{Ell}$ *such that* $\Phi(\gamma)$ *represents* $\mathrm{Ell}(C^*(\gamma))$. □

### 7.3. Comparing complexities of analytic equivalence relations

In this section we introduce some important classes of analytic equivalence relations.

7.3.1. *Relation* $E_0$ On $\{0,1\}$ define $x \mathrel{\mathrm{E}_0} y$ if $(\forall^\infty n) x(n) = y(n)$. This equivalence relation is $F_\sigma$ and a simple Baire category argument shows that it is not smooth. By [40], $E_0$ is the minimal non-smooth Borel equivalence relation. (It is, however, not the minimal non-smooth analytic equivalence relation; see Exercise 7.4.3.) It should be noted that the combinatorial essence of this result, called *Glimm–Effros dichotomy*, first appeared in Glimm's [38] as a device for embedding $M_{2^\infty}$ into C*-algebras with non-smooth dual (non-type I C*-algebras) as a subquotient.

7.3.2. *Essentially countable equivalence relations* A Borel equivalence relation all of whose classes are countable is said to be *countable*. A Borel equivalence relation which is Borel-reducible to a countable equivalence relation is *essentially countable*. By the above, $E_0$ is the $\leq_B$-minimal essentially countable equivalence relation. Orbit equivalence relation of the shift action of the free group on infinitely many generators on its power-set the $\leq_B$-maximal essentially countable equivalence relation and by a result of Adams and Kechris the $\leq_B$ ordering on essentially countable equivalence relations has a very rich structure ([1]).

7.3.3. *Countable structures* An equivalence relation $E$ is *classifiable by countable structures* if $E$ is Borel-reducible to the isomorphism relation of countable structures in some countable signature (see §7.1.1). By a result of Hjorth–Kechris–Louveau, this is equivalent to $E$ being Borel-reducible to an orbit equivalence relation of a continuous action of a closed subgroup of $S_\infty$ on $\mathcal{P}(\mathbb{N})$.

The $\leq_B$-maximal analytic equivalence relation in this class is the graph isomorphism relation.

7.3.4. *Orbit equivalence relations* If $G$ is a Polish group that acts continuously on a Polish space $X$, then $E^X_G$ is the orbit equivalence relation,

$$x \mathrel{\mathrm{E}^X_G} y \text{ iff } (\exists g \in G) g.x = y.$$

Being a projection of the closed set $\{(x, y, g) : g.x = y\}$, it is clearly analytic. By a result of Becker and Kechris ([3]), considering Borel actions instead of continuous ones does not result in a larger class of equivalence relations.

This class also has the $\leq_B$-maximal equivalence relation. It is the shift action of the isometry group $G$ of the Urysohn space on the Effros Borel space of its closed subsets, denoted $E_{G_\infty}^{X_\infty}$.

7.3.5. *Turbulence* Being classifiable by countable structures is strictly weaker than being below a Polish group action. This is a result of Hjorth ([43]) and it utilizes the notion of *turbulence*, a dynamical property of group action. This is the main tool used in §8.1.

7.3.6. *The dark side* On $[0,1]^{\mathbb{N}}$ define $x \mathrel{\mathrm{E}_1} y$ if $(\forall^\infty n) x(n) = y(n)$. This equivalence relation is $F_\sigma$. By [50], $E_1$ is an immediate $\leq_B$ successor of $E_0$ and it is not Borel-reducible to any orbit equivalence relation of a Polish group action. Therefore being 'above $E_1$' is a measure of an equivalence relation not being simply classifiable.

In [72] Rosendal isolated the $\leq_B$-maximal $K_\sigma$ equivalence relation, $E_{K_\sigma}$. Its underlying space is the space of nondecreasing functions in $\mathbb{N}^{\mathbb{N}}$ and we let $f \mathrel{\mathrm{E}_{K_\sigma}} g$ if and only if there is $n$ such that $f(j) \leq g(j)+n$ and $g(j) \leq f(j)+n$ for all $n$.

There exists a maximal analytic equivalence relation $E_{\Sigma^1_1}$, isolated in [57]. Both bi-embeddability of separable Banach spaces ([57]) and the isomorphism of separable Banach spaces ([35]) are bireducible with the maximal analytic equivalence relation.

## 7.4. *Exercises*

**7.4.1.** Compare classification of rank 1 torsion-free abelian groups (Example 7.3) with the classification of UHF algebras (§4.1) and nonunital separable direct limits of full matrix algebras (Exercise 4.3.7).

**7.4.2.** Prove that both the isomorphism and bi-embeddability of C*-algebras are analytic subsets of the square of Kechris's space $\Gamma$ (§7.1.5).

**7.4.3.** On the space of compact subspaces of the Hilbert cube defined in §7.1.2 consider the equivalence relation $x \mathrel{\mathrm{E}} y$ iff either both $x$ and $y$ are uncountable or both $x$ and $y$ are countable and homeomorphic. Then show that this relation is neither smooth nor $\geq_B E_0$.

(Hint: Count the equivalence classes.)

The following exercise gives a Borel space of UHF algebras.

**7.4.4.** Prove that there exists a universal separable UHF algebra $Q$. Also prove that not all unital infinite-dimensional subalgebras of this algebra are UHF, but its UHF subalgebras form a Borel subset of $F(Q)$.

**7.4.5.** Do the nonunital case of Exercise 7.4.4.

## 8. Estimating the Complexity of the Isomorphism of C*-Algebras

Upper and lower bounds for the complexity of the isomorphism relation for C*-algebras were proved in [33].

### 8.1. *Turbulence: A lower bound for complexity*

Dynamical properties of Polish group actions affect the complexity of the orbit equivalence relation. We shall use the following classical result as a warmup.

**Proposition 8.1.** *Assume $G \curvearrowright X$ is a continuous action of a Polish group on a Polish space such that*

*(1) every orbit is dense, and*
*(2) every orbit is meager.*

*Then the orbit equivalence relation is not smooth.*

*Proof.* Assume $f\colon X \to \mathbb{R}$ is a Borel map such that $x \mathrm{E}_G^X y$ implies $f(x) = f(y)$. Let $Y$ be a dense $G_\delta$ subset of $X$ such that the restriction of $f$ to $Y$ is continuous. Since $x \mapsto g.x$ is a homeomorphism for all $g \in G$, the set $\{(g,x) : g.x \in Y\}$ is by Kuratowski–Ulam theorem comeager in $G \times X$. Therefore there exists $x \in X$ such that $g.x \in Y$ for comeager many $g \in G$. But this implies that the intersection of the orbit of $x$ with $Y$ is dense in $Y$. Since $f$ is constant on this set, it is by the continuity constant on $Y$. Since all orbits are meager, $f$ cannot be a reduction of $E_G^X$ to $=_{\mathbb{R}}$. □

Let $G \curvearrowright X$ be a continuous action of a Polish group on a Polish space. Fix $x \in X$, a symmetric open neighborhood $U$ of $e_G$ and an open neighbourhood $V$ of $x$. On $V$ define a graph by letting $\{x, y\}$ be an edge if $x \neq y$ and there exists $g \in U$ such that $g.x = y$. Since $U = U^{-1}$, this relation is symmetric. Let $\mathcal{O}(x, U, V)$ be the connected component of $x$ in this graph. Then $\mathcal{O}(x, U, V)$ is the set of points that can be reached from $x$ by taking small steps (smallness being measured by $U$) while staying inside $V$.

**Definition 8.2** (Hjorth, [43]). An action $G \curvearrowright X$ as above is *turbulent* if

(1) every orbit is dense,
(2) every orbit is meager, and
(3) the closure of every local orbit of every $x \in X$ has a nonempty interior.

In the presence of (1) and (2), (3) above is equivalent to the assertion that the closure of every local orbit intersects every $G$-orbit (exercise!). There are some other equivalent reformulations of the definition, and for all purposes it suffices to assume that the set of points $x$ satisfying (3) is comeager (such actions are *generically turbulent*).

**Theorem 8.3** (Hjorth, [43]). *If the action $G \curvearrowright X$ is turbulent then $E_G^X$ is not classifiable by countable structures.* □

**Example 8.4.** (1) Consider $c_0$ as an additive group. It is a Polish group with respect to its norm topology. Then the shift action of $\ell_2$ on $\mathbb{R}^{\mathbb{N}}$ is turbulent.

In order to prove this, fix $x = (x_n), U, V$ as above and fix $y \in \mathbb{R}^{\mathbb{N}}$. We shall prove that the orbit of $y$ intersects the closure of $\mathcal{O}(x, U, V)$. Then there exist $\varepsilon > 0$ and $k$ such that

$$U \supseteq \{(g_n) \in c_0 : |g_n| < \varepsilon \text{ for all } n\}$$

and

$$V \supseteq \{(z_n) \in \mathbb{R}^{\mathbb{N}} : |x_n - z_n| < \varepsilon \text{ for all } n < k\}.$$

Let $z = (z_n) \in \mathbb{R}^{\mathbb{N}}$ be such that $z_n = x_n$ for $n < k$ and $z_n = y_n$ for $n \geq k$. Then $z \in V \cap [y]$. We shall prove that $z \in \overline{\mathcal{O}(x, U, V)}$. Fix $m \in \mathbb{N}$ and let $K = \max_{n \leq m} |x_n - z_n|$. If $j > K/\varepsilon$, then we can find a sequence $x = z^0, z^1, \ldots, z^j$ such that the first $m$ coordinates of $z^j$ coincide with the first $m$ coordinates of $z$ and $z^i - z^{i+1} \in U$ for all $i < j$. Therefore all $z^i$ belong to $\mathcal{O}(x, U, V)$ and since $m$ was arbitrary $z$ is an accumulation point of $\mathcal{O}(x, U, V)$.

(2) The above proof shows that the action of any classical Banach space $\ell_p$, for $p \geq 1$ on $\mathbb{R}^{\mathbb{N}}$ is turbulent,

(3) The following example will be used later. Consider the Polish group $G = \mathbb{Z}^{\mathbb{N}}$ and let

$$G_0 = \{g \in G : \lim_{n \to \infty} \frac{g(n)}{n} = 0\}.$$

Then $G_0$ is a Polish subgroup of $G$, and the coset action is turbulent.

Here is our main application of turbulence.

**Theorem 8.5** (Farah–Toms–Törnquist, [33]). *The isomorphism relation $\cong_{AI}$ of simple, separable, unital AI algebras has the following two properties:*

*(1) it is not classifiable by countable structures.*
*(2) for any countable signature $\mathcal{L}$ the isomorphism relation of countable $\mathcal{L}$-models is Borel-reducible to it.*

The proof of this theorem proceeds in two steps, via using $\mathcal{H}^2$, the homeomorphism relation of compact subsets of $[0,1]^2$: (i) showing that $\mathcal{H}^2$ has these properties (Lemma 8.6) and (ii) $\mathcal{H}^2 \leq_B \cong_{AI}$ (Lemma 8.9). A curious feature of (ii) is that we shall use classification result to prove a non-classification result. The following is a slight improvement of a result of Hjorth used in [43, 4.21].

**Lemma 8.6.** *(1) The homeomorphism relation of compact subspaces of $[0,1]^2$ is not classifiable by countable structures.*
*(2) If $\mathcal{L}$ is any countable signature then the isomorphism relation of countable $\mathcal{L}$-models is Borel-reducible to the homeomorphism of compact subsets of $[0,1]^2$.*

*Proof.* (1) It is notationally convenient to work with $[-1,1]^2$ instead of $[0,1]^2$.

Let $G = \mathbb{Z}^{\mathbb{N}}$. Note that $G$ is a Polish group when given the product group structure and product topology. We let

$$G_0 = \{g \in G : \lim_{n\to\infty} \frac{g(n)}{n} = 0\}.$$

By [43, 4.16] $G_0$ acts turbulently on $G$ by translation. Let

$$T_1 = \{(x,y) \in [-1,1] \times [0,1] : |x| \leq y\}$$

and for $n \in \mathbb{N}$, let in general

$$T_n = \bigcup_{k=1}^{n} [(2k-2,0) + T_1].$$

Then the $T_n$ are compact and connected and for $m \neq n$ we have $T_m \not\cong_{\mathrm{hom}} T_n$. Fix an order-preserving homeomorphism $f : \mathbb{R} \to (-1,1)$ such that $f(0) = 0$. For each $m \in \mathbb{Z}$ and $n, k \in \mathbb{N}$ and let

$$J_{m,n,k} \subseteq [\frac{1}{2n-1}, \frac{1}{2n} - \frac{|m|}{2n(2n-1)(|m|+1)}] \times [f\left(\frac{2m-1}{n}\right), f\left(\frac{2m}{n}\right)]$$

be a closed set homeomorphic to $T_k$. Note that $J_{m,n,k} \subseteq [\frac{1}{2n-1}, \frac{1}{2n}] \times [-1,1]$ and that for all $n$,

$$\lim_{m\to\pm\infty} \sup\{\operatorname{diam}(J_{m,n,k}) : k \in \mathbb{N}\} \to 0.$$

Fix a bijection $\varphi : \mathbb{Z}\times\mathbb{N} \to \mathbb{N}$ and define for each $g \in G$ a set $K(g) \subseteq [-1,1]^2$ by

$$X(g) = \left(\bigsqcup_{m,n} J_{m,n,\varphi(m+g(n),n)}\right) \sqcup (\partial([0,1] \times [-1,1]) \cup [-1,0] \times \{0\}). \quad (*)$$

It is easy to see that $X(g)$ is compact for all $g \in G$, and that the map

$$G \to K([-1,1]^2) : g \mapsto X(g)$$

is Borel, where $K([-1,1]^2)$ denotes the compact hyperspace of $[-1,1]^2$. Note that $(*)$ provides a decomposition of each $X(g)$ into mutually non-homeomorphic connected components.

By the $(m,n)$ component of $X(g)$ we mean the set

$$J_{m,n,\varphi(m+g(n),n)},$$

and by the $n$-th column of $X(g)$ we mean the set

$$X(g)_n = \bigsqcup_m J_{m,n,\varphi(m+g(n),n)}.$$

It is clear that the set of components of the $n$-th column

$$\{J_{m,n,\varphi(m+g(n),n)} : m \in \mathbb{N}\}$$

does not depend on $g \in G$. On the other hand, if $n_0 \neq n_1$, then the components of $X(g)_{n_0}$ are not homeomorphic to any of the components of $X(g)_{n_1}$. Therefore any homeomorphism from $X(g)$ to $X(h)$ must map $X(g)_n$ to $X(h)_n$, for all $n \in \mathbb{N}$. Moreover, any homeomorphism between $X(g)$ and $X(h)$ must fix the point $(0,0)$.

**Claim 8.7.** *For all $g \in G$, if $(x_n, y_n) \in \bigsqcup_{m,n} J_{m,n,\varphi(m+g(n),n)}$ is a sequence such that $(x_n, y_n) \to (0,0)$ and $x_n \in [\frac{1}{2n-1}, \frac{1}{2n}]$ for all $n$, then the function $\theta : \mathbb{N} \to \mathbb{Z}$ defined by*

$\theta(n) = k \iff$ *the connected component of $(x_n, y_n)$ isomorphic to $T_{\varphi(k,n)}$*

*satisfies $\theta - g \in G_0$.*

*Proof.* For each $n \in \mathbb{N}$, let $m_n \in \mathbb{Z}$ be such that

$$y_n \in [f\left(\frac{2m_n - 1}{n}\right), f\left(\frac{2m_n}{n}\right)].$$

Since $y_n \to 0$ it follows that $f(\frac{2m_n}{n}) \to 0$, and so $\frac{m_n}{n} \to 0$. By definition of $X(g)$ the connected component in which $(x_n, y_n)$ lies in is isomorphic to $T_{\varphi(m_n+g(n),n)}$, and so

$$\theta(n) = m_n + g(n).$$

Thus $\theta - g \in G_0$. □

**Claim 8.8.** *For all $g, h \in G$, $X(g)$ is homeomorphic $X(h)$ if and only if $g - h \in G_0$.*

*Proof.* Suppose first that $X(g)$ and $X(h)$ are homeomorphic, and let $\hat{\pi}$ : $X(g) \to X(h)$ witness this. Fix a sequence $(x'_n, y'_n) \in X(g)$ such that $(x'_n, y'_n) \in J_{0,n,\varphi(g(n),n)}$ for all $n \in \mathbb{N}$. Then $(x'_n, y'_n) \to (0,0)$, and so since $\hat{\pi}(0,0) = (0,0)$ we have $(x_n, y_n) = \hat{\pi}(x'_n, y'_n) \to 0$. Since $X(g)_n$ is mapped to $X(h)_n$ if holds for all $n$ that $x_n \in [\frac{1}{2n-1}, \frac{1}{2n}]$. Moreover,

$$g(n) = k \iff \text{the connected component of } (x_n, y_n) \text{ isomorphic to } T_{\varphi(k,n)}$$

and so $g - h \in G_0$ by Claim 8.7.

Suppose conversely that $z = g - h \in G_0$. Then define $\pi : X(g) \to X(h)$ by letting

$$\pi \restriction \partial([0,1] \times [-1,1]) \cup [-1,0] \times \{0\} = \text{id},$$

and for each $n \in \mathbb{N}$ and $m \in \mathbb{Z}$ letting $\pi \restriction J_{m,n,\varphi(m+g(n),n)}$ be a homeomorphism

$$J_{m,n,\varphi(m+g(n),n)} \to J_{m+z(n),n,\varphi(m+z(n)+h(n),n)}.$$

To see that $\pi$ is a homeomorphism it suffices to see that $\pi$ is continuous, since it is clearly 1-1 and onto. To see this, for each $x \in \bigsqcup_{m,n} J_{m,n,\varphi(m+g(n),n)}$ let $m_x \in \mathbb{Z}$ and $n_x \in \mathbb{N}$ be such that

$$x \in [\frac{1}{2n_x - 1}, \frac{1}{2n_x}] \times [f\left(\frac{2m_x - 1}{n_x}\right), f\left(\frac{2m_x}{n_x}\right)].$$

Then by the definition of $\pi$ we have

$$\begin{aligned}d(x,\pi(x)) \le\ & d\left(f\left(\frac{2m_x}{n_x}\right), f\left(\frac{2(m_x+z(n_x))}{n_x}\right)\right)\\ &+ d\left(f\left(\frac{2m_x-1}{n_x}\right), f\left(\frac{2m_x}{n_x}\right)\right)\\ &+ d\left(\frac{2m_x-1}{n_x}, \frac{2m_x}{n_x} - \frac{|m_x|}{2n_x(2n_x-1)(|m_x|+1)}\right)\\ &+ d\left(\frac{2(m_x+z(n_x))-1}{n_x}, \frac{2(m_x+z(n_x))}{n_x}\right.\\ &\qquad \left. - \frac{|m_x+z(n_x)|}{2n_x(2n_x-1)(|m_x+z(n_x)|+1)}\right)\end{aligned}$$

which shows that $d(x,\pi(x)) \to 0$ as $x \to \partial([0,1]\times[-1,1])$. Thus $\pi$ is continuous. □

(2) follows from the argument suggested in [43, 4.22] but I shall provide a slightly different proof. By [36], it suffices to reduce the isomorphism of countable graphs. With $[\mathbb{Z}]^2 = \{(m,n) : m < n\}$ fix a bijection $\phi\colon [\mathbb{Z}]^2 \to \mathbb{N}$. Let $E_{m,n}$ be the union of straight lines in $\mathbb{R}^3$ connecting $(m,0,0)$ with $(0,\phi(m,n),\phi(m,n))$ and $(0,\phi(m,n),\phi(m,n))$ with $(n,0,0)$. These lines are pairwise disjoint and $\bigcup_{m,n} E_{m,n}$ is closed.

Let $L_n$ be union of three straight lines connecting $(m,0,0)$ and $(m,0,-1)$, $(m,0,0)$ and $(m,-1,-1)$ and $(m,0,0)$ and $(m,-1,0)$ for all $n$.

To a countable graph $G = (\mathbb{N},\mathcal{E})$ associate the set

$$X(G) = \bigcup_{n\in\mathbb{Z}} L_n \cup \bigcup_{\{m,n\}\in\mathcal{E}} L_{m,n}.$$

This map is clearly continuous. If graphs $G$ and $H$ are isomorphic, then $X(G)$ and $X(H)$ are clearly isomorphic. Now assume $X(G)$ and $X(H)$ are homeomorphic. Homeomorphism has to send each endpoint of the form $(m,0,-1)$, $(m,-1,-1)$, or $(m,-1,0)$ to a point of the same form. Therefore each $(m,0,0)$ goes to some $(f(m),0,0)$ and It is then easy to show that $f$ is an isomorphism between $G$ and $H$. □

The following lemma completes the proof of Theorem 8.5.

**Lemma 8.9.** *There is a Borel map $\Phi$ from $\mathcal{K}([0,1]^{\mathbb{N}})$, the space of compact subspaces of $[0,1]^{\mathbb{N}}$, to $\Gamma$ such that each $A(K) = C^*(\Phi(K))$ is a simple, unital AI algebra and $A(K) \cong A(L)$ if and only if $K$ and $L$ are homeomorphic.*

*Proof.* In [78] Thomsen constructed (among other things) for every compact metric $K$ algebra $A(K)$ such that

(1) $A(K)$ is a simple, unital AI algebra,
(2) $K_0(A(K)) = (\mathbb{Q}, \mathbb{Q}^+, [1])$,
(3) $K_1(A(K)) = \{0\}$,
(4) $T(A(K))$ is affinely homeomorphic to $\mathcal{P}(K)$, the space of Borel probability measures on $K$.

In [33] it was demonstrated that Thomsen's construction can be performed effectively. Clearly, if $K$ is not homeomorphic to $L$ then $A(K)$ and $A(L)$ are not isomorphic. For the converse we use a result of Elliott who proved that simple, separable AI algebras are classified by their Elliott invariant (see [70]). (Since $(\mathbb{Q}, \mathbb{Q}^+, [1])$ has the unique state, the pairing function $\rho$ is uniquely determined in each $A(K)$.) □

### 8.2. *Below a group action: An upper bound for complexity*

While the isomorphism of separable Banach spaces is the $\leq_B$ maximal analytic equivalence relation ([35]), the isomorphism of von Neumann factors is reducible to an orbit equivalence relation of a Polish group action (classical, see e.g., [74]). This is not surprising since Banach spaces are much wilder objects than von Neumann algebras (note, however, that the isometry relation of separable Banach spaces is reducible to an orbit equivalence relation of a Polish group action, by [57]).

C*-algebras are not as wild as arbitrary Banach spaces and not as well-behaved as von Neumann algebras, and one can ask what is the complexity of the isomorphism of separable C*-algebras. While [33] left the general problem open, it did show that the isomorphism of algebras relevant to Elliott's program is below a group action. The proof, however, took a detour and used some of the deepest results on the structure of separable nuclear C*-algebras. This detour gives us an excuse to introduce a fascinating object.

### 8.3. *Cuntz algebra $\mathcal{O}_2$*

Let $H$ be a separable complex Hilbert space with orthonormal basis $e_n$, for $n \in \mathbb{N}$. The equations

$$s(e_n) = e_{2n} \qquad t(e_n) = e_{2n+1} \text{ for all } n$$

uniquely define linear operators $s$ and $t$. We have that

$$s^*s = 1, \qquad t^*t = 1, \text{ and } ss^* + tt^* = 1 \tag{8.1}$$

since $ss^*$ and $tt^*$ are mutually orthogonal projections.

Consider the C*-algebra generated by $s$ and $t$. The unit in this algebra satisfies the following definition (note that the property of $p$ is computed relative to the ambient C*-algebra $A$, and recall that for projections $p$ and $q$ we write $p \leq q$ if $pq = p$).

**Definition 8.10.** A projection $p$ in $A$ is *properly infinite* if there are projections $q \sim p$ and $r \sim p$ such that $q + r \leq p$.

The first separable simple C*-algebra in which the unit is properly infinite was constructed by Dixmier. He considered $C^*(s,t)/J$ with the above $s$ and $t$, where $J$ is the maximal ideal of $C^*(s,t)$. However, moding out by $J$ was not necessary.

**Theorem 8.11** (Cuntz, [10]). *If $s$ and $t$ are isometries satisfying* (8.1) *then the algebra $C^*(s,t)$ is simple. Moreover, any two algebras generated in this way are isomorphic.* □

This algebra generated by $s$ and $t$ is denoted by $\mathcal{O}_2$ and we shall spend some time investigating its properties.

**Lemma 8.12.** *If $s$ and $t$ satisfy* (8.1) *then $\mathcal{O}_2$ is the closed linear span of all monomials of the form $\prod_{i \leq m} x_i \prod_{j \leq n} y_j^*$, where $\{x_j, y_i\} \subseteq \{s,t\}$.*

*Proof.* We need to show that any monomial in $s, t, s^*$ and $t^*$ is either 0 or equal to a monomial of the form $\prod_{i \leq m} x_i \prod_{j \leq n} y_j^*$, where $\{x_j, y_i\} \subseteq \{s,t\}$. Since $s^*s = t^*t = 1$, it will suffice to prove that $s^*t = t^*s = 0$. But by (8.1) we have $ss^*t + tt^*t = t$, and therefore $ss^*t = 0$. By multiplying by $s^*$ on the left we obtain $s^*t = 0$. A proof of $t^*s = 0$ is similar. □

The structure of $\mathcal{O}_2$ is discussed below in Exercises 8.4.1 and 8.4.2.

Every separable simple C*-algebra, including $\mathcal{O}_2$, has outer automorphisms. The following lemma shows that automorphisms of $\mathcal{O}_2$ are coded by its unitaries. However, not only that $\alpha_u$ is distinct from $\operatorname{Ad} u$, but moreover $u \mapsto \alpha_u$ is not an action of the unitary group $\mathcal{O}_2$ on $\mathcal{O}_2$.

**Lemma 8.13.** *Every automorphism $\alpha$ of $\mathcal{O}_2$ is determined by*

$$\alpha(s) = us \text{ and } \alpha(t) = ut$$

*for some unitary $u$. Conversely, every unitary $u \in \mathcal{O}_2$ uniquely determines an automorphism $\alpha_u$ of $\mathcal{O}_2$ such that $\alpha_u(s) = us$ and $\alpha_u(t) = ut$.*

*Proof.* Let $u = \alpha(s)s^* + \alpha(t)t^*$. We claim that $u$ is a unitary. By applying Lemma 8.12 we have $uu^* = \alpha(s)\alpha(s)^* + \alpha(t)\alpha(t)^* = \alpha(ss^* + tt^*) = 1$. Similarly, $u^*u = 1$ and therefore $u$ is a unitary. Using Lemma 8.12 again we get $us = \alpha(s)$ and $ut = \alpha(t)$. □

More information on this fascinating object, as well as its relatives $\mathcal{O}_n$ for $n \in \{2, 3, \ldots, \infty\}$, including the proofs of the results below, can be found in [10] and in [70]. The definition of approximate unitary equivalence was given in the paragraph before Theorem 5.9.

**Theorem 8.14.** *Every endomorphism $\beta\colon \mathcal{O}_2 \to \mathcal{O}_2$ is approximately unitarily equivalent to the identity map.* □

By the following remarkable result $\mathcal{O}_2$ tensorially absorbs exactly the algebras that are subject of the Elliott's program. Its precursor, the proof that $\mathcal{O}_2 \otimes \mathcal{O}_2$ is isomorphic to $\mathcal{O}_2$, was proved by Elliott using Theorem 8.14 and a clever use of ultrapowers and an approximate intertwining argument (see [70]). The difficult proof of Theorem 8.15 proof can be found in [53] or (with fewer details) in [70].

**Theorem 8.15** (Kirchberg). *Every separable nuclear unital $C^*$-algebra is isomorphic to a unital subalgebra of $\mathcal{O}_2$.*

*Also, $A$ is separable, nuclear, simple and unital $C^*$-algebra if and only if $A \otimes \mathcal{O}_2 \cong \mathcal{O}_2$.* □

Unlike the case of Thomsen's theorem (see the proof of Theorem 8.5), we were unable to find a direct Borel proof of Kirchberg's theorem. Its known proofs are not taking place in the 'Borel world' but are instead considering embeddings into the corona of the stabilization of $\mathcal{O}_2$. Nevertheless, Kirchberg's theorem can be used to prove its Borel version (cf. Exercise 8.4.6).

I can now give a rough sketch of the proof of an upper bound for the complexity of the isomorphism relation. By $\mathrm{SA}(\mathcal{O}_2)$ we denote the space of subalgebras of $\mathcal{O}_2$, with respect to the Effros Borel structure.

**Theorem 8.16** (Farah–Toms–Törnquist, [33]). *The isomorphism relation of separable, simple, unital and nuclear $C^*$-algebras is Borel-reducible to an orbit equivalence relation of a Polish group action.*

*Sketch of a proof.* More precisely, the set $\Gamma_N = \{\gamma \in \Gamma : C^*(\gamma)$ is simple, unital and nuclear$\}$ is Borel. There is a Borel reduction $\Phi\colon N \to \mathrm{SA}(\mathcal{O}_2)$ such that $C^*(\gamma_1) \cong C^*(\gamma_2)$ if and only if $\Phi(\gamma_1)\, E^{\mathrm{SA}(\mathcal{O}_2)}_{\mathrm{Aut}(\mathcal{O}_2)}\, \Phi(\gamma_2)$.

By a Borel version of Kirchberg's Theorem 8.15 proved in [33] there is a Borel map $\Phi$ from $N$ into $\mathrm{SA}(\mathcal{O}_2)$ such that $A = \Phi(\gamma)$ is a unital subalgebra of $\mathcal{O}_2$ isomorphic to $C^*(\gamma)$ with the property that the *relative commutant* of $A$ in $\mathcal{O}_2$,

$$B = \{b \in \mathcal{O}_2 : ab = ba \text{ for all } a \in A\}$$

is isomorphic to $\mathcal{O}_2$ and $C^*(A, B)$ is both equal to $\mathcal{O}_2$ and is naturally isomorphic to $A \otimes \mathcal{O}_2$.

Clearly we only need to check that if $C^*(\gamma_1) \cong C^*(\gamma_2)$ then there is $\alpha \in \mathrm{Aut}(\mathcal{O}_2)$ that sends $\Phi(\gamma_1)$ onto $\Phi(\gamma_2)$. Fix an isomorphism $\alpha_0$ between $\Phi(\gamma_1)$ and $\Phi(\gamma_2)$. Let $B_1$ and $B_2$ be the relative commutants of these algebras. Since they are both isomorphic to $\mathcal{O}_2$, we can fix an isomorphism $\alpha_2 \colon B_1 \to B_2$. Then $\alpha = \alpha_1 \otimes \alpha_2$ is as required. □

### 8.4. *Exercises*

**8.4.1.** Let $s$ and $t$ be the generators of $\mathcal{O}_2$. For $n \in \mathbb{N}$ let $\mathcal{F}_n$ be the linear span of $\prod_{j\leq n} x_j \prod_{j\leq n} y_j^*$, where $\{x_j, y_j\} \subseteq \{s, t\}$. Prove that $\mathcal{F}_n$ is isomorphic to $M_{2^n}(\mathbb{C})$.

(Hint: First consider $n = 1$ and check that $ss^*$, $tt^*$, $st^*$ and $ts^*$ are the matrix units in $M_2(\mathbb{C})$ (cf. Exercise 2.6.7). Then go up.)

**8.4.2.** Using notation of the previous exercise, show that the 'balanced' products $\prod_{j\leq n} x_j \prod_{j\leq n} y_j^*$, for $n \in \mathbb{N}$ and $\{x_j, y_j\} \subseteq \{s, t\}$ generate a subalgebra isomorphic to the CAR algebra.

Then prove that $\mathcal{O}_2$ is generated by the CAR algebra $A$ and a partial isometry $s$ such that $a \mapsto sas^*$ is an endomorphism sending $A$ onto a corner $pAp$ where $p \in A$ is a projection of trace $1/2$.

**8.4.3.** Show that all nonzero projections in $\mathcal{O}_2$ are Murray-von Neumann equivalent and that $K_0(\mathcal{O}_2)$ is trivial.

**8.4.4.** Prove that $\mathcal{O}_2$ has no normalized traces.

(Hint: $1_{\mathcal{O}_2}$ is properly infinite.)

One can prove that $K_1(\mathcal{O}_2)$ is also trivial, and therefore by the previous two exercises the Elliott invariant of $\mathcal{O}_2$ is equal to the Elliott invariant of the C*-algebra $\{0\}$ (assuming that we accept the latter as a C*-algebra). Compare this with Theorem 8.15 and think $0 \cdot A = 0$.

**8.4.5.** Let $A_n$, $F_{mn}$, for $m \leq n$, $m, n \in \mathbb{N}$, be a directed unital system of C*-algebras all of them isomorphic to $\mathcal{O}_2$. Show that the direct limit is isomorphic to $\mathcal{O}_2$.

(Hint: Theorem 8.14 and approximate intertwining.)

I do not know what happens if one considers a transfinite direct limit of an $\aleph_1$-directed sequence of copies of $\mathcal{O}_2$, or whether such algebra of density character $\aleph_1$ is uniquely determined.[e] I also do not know whether the analogue of Kirchberg's embedding theorem (see Theorem 8.15) holds for nuclear, or exact, C*-algebras of density character $\aleph_1$.

Note that the conclusion of the following exercise is strictly weaker than what is required in the proof of Theorem 8.16.

**8.4.6.** Using notation from the proof of Theorem 8.16, prove that there exists function $\Phi\colon N \to \mathrm{SA}(\mathcal{O}_2)$ that is C-measurable.

(Hint: Use Jankov, von Neumann selection theorem.)

## 9. Concluding Remarks

### 9.1. *The Borel-reducibility diagram*

Figure 13 summarizes some of the known Borel reductions, concentrating on results relevant to operator algebras.[f]

All classes of C*-algebras occurring in the diagram are separable and unital (unless otherwise specified). Unless otherwise specified, the equivalence relation on a given class is the isomorphism relation. The bi-reducibility between the isomorphism for UHF algebras and bi-embeddability of UHF algebras is an immediate consequence of Exercise 4.3.9, or rather of its (straightforward) Borel version. Classification of compact metric spaces up to isometry is due to Gromov. Borel bi-reducibility between abelian C*-algebras and compact metric spaces was proved in [32]. Borel reductions from compact metric spaces to Choquet simplexes to simple AI algebras (as well as the definition of the latter) are given in [33]. A Borel-reduction of Choquet simplexes to the isometry of Banach spaces is given by sending a Choquet simplex $K$ to the Banach space of affine functions on $K$.

A Borel version of Elliott's reduction of simple AI algebras to Elliott invariant follows from Elliott's classification result and the fact that the computation of the Elliott invariant is Borel was proved in [32] (see Lemma 8.9). Equireducibility of the maximal orbit equivalence relation of a Polish group action with the isometry of Polish spaces and the isometry of Banach spaces was proved in [8] and [61], respectively.

[e] Added on April 20, 2013: Now I do know. It is not. The tensor product of the algebra constructed in [26] with $\mathcal{O}_2$ is not isomorphic to $\bigotimes_{\aleph_1} \mathcal{O}_2$.

[f] Added in July 2013. Recently Marcin Sabok proved that the affine homeomorphism of metrizable Choquet simplexes is the maximal orbit equivalence relation ([73]).

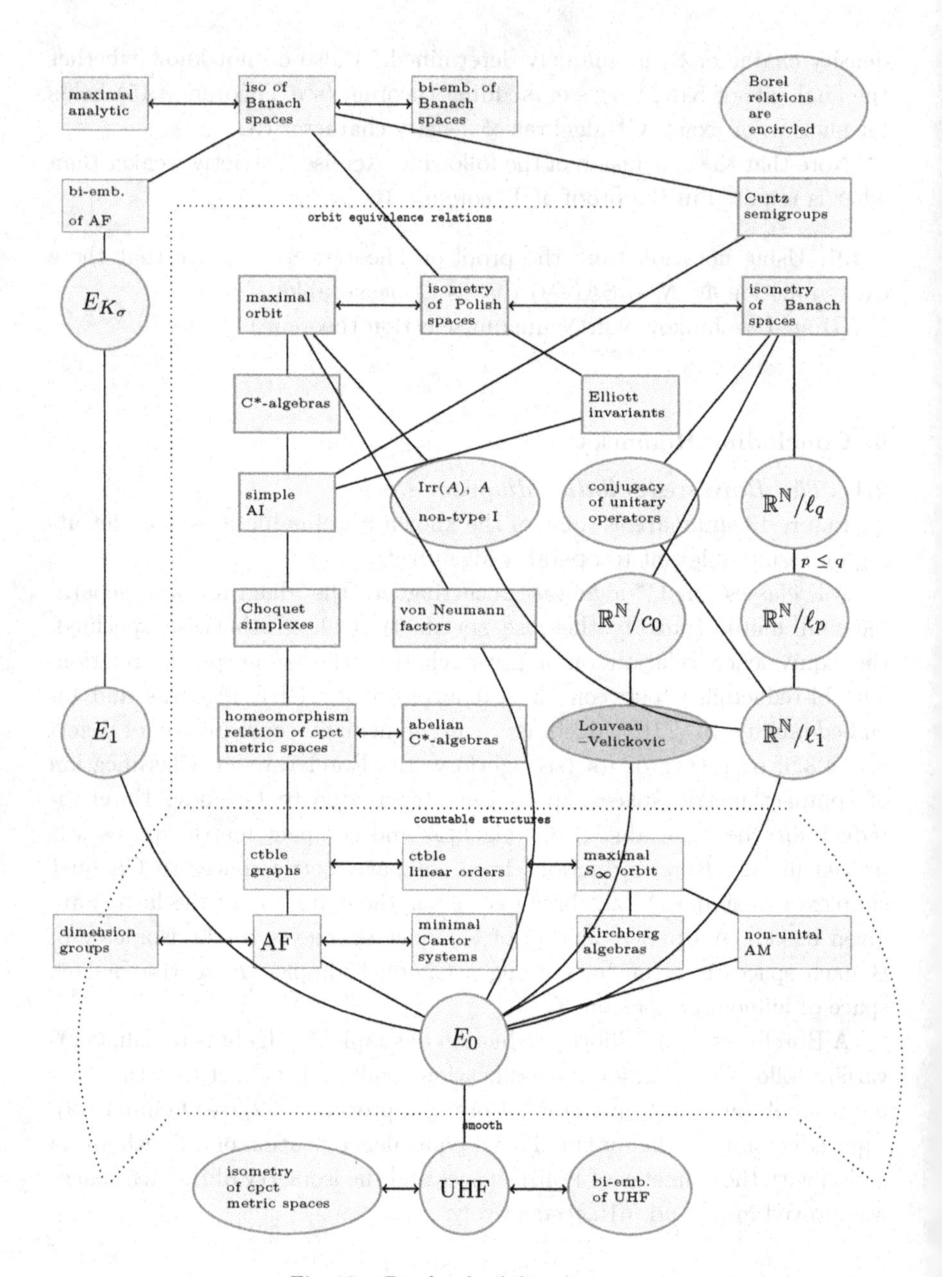


Fig. 13. Borel reducibility diagram.

The reduction of the Elliott invariant to the maximal orbit equivalence relation, as well as the facts about the Cuntz semigroup, are proved in [32].

Dougherty–Hjorth ([14]) proved that $\mathbb{R}^{\mathbb{N}}/\ell_p \leq_B \mathbb{R}^{\mathbb{N}}/\ell_q$ if and only if $p \leq q$, and Hjorth proved ([42]) that the translation action of $\ell_1$ on $\mathbb{R}^{\mathbb{N}}$ is a minimal turbulent action. On the other hand, I proved ([25]) that there is no minimal turbulent action Borel-reducible to $\mathbb{R}^{\mathbb{N}}/c_0$ and that the complicated structure of turbulent equivalence relations constructed in [58] occurs below orbit equivalence relation of any turbulent action Borel-reducible to $\mathbb{R}^{\mathbb{N}}/c_0$. Also, $\mathbb{R}^{\mathbb{N}}/c_0$ and $\mathbb{R}^{\mathbb{N}}/\ell_1$ are $\leq_B$-incomparable (see [25]). A class of $2^{\aleph_0}$ many $F_\sigma$ equivalence relations not classifiable by countable structures that are *incompatible*, in the sense that any relation Borel-reducible to any two of them is classifiable by countable structures, was constructed in [24], but I could not fit this family into the picture.

The fact that $\mathbb{R}^{\mathbb{N}}/\ell_1$ is Borel-reducible to the conjugacy of unitary operators on $\mathcal{B}(H)$ was proved in [51]. As Todor Tsankov and Alain Louveau pointed out, Dellacherie proved that the conjugacy relation of unitary operators in $\mathcal{B}(H)$ is Borel. Characterization of $\leq_B$-maximal isomorphism relation of countable structures as the graph isomorphism or isomorphism of countable linear orderings is in [36] (and so is the original definition of $\leq_B$!). AF algebras are classifiable by countable structures by Elliott's Theorem 5.8 and the fact that the computation of $K_0$ is Borel (Lemma 7.9). The fact that the topological orbit equivalence of minimal Cantor systems is equireducible with AF algebras was proved in [37], and that these are equivalent with the isomorphism of dimension groups is in [16].

Classification of Kirchberg algebras by countable structures is a consequence of Kirchberg–Phillips theorem ([65], [54]) and Borel-computability of the Elliott invariant. Classification of separable, but not necessarily unital direct limits of full matrix algebras (i.e., non-unital AM algebras) by countable structures is given in [13]; see also Exercise 4.3.7 .

Non-classifiability of the isomorphism relation in every major class of von Neumann factors that was not already classified is given in [74] and [75]; see also [60].

Since the maximal isomorphism relation for countable structures is not Borel ([43]), it is not Borel-reducible to any Borel equivalence relation. This in particular implies that the conjugacy of unitary operators is strictly $\leq_B$-below the isometry of Polish space.

By Irr($A$) we denote the space of irreducible representations of a C*-algebra $A$, or equivalently, the space of pure states of $A$ up to the unitary conjugacy. This relation satisfies a dichotomy result: it is either smooth (when $A$ is a type I C*-algebra, by the definition of type I algebra) or turbulent was proved independently in [52] and [27], and the latter proof gives that this relation Borel-reduces $\mathbb{R}^{\mathbb{N}}/\ell_1$ (and not $\mathbb{R}^{\mathbb{N}}/\ell_2$, as stated in this paper). The fact that this relation is Borel (actually $F_\sigma$) follows from a result of Glimm and Kadison (see [27]).

Bi-embeddability of AF algebras is proved to be above $E_{K_\sigma}$ in [33, §8]. The isomorphism of separable Banach spaces is the complete analytic equivalence relation by [35]. Several results on non-classifiability of automorphisms of C*-algebras with respect to unitary equivalence were obtained in [59].

Last, but not least, the isomorphism relation of separable C*-algebras is reducible to an orbit equivalence relation of a Polish group action [19]. This result (stated as an open problem in my last lecture) was obtained in email correspondence between the authors of [19] in the wake of the June 2012 BIRS meeting on applications of descriptive set theory to functional analysis.

Some of the results presented in Figure 13 may not be optimal. For example, it is not known whether the homeomorphism of compact metric spaces is equireducible with the maximal orbit equivalence relation. See also [73].

### 9.2. *Selected open problems*

The following problem was suggested by N. Christopher Phillips.

**Question 9.1.** *Not necessarily self-adjoint, norm-closed algebras are well-studied (see e.g., [66]). Finite-dimensional algebras in this class are not necessarily direct sums of matrix algebras, and there is no simple classification of such algebras. Is there a descriptive set-theoretic explanation for this? For example, are these algebras classifiable by countable structures?*

Let $H$ be the class of homogeneous algebras of the form $C(X, M_n(\mathbb{C}))$ for $X$ compact metric and form the class AH using H as the building blocks (Definition 6.1). By [11], not every direct limit of AH algebras is AH.

**Question 9.2.** *Let $\alpha$ be the minimal ordinal such that the class of algebras obtained by closing $H$ as above under direct limits $\alpha$ times is closed under direct limits. By the above, $\alpha \geq 2$. What is the value of $\alpha$?*

(Warning: I do not have an impression that the C*-algebraists are too keen to find an answer to Question 9.2.)

**Question 9.3.** *For $2 \leq m \leq \infty$ let $\mathrm{E}_m$ be the homeomorphism relation of compact subspaces of $[0,1]^m$. Clearly $\mathrm{E}_n \leq_B \mathrm{E}_{n+1} \leq_B \mathrm{E}_\infty$ for all $n$ and $\mathrm{E}_\infty \leq E_{G_\infty}^{X_\infty}$. Does any of the converses hold? In particular, is $\mathrm{E}_2$ the $\leq_B$-maximal orbit equivalence relation of a Polish group action?*

The following important question is vague since it is really a large number of questions in one.

**Question 9.4.**[g] *How does the complexity of the isomorphism relation of a class of C*-algebras change as the class increases? In particular, is the isomorphism relation of all C*-algebras Borel-reducible to the isomorphism relation of nuclear C*-algebras, or to the isomorphism relation of simple C*-algebras, or to the isomorphism relation of simple nuclear C*-algebras?*

**Question 9.5.** *What is the complexity of the isomorphism relation of (nuclear, simple) C*-algebras with the same fixed value of the Elliott invariant?*

In [79] and [81] Toms constructed infinitely, and then continuum many, nonisomorphic nuclear simple C*-algebras with the same Elliott invariant.

The present paragraph contains a vague suggestion instead of a question or a problem. By the basic theory of vector bundles ([41]), if $2\dim(X) < n$ then $C(X, M_n(\mathbb{C}))$ contains only trivial projections (cf. §6.0.1). Therefore counterexamples to Elliott's conjecture (e.g., [80]) are AH algebras that are direct limits of direct sums of algebras $C(X, M_n(\mathbb{C}))$ where $\dim(X) > n$. On the other hand, in a technical tour de force the AH algebras with 'slow dimension growth' were classified by Elliott's invariant ([23]). One could therefore speculate that the complexity of the isomorphism relation for AH algebras can be tied with very fast growing functions, especially because the latter were a source of interesting logical result for decades. George Elliott suggested that the substantial paper of Rieffel ([69]) could be a useful source for this project.

Many classification results have matching range of invariant results and it would be good to have their Borel versions. We have already mentioned [78] whose partial Borel version was used in Lemma 8.9 and the Effros–Handelman–Shen result that every dimension group corresponds to an AF algebra ([16]). A Borel version of this result is not difficult to prove and it was used implicitly above. In [77] it was proved that a pair of countable abelian groups appears as the Elliott invariant of a Kirchberg algebra if and only if the second group is free. It would also be good to have a Borel version

[g] By Sabok's results ([73]), all of these classes have the same Borel complexity.

of the result of [6], that the Cuntz semigroup is recovered functorially from the Elliott invariant for a large class of C*-algebras.

**Question 9.6.** *Do the above range of invariant results have Borel versions?*

It would be nice to have a general selection theorem that provides such results automatically. However, the obvious strategy falls short of obtaining Borel maps (see Exercise 8.4.6). See, however, [19, Theorem A] and [32, Lemma 6.2] for some partial results.

As pointed out before, the proof of Kirchberg's Theorem makes a detour into the nonseparable world. The following question may have an interesting answer.

**Problem 9.7.** What is the strength of Kirchberg's Theorem 8.15 in the sense of reverse mathematics?

It is not known whether nuclear simple separable C*-algebras of real rank zero (see Exercise 3.5.9) are classifiable by $K_0$ and $K_1$. An obvious set-theoretic take on this problem is the following.

**Problem 9.8.** Can separable nuclear C*-algebras of real rank zero be classified by countable structures? What about separable nuclear simple C*-algebras or real rank zero?

It is known that $K_0$ and $K_1$ do not classify non-nuclear separable C*-algebras of real rank zero (a Löwenheim–Skolem type argument is given in [65]).

A distinguishing feature of Elliott's view of the classification program of C*-algebras is its functoriality. This does not seem to be captured by the present theory of Borel equivalence relations and the following problem is an attempt to remedy this situation. Polish groupoids were defined in [67] and one can similarly define a 'Polish category' whose objects are metric structures based on Polish spaces. Some preliminary results on the following problem were obtained by the author in a joint work with S. Coskey, G. Elliott, and M. Lupini.

**Problem 9.9.** Develop Borel reduction theory for Polish groupoids, and more generally for Polish categories.

## References


1. S. Adams and A.S. Kechris, *Linear algebraic groups and countable Borel equivalence relations*, J. Amer. Math. Soc. **13** (2000), no. 4, 909–943.
2. W. Arveson, *A short course on spectral theory*, Graduate Texts in Mathematics, vol. 209, Springer-Verlag, New York, 2002.

3. H. Becker and A.S. Kechris, *The descriptive set theory of Polish group actions*, Cambridge University Press, 1996.
4. B. Blackadar, *Operator algebras*, Encyclopaedia of Mathematical Sciences, vol. 122, Springer-Verlag, Berlin, 2006, Theory of $C^*$-algebras and von Neumann algebras, Operator Algebras and Non-commutative Geometry, III.
5. N. Brown and N. Ozawa, *$C^*$-algebras and finite-dimensional approximations*, Graduate Studies in Mathematics, vol. 88, American Mathematical Society, Providence, RI, 2008.
6. N.P. Brown, F. Perera, and A.S. Toms, *The Cuntz semigroup, the Elliott conjecture, and dimension functions on $C^*$-algebras*, J. Reine Angew. Math. **621** (2008), 191–211.
7. K. Carlson, E. Cheung, I. Farah, A. Gerhardt-Bourke, B. Hart, L. Mezuman, N. Sequeira, and A. Sherman, *Omitting types and AF algebras*, Arch. Math. Logic **53** (2014), no. 1–2, 157–169.
8. J.D. Clemens, S. Gao, and A.S. Kechris, *Polish metric spaces: their classification and isometry groups*, Bull. Symbolic Logic **7** (2001), no. 3, 361–375.
9. K.T. Coward, G.A. Elliott, and C. Ivanescu, *The Cuntz semigroup as an invariant for $C^*$-algebras*, J. Reine Angew. Math. **623** (2008), 161–193.
10. J. Cuntz, *Simple $C^*$-algebras generated by isometries*, Comm. Math. Phys. **57** (1977), no. 2, 173–185.
11. M. Dădărlat and S. Eilers, *Approximate homogeneity is not a local property*, J. Reine Angew. Math. **507** (1999), 1–13.
12. K.R. Davidson, *$C^*$-algebras by example*, Fields Institute Monographs, vol. 6, American Mathematical Society, Providence, RI, 1996.
13. J. Dixmier, *On some $C^*$-algebras considered by Glimm*, J. Functional Analysis **1** (1967), 182–203.
14. R. Dougherty and G. Hjorth, *Reducibility and nonreducibility between $l^p$ equivalence relations*, Trans. Amer. Math. Soc. **351** (1999), no. 5, 1835–1844.
15. E.G. Effros, *Classifying the unclassifiables*, Group representations, ergodic theory, and mathematical physics: a tribute to George W. Mackey, Contemp. Math., vol. 449, Amer. Math. Soc., Providence, RI, 2008, pp. 137–147.
16. E.G. Effros, D.E. Handelman, and C.L. Shen, *Dimension groups and their affine representations*, Amer. J. Math. **102** (1980), no. 2, 385–407.
17. E.G. Effros and Z.-J. Ruan, *Operator spaces*, London Mathematical Society Monographs. New Series, vol. 23, The Clarendon Press Oxford University Press, New York, 2000.
18. G.A. Elliott, *On the classification of inductive limits of sequences of semisimple finite-dimensional algebras*, J. Algebra **38** (1976), no. 1, 29–44.
19. G.A. Elliott, I. Farah, V. Paulsen, C. Rosendal, A.S. Toms, and A. Törnquist, *The isomorphism relation of separable C*-algebras*, Math. Res. Letters (to appear), preprint, arXiv 1301.7108.
20. G.A. Elliott and A.S. Toms, *Regularity properties in the classification program for separable amenable $C^*$-algebras*, Bull. Amer. Math. Soc. **45** (2008), no. 2, 229–245.
21. G.A. Elliott, *A classification of certain simple $C^*$-algebras*, Quantum and non-commutative analysis (Kyoto, 1992), Math. Phys. Stud., vol. 16, Kluwer Acad. Publ., Dordrecht, 1993, pp. 373–385.

22. ______, *The classification problem for amenable $C^*$-algebras*, Proceedings of the International Congress of Mathematicians, Vol. 1, 2 (Zürich, 1994) (Basel), Birkhäuser, 1995, pp. 922–932.
23. G.A. Elliott, G. Gong, and L. Li, *On the classification of simple inductive limit $C^*$-algebras. II. The isomorphism theorem*, Invent. Math. **168** (2007), no. 2, 249–320.
24. I. Farah, *Basis problem for turbulent actions I: Tsirelson submeasures*, Proceedings of XI Latin American Symposium in Mathematical Logic, Merida, July 1998, Annals of Pure and Applied Logic, vol. 108, 2001, pp. 189–203.
25. ______, *Basis problem for turbulent actions II: $c_0$-equalities*, Proceedings of the London Mathematical Society **82** (2001), 1–30.
26. ______, *Graphs and CCR algebras*, Indiana Univ. Math. Journal **59** (2010), 1041–1056.
27. ______, *A dichotomy for the Mackey Borel structure*, Proceedings of the 11th Asian Logic Conference In Honor of Professor Chong Chitat on His 60th Birthday (Yang Yue et al., eds.), 2011, pp. 86–93.
28. I. Farah, B. Hart, and D. Sherman, *Model theory of operator algebras I: Stability*, Bull. London Math. Soc. **45** (2013), 825–838.
29. ______, *Model theory of operator algebras II: Model theory*, Israel J. Math. (to appear), arXiv:1004.0741.
30. I. Farah and T. Katsura, *Nonseparable UHF algebras I: Dixmier's problem*, Adv. Math. **225** (2010), no. 3, 1399–1430.
31. ______, *Nonseparable UHF algebras II: Classification*, Math. Scand. (to appear), preprint arXiv:1301.6152.
32. I. Farah, A.S. Toms, and A. Törnquist, *The descriptive set theory of $C^*$-algebra invariants*, Int. Math. Res. Notices **22** (2013), 5196–5226, Appendix with Caleb Eckhardt.
33. ______, *Turbulence, orbit equivalence, and the classification of nuclear $C^*$-algebras*, J. Reine Angew. Math. **688** (2014), 101–146.
34. I. Farah and E. Wofsey, *Set theory and operator algebras*, Appalachian set theory 2006-2010 (J. Cummings and E. Schimmerling, eds.), Cambridge University Press, 2013, pp. 63–120.
35. V. Ferenczi, A. Louveau, and C. Rosendal, *The complexity of classifying separable Banach spaces up to isomorphism*, J. Lond. Math. Soc. (2) **79** (2009), no. 2, 323–345.
36. H. Friedman and L. Stanley, *A Borel reducibility theory for classes of countable structures*, The Journal of Symbolic Logic **54** (1989), 894–914.
37. T. Giordano, I.F. Putnam, and C.F. Skau, *Topological orbit equivalence and $C^*$-crossed products*, J. Reine Angew. Math. **469** (1995), 51–111.
38. J. Glimm, *Type I $C^*$-algebras*, Ann. of Math. (2) **73** (1961), 572–612.
39. J.G. Glimm, *On a certain class of operator algebras*, Trans. Amer. Math. Soc. **95** (1960), 318–340.
40. L.A. Harrington, A.S. Kechris, and A. Louveau, *A Glimm–Effros dichotomy for Borel equivalence relations*, Journal of the American Mathematical Society **4** (1990), 903–927.
41. A. Hatcher, *Vector bundles and K-theory*, 2003.

42. G. Hjorth, *Actions by classical Banach spaces*, The Journal of Symbolic Logic **65** (2000), 392–420.
43. ______, *Classification and orbit equivalence relations*, Mathematical Surveys and Monographs, vol. 75, American Mathematical Society, 2000.
44. ______, *Borel equivalence relations*, Handbook of set theory (M. Foreman and A. Kanamori, eds.), Springer, 2010, pp. 297–332.
45. X. Jiang and H. Su, *On a simple unital projectionless $C^*$-algebra*, Amer. J. Math **121** (1999), 359–413.
46. V.F.R. Jones, *Von Neumann algebras*, 2010, lecture notes, http://math.berkeley.edu/~vfr/.
47. M. Junge and G. Pisier, *Bilinear forms on exact operator spaces and $B(H) \otimes B(H)$*, Geom. Funct. Anal. **5** (1995), no. 2, 329–363.
48. A.S. Kechris, *Classical descriptive set theory*, Graduate texts in mathematics, vol. 156, Springer, 1995.
49. ______, *The descriptive classification of some classes of $C^*$-algebras*, Proceedings of the Sixth Asian Logic Conference (Beijing, 1996), World Sci. Publ., River Edge, NJ, 1998, pp. 121–149.
50. A.S. Kechris and A. Louveau, *The structure of hypersmooth Borel equivalence relations*, Journal of the American Mathematical Society **10** (1997), 215–242.
51. A.S. Kechris and N.E. Sofronidis, *A strong generic ergodicity property of unitary and self-adjoint operators*, Ergodic Theory Dynam. Systems **21** (2001), no. 5, 1459–1479.
52. D. Kerr, H. Li, and M. Pichot, *Turbulence, representations, and trace-preserving actions*, Proc. Lond. Math. Soc. (3) **100** (2010), no. 2, 459–484.
53. E. Kirchberg and N.C. Phillips, *Embedding of exact $C^*$-algebras in the Cuntz algebra $\mathcal{O}_2$*, J. reine angew. Math. **525** (2000), 17–53.
54. E. Kirchberg, *Exact* C**-algebras, tensor products, and the classification of purely infinite algebras*, Proceedings of the International Congress of Mathematicians, Vol. 1, 2 (Zürich, 1994) (Basel), Birkhäuser, 1995, pp. 943–954.
55. H. Lin, *Classification of simple $C^*$-algebras of tracial topological rank zero*, Duke Math. J. **125** (2004), no. 1, 91–119.
56. T.A. Loring, *Lifting solutions to perturbing problems in $C^*$-algebras*, Fields Institute Monographs, vol. 8, American Mathematical Society, Providence, RI, 1997.
57. A. Louveau and C. Rosendal, *Complete analytic equivalence relations*, Trans. Amer. Math. Soc. **357** (2005), no. 12, 4839–4866.
58. A. Louveau and B. Velickovic, *A note on Borel equivalence relations*, Proceedings of the American Mathematical Society **120** (1994), 255–259.
59. M. Lupini, *Unitary equivalence of automorphisms of separable $C^*$-algebras*, arXiv preprint arXiv:1304.3502 (2013).
60. M. Lupini and A. Törnquist, *Set theory and von Neumann algebras*, Appalachian set theory 2006-2010 (J. Cummings and E. Schimmerling, eds.), Cambridge University Press, to appear.
61. J. Melleray, *Computing the complexity of the relation of isometry between separable Banach spaces*, MLQ Math. Log. Q. **53** (2007), no. 2, 128–131.
62. G.J. Murphy, *$C^*$-algebras and operator theory*, Academic Press Inc., Boston, MA, 1990.

63. V. Paulsen, *Completely bounded maps and operator algebras*, Cambridge Studies in Advanced Mathematics, vol. 78, Cambridge University Press, Cambridge, 2002.
64. G.K. Pedersen, *Analysis now*, Graduate Texts in Mathematics, vol. 118, Springer-Verlag, New York, 1989.
65. N.C. Phillips, *A classification theorem for nuclear purely infinite simple $C^*$-algebras*, Doc. Math. **5** (2000), 49–114 (electronic).
66. S.C. Power, *Limit algebras: an introduction to subalgebras of $C^*$-algebras*, Pitman Research Notes in Mathematics Series, vol. 278, Longman Scientific & Technical, Harlow, 1992.
67. A. Ramsay, *The Mackey-Glimm dichotomy for foliations and other Polish groupoids*, Journal of Functional Analysis **94** (1990), no. 2, 358–374.
68. M. Rørdam, F. Larsen, and N.J. Laustsen, *An introduction to K-theory for $C^*$ algebras*, London Mathematical Society Student Texts, no. 49, Cambridge University Press, 2000.
69. M.A. Rieffel, *Projective modules over higher-dimensional noncommutative tori*, Canad. J. Math. **40** (1988), no. 2, 257–338.
70. M. Rørdam, *Classification of nuclear $C^*$-algebras*, Encyclopaedia of Math. Sciences, vol. 126, Springer-Verlag, Berlin, 2002.
71. ———, *A simple $C^*$-algebra with a finite and an infinite projection*, Acta Math. **191** (2003), 109–142.
72. C. Rosendal, *Cofinal families of Borel equivalence relations and quasiorders*, Journal of Symbolic Logic **70** (2005), 1325–1340.
73. M. Sabok, *Completeness of the isomorphism problem for separable $C^*$-algebras*, arXiv preprint arXiv:1306.1049 (2013).
74. R. Sasyk and A. Törnquist, *Borel reducibility and classification of von Neumann algebras*, Bulletin of Symbolic Logic **15** (2009), no. 2, 169–183.
75. ———, *Turbulence and Araki-Woods factors*, J. Funct. Anal. **259** (2010), no. 9, 2238–2252.
76. P. Skoufranis, *Operator theory notes*, notes available at http://www.math.ucla.edu/%7Epskoufra/OperatorAlgebras.html, 2012.
77. W. Szymański, *The range of K-invariants for $C^*$-algebras of infinite graphs*, Indiana Univ. Math. J. **51** (2002), no. 1, 239–249.
78. K. Thomsen, *Inductive limits of interval algebras: the tracial state space*, Amer. J. Math. **116** (1994), no. 3, 605–620.
79. A.S. Toms, *An infinite family of non-isomorphic $C^*$-algebras with identical K-theory*, Trans. Amer. Math. Soc. **360** (2008), no. 10, 5343–5354.
80. ———, *On the classification problem for nuclear $C^*$-algebras*, Ann. of Math. (2) **167** (2008), no. 3, 1029–1044.
81. ———, *Comparison theory and smooth minimal $C^*$-dynamics*, Comm. Math. Phys. **289** (2009), no. 2, 401–433.
82. A.S. Toms and W. Winter, *The Elliott conjecture for Villadsen algebras of the first type*, J. Funct. Anal. **256** (2009), no. 5, 1311–1340.
83. J. Villadsen, *Simple $C^*$-algebras with perforation*, J. Funct. Anal. **154** (1998), no. 1, 110–116.
84. N. Weaver, *Set theory and $C^*$-algebras*, Bull. Symb. Logic **13** (2007), 1–20.

85. C. Winslow and U. Haagerup, *The Effros–Maréchal topology in the space of von Neumann algebras*, American Journal of Mathematics **120** (1998), 567–617.
86. W. Winter, *Decomposition rank and $\mathcal{Z}$-stability*, Invent. Math. **179** (2010), 229–301.
87. ———, *Nuclear dimension and $\mathcal{Z}$-stability of pure $C^*$-algebras*, Invent. Math. **187** (2012), 259–342.
88. ———, *Regularity of nuclear $C^*$-algebras*, CBMS conference notes, to appear.

# SUBCOMPLETE FORCING AND $\mathcal{L}$-FORCING

Ronald Jensen
*Humboldt-Universität zu Berlin, Institut für Mathematik*
*Sitz: Rudower Chaussee 25, D-10099 Berlin, Germany*
*jensen@mathematik.hu-berlin.de*

In his book **Proper Forcing** (1982) Shelah introduced three classes of forcings (complete, proper, and semi-proper) and proved a strong iteration theorem for each of them: The first two are closed under countable support iterations. The latter is closed under revised countable support iterations subject to certain standard restraints. These theorems have been heavily used in modern set theory. For instance using them, one can formulate "forcing axioms" and prove them consistent relative to a supercompact cardinal. Examples are PFA, which says that Martin's axiom holds for proper forcings, and MM, which says the same for semiproper forcings. Both these axioms imply the negation of CH. This is due to the fact that some proper forcings add new reals. Complete forcings, on the other hand, not only add no reals, but also no countable sets of ordinals. Hence they cannot change a cofinality to $\omega$. Thus none of these theories enable us e.g. to show, assuming CH, that Namba forcing can be iterated without adding new reals.

More recently we discovered that the three forcing classes mentioned above have natural generalizations which we call "subcomplete", "subproper" and "semi-subproper". It turns out that each of these is closed under Revised Countable Support (RCS) iterations subject to the usual restraints.

The first part of our lecture deals with subcomplete forcings. These forcings do not add reals. Included among them, however, are Namba forcing, Prikry forcing, and many other forcings which change cofinalities. This gives a positive solution to the above mentioned iteration problem for Namba forcing. Using the iteration theorem one can also show that the **Subcomplete Forcing Axiom** (SCFA) is consistent relative to a supercompact cardinal. It has some of the more striking consequences of MM but is compatible with CH (and in fact with $\diamondsuit$).

(Note: Shelah was able to solve the above mentioned iteration problem for Namba forcing by using his ingenious and complex theory of

"I-condition forcing". The relationship of I-condition forcing to subcomplete forcing remains a mystery. There are, however, many applications of subcomplete forcing which have not been replicated by I-condition forcing.)

In the second part of the lecture, we give an introduction to the theory of "$\mathcal{L}$-forcings". We initially developed this theory more than twenty years ago in order to force the existence of new reals. More recently, we discovered that there is an interesting theory of $\mathcal{L}$-forcings which do **not** add reals. (In fact, if we assume CH $+2^{\omega_1} = \omega_2$, then Namba forcing is among them.) Increasingly we came to feel that there should be a "natural" iteration theorem which would apply to a large class of these forcings. This led to the iteration theorem for subcomplete forcing.

Combining all our methods, we were then able to prove:

(1) Let $\kappa$ be a strongly inaccessible cardinal. Assume CH. There is a subcomplete forcing extension in which $\kappa$ becomes $\omega_2$ and every regular cardinal $\tau \in (\omega_1, \kappa)$ acquires cofinality $\omega$.

(2) Let $\kappa$ be as above, where GCH holds below $\kappa$. Let $A \subset \kappa$. There is a subcomplete forcing extension in which:

- $\kappa$ becomes $\omega_2$;
- If $\tau \in (\omega_1, \kappa) \cap A$ is regular, then it acquires cofinality $\omega$;
- If $\tau \in (\omega_1, \kappa) \setminus A$ is regular, then it acquires cofinality $\omega_1$.

We will not be able to fully prove these theorems in our lectures, but we hope to develop some of the basic methods involved.

## Contents

## 0. Preliminaries

ZF$^-$ ("ZF without power set") consists of the axioms of extensionality and foundation together with:

(1) $\emptyset$, $\{x, y\}$, $\bigcup x$ are sets.
(2) (Axiom of Subsets or "Aussonderungsaxiom")
$x \cap \{z \mid \varphi(z)\}$ is a set.
(3) (Axiom of Collection)
$\bigwedge x \bigvee y\, \varphi(x, y) \to \bigwedge u \bigvee v \bigwedge x \in u \bigvee y \in v\, \varphi(x, y)$.
(4) (Axiom of Infinity)
$\omega$ is a set.

**Note** (3) implies the usual replacement axiom, but cannot be derived from it without the power set axiom.

ZFC$^-$ is ZF$^-$ together with the strong form of the axiom of choice:

(5) Every set is enumerable by an ordinal.

**Note** The power set axiom is required to derive (5) from the weaker forms of choice.

The *Levy hierarchy* of formulae is defined in the usual way:
$\Sigma_0$ formulae are the formulae containing only bounded quantification – i.e. $\Sigma_0$ = the smallest set of formulae containing the primitive formulae and closed under sentential operations and bounded quantification:

$$\bigwedge x \in y\, \varphi, \quad \bigvee x \in y\, \varphi$$

(where $\bigwedge x \in y\, \varphi = \bigwedge x(x \in y \to \varphi)$ and $\bigvee x \in y\, \varphi = \bigvee x(x \in y \wedge \varphi)$). (In some contexts it is useful to introduce bounded quantifiers as primitive signs rather than defined operations.)

We set: $\Pi_0 = \Sigma_0$. $\Sigma_{n+1}$ formulae are then the formulae of the form $\bigvee x\, \varphi$, where $\varphi$ is $\Pi_n$. Similarly $\Pi_{n+1}$ formulae have the form $\bigwedge x\, \varphi$, where $\varphi$ is $\Sigma_n$.

A relation $R$ on the model $\mathfrak{A}$ is called $\Sigma_n(\mathfrak{A})$ $(\Pi_n(\mathfrak{A}))$ iff it is definable over $\mathfrak{A}$ by a $\Sigma_n$ $(\Pi_n)$ formula.

$R$ is $\Sigma_n(\mathfrak{A})$ $(\Pi_n(\mathfrak{A}))$ *in the parameters* $p_1, \ldots, p_m$ iff it is $\Sigma_n(\Pi_n)$ definable in the parameters $p_1, \ldots, p_n \in \mathfrak{A}$. It is $\underline{\Sigma}_n(\mathfrak{A})$ $(\underline{\Pi}_n(\mathfrak{A}))$ iff it is $\Sigma_n$ $(\Pi_n)$ definable in some parameters. It is $\Delta_n(\mathfrak{A})$ iff it is $\Sigma_n(\mathfrak{A})$ and $\Pi_n(\mathfrak{A})$.

⋆ ⋆ ⋆ ⋆ ⋆

$\overline{\overline{x}}$ or card$(x)$ denotes the cardinality of $x$. (We reserve the notation $|x|$ for other uses.)

If $r$ is a well ordering or a set of ordinals, then $\mathrm{otp}(r)$ denotes its order type. $\mathrm{crit}(f)$ is the critical point of the function $f$ (i.e. $\alpha = \mathrm{crit}(f) \leftrightarrow (f \upharpoonright \alpha = \mathrm{id} \wedge f(\alpha) > \alpha)$.

$F''A$ is the image of $A$ under the function (or relation) $F$.

$\mathrm{rng}(R)$ is the range of the relation $R$.

$\mathrm{dom}(R)$ is the domain of the relation $R$.

$\mathrm{TC}(x)$ is the transitive closure of $x$, $H_\alpha = \{x \mid \overline{\overline{\mathrm{TC}(x)}} < \alpha\}$.

### *Boolean Algebras and Forcing*

The theory of forcing can be developed using "sets of conditions" or complete Boolean algebras. The former is most useful when we attempt to devise a forcing for a specific end. The latter is more useful when we deal with the general theory of forcing, as in the theory of iterated forcing. We adopt here an integrated approach which begins with Boolean algebras. By a Boolean algebra we mean a partial ordering $\mathbb{B} = \langle |\mathbb{B}|, c_{\mathbb{B}} \rangle$ with maximal and minimal elements 0, 1, lattice operations $\cap$, $\cup$ defined by:

$$a \subset (b \cap c) \longleftrightarrow (a \subset b \wedge a \subset c)$$

$$(b \cup c) \subset a \longleftrightarrow (b \subset a \wedge c \subset a)$$

and a complement operation $\neg$ defined by:

$$a \subset \neg b \longleftrightarrow a \cap b = 0,$$

satisfying the usual Boolean equalities. We call $\mathbb{B}$ a *complete Boolean algebra* if, in addition, for each $X \subset \mathbb{B}$ there are operations $\bigcap^{\mathbb{B}} X$, $\bigcup^{\mathbb{B}} X$ defined by:

$$a \subset \bigcap{}^{\mathbb{B}} X \longleftrightarrow \bigwedge b \in X\; a \subset b,$$

$$\bigcup{}^{\mathbb{B}} X \subset a \longleftrightarrow \bigwedge b \in X\; b \subset a,$$

s.t.

$$a \cap \bigcup_{b \in I} b = \bigcup_{b \in I} (a \cap b), \qquad a \cup \bigcap_{b \in I} = \bigcap_{b \in I} (a \cap b).$$

We shall generally write 'BA' for 'Boolean algebra'.

We write $\mathbb{A} \subseteq \mathbb{B}$ to mean that $\mathbb{A}$, $\mathbb{B}$ are BA's, $\mathbb{A}$ is complete, and $\mathbb{A}$ is completely contained in $\mathbb{B}$ – i.e.

$$\bigcap{}^{\mathbb{B}} X = \bigcap{}^{\mathbb{A}} X, \quad \bigcup{}^{\mathbb{B}} X = \bigcup{}^{\mathbb{A}} X \text{ for } X \subset \mathbb{A}.$$

If $\mathbb{A} \subseteq \mathbb{B}$ and $b \in \mathbb{B}$, we define $h(b) = h_{\mathbb{A},\mathbb{B}}(b)$ by: $h(b) = \bigcap\{a \in \mathbb{A} \mid b \subset a\}$. Thus:

- $h(\bigcup_i b_i) = \bigcup_i h(b_i)$ if $b_i \in \mathbb{B}$ for $i \in I$.
- $h(a \cap b) = a \cap h(b)$ if $a \in \mathbb{A}$.
- $b = 0 \leftrightarrow h(b) = 0$ for $b \in \mathbb{B}$.

⋆ ⋆ ⋆ ⋆ ⋆

If $\mathbb{B}$ is a complete BA, we can form the canonical *maximal* $\mathbb{B}$*-valued model* $\mathbf{V}^{\mathbb{B}}$. The elements of $\mathbf{V}^{\mathbb{B}}$ are called *names* and there is a valuation function $\varphi \to [[\varphi]]_{\mathbb{B}}$ attaching to each statement $\varphi = \varphi'(t_1, \ldots, t_n)$ a truth value in $\mathbb{B}$. (Here $\varphi$ is a ZFC formula and $t_1, \ldots, t_n$ are names.) All axioms of ZFC have truth value 1 (assuming ZFC). The sentential connectives are interpreted by:

$$[[\varphi \wedge \psi]] = [[\varphi]] \cap [[\psi]]; \quad [[\varphi \vee \psi]] = [[\varphi]] \cup [[\psi]];$$
$$[[\varphi \to \psi]] = [[\varphi]] \Rightarrow [[\psi]], \text{ where } (a \Rightarrow b) =_{\mathrm{Df}} \neg a \cup b;$$
$$[[\neg\varphi]] = \neg[[\varphi]].$$

The quantifiers are interpreted by:

$$[[\bigwedge v\, \varphi(v)]] = \bigcap_{x \in \mathbf{V}^{\mathbb{B}}} [[\varphi(x)]], \quad [[\bigvee v\, \varphi(v)]] = \bigcup_{x \in \mathbf{V}^{\mathbb{B}}} [[\varphi(x)]].$$

If $u \subset \mathbf{V}^{\mathbb{B}}$ is a set and $f : u \to \mathbb{B}$, then there is a name $x \in \mathbf{V}^{\mathbb{B}}$ s.t.

$$[[y \in x]] = \bigcup_{z \in u} [[y = z]] \cap f(x)$$

for all $y \in \mathbf{V}^{\mathbb{B}}$. Conversely, for each $x \in \mathbf{V}^{\mathbb{B}}$ there is a set $u_x \subset \mathbf{V}^{\mathbb{B}}$ s.t.

$$[[y \in x]] = \bigcup_{z \in u_x} [[y = z]] \cap [[z \in x]].$$

We can, in fact, arrange things s.t. $\{\langle z, x\rangle \mid z \in u_x\}$ is a well founded relation. If $U \subset \mathbf{V}^{\mathbb{B}}$ is a class $j$ and $A : U \to \mathbb{B}$, we may add to the language a predicate $\mathring{A}$ interpreted by: $[[\mathring{A}x]] = \bigcup_{z \in u} [[x = z]] \cap A(z)$. We inductively define for each $x \in \mathbf{V}$ a name $\check{x}$ by:

$$[[y \in \check{x}]] = \bigcup_{z \in x} [[y = \check{z}]],$$

and a predicate $\check{\mathbf{V}}$ by:

$$[[y \in \check{\mathbf{V}}]] = \bigcup_{\tau \in \mathbf{V}} [[y = \check{z}]].$$

If $\sigma : \mathbb{A} \overset{\sim}{\leftrightarrow} \mathbb{B}$ is an isomorphism, then we can define an injection $\sigma^* : \mathbf{V}^{\mathbb{A}} \to \mathbf{V}^{\mathbb{B}}$ as follows: Let $R = \{\langle z, x\rangle \mid z \in u_x\}$ be the above mentioned well founded relation for $\mathbf{V}^{\mathbb{A}}$. By $R$-induction we define $\sigma^*(x)$, picking $\sigma^*(x)$ to be a $w \in \mathbf{V}^{\mathbb{B}}$ s.t.

$$[[y \in w]]_{\mathbb{B}} = \bigcup_{z \in u_x} [[y = \sigma^*(z)]]_{\mathbb{B}} \cap \sigma([[z \in x]]_{\mathbb{A}}).$$

Then:

$$(1) \quad \sigma([[\varphi(\vec{x})]]_{\mathbb{A}}) = [[\varphi(\sigma^*(\vec{x}))]]_{\mathbb{B}}$$

for all ZFC formulae and all $x_1, \ldots, x_n \in \mathbf{V}^{\mathbb{A}}$. If $\sigma : \mathbb{A} \to \mathbb{B}$ is a complete embedding (i.e. $\sigma : \mathbb{A} \overset{\sim}{\leftrightarrow} \mathbb{A}' \subseteq \mathbb{B}$ for some $\mathbb{A}'$), then $\sigma^*$ can be defined the same way, but (1) then holds only for $\Sigma_0$ formulae. In such contexts it is often useful to take $\mathbf{V}^{\mathbb{B}}$ as a $\mathbb{B}$-valued *identity model*, meaning that

$$[[x = y]] = 1 \longrightarrow x = y \quad \text{for } x, y \in \mathbf{V}^{\mathbb{B}}.$$

(If $\mathbf{V}^{\mathbb{B}}$ does not already have this property, we can attain it by factoring.) If $\sigma : \mathbb{A} \overset{\sim}{\leftrightarrow} \mathbb{B}$ and $\mathbf{V}^{\mathbb{A}}$, $\mathbf{V}^{\mathbb{B}}$ are identity models, then $\sigma^*$ is bijective (and is, in fact, an isomorphism of $\langle \mathbf{V}^{\mathbb{A}}, I^{\mathbb{A}}, E^{\mathbb{A}} \rangle$ onto $\langle \mathbf{V}^{\mathbb{B}}, I^{\mathbb{B}}, E^{\mathbb{B}} \rangle$, where $I = (x, y) = [[x = y]]$, $E(x, y) = [[x \in y]]$). Another advantage of identity models in that $\{z \mid [[z \in x]] = 1\}$ is then a set, rather than a proper class.

There are many ways to construct a maximal $\mathbb{B}$-valued model $\mathbf{V}^{\mathbb{B}}$ and we can take its elements as being anything we want. Noting that $\mathbb{A} \subseteq \mathbb{B}$ means that $\mathrm{id} \restriction \mathbb{A}$ is a complete embedding, it is useful, when dealing with such a pair $\mathbb{A}$, $\mathbb{B}$, to arrange that $\mathbf{V}^{\mathbb{A}} \subset \mathbf{V}^{\mathbb{B}}$ and $(\mathrm{id} \restriction \mathbb{A})^* = \mathrm{id} \restriction \mathbf{V}^{\mathbb{A}}$. (We express this by: $\mathbf{V}^{\mathbb{A}} \subseteq \mathbf{V}^{\mathbb{B}}$.)

The *forcing relation* $\Vdash_{\mathbb{B}}$ is defined by:

$$b \Vdash \varphi \longleftrightarrow_{\mathrm{Df}} (b \neq 0 \wedge b \subset [[\varphi]]).$$

We also set: $\Vdash \varphi \leftrightarrow_{\mathrm{Df}} [[\varphi]] = 1$. Now suppose that $W$ is an inner model of ZF and $\mathbb{B} \in W$ is complete in the sense of $W$. We can form $W^{\mathbb{B}}$ internally in $W$, and it turns out that all ZF axioms are true in $W^{\mathbb{B}}$. (If $W$ satisfies ZFC, then ZFC holds in $W^{\mathbb{B}}$.) $W$ could also be a set rather than a class. If $W$ is only a model of $\mathrm{ZF}^-$, we can still form $W^{\mathbb{B}}$, which will then model $\mathrm{ZF}^-$ (or $\mathrm{ZFC}^-$ if $W$ models $\mathrm{ZFC}^-$). (In this case, however, we may not be able – internally in $W$ – to factor $W^{\mathbb{B}}$ to an identity model.)

We say that $G \subset \mathbb{B}$ is $\mathbb{B}$-*generic over* $W$ iff $G$ is an ultrafilter on $\mathbb{B}$ which respects all intersections and unions of $X \subset \mathbb{B}$ s.t. $X \in W$ – i.e.

$$\bigcap x \in G \longleftrightarrow \bigwedge b \in x \; b \in G, \quad \bigcup x \in G \longleftrightarrow \bigvee b \in X \; b \in G.$$

If $G$ is generic, we can form the *generic extension* $W[G]$ of $W$ by:

$$W[G] = \{x^G \mid x \in W^G\}, \text{ where } x^G = \{z^G \mid z \in u_x \wedge [[z \in x]] \in G\}.$$

Then $W \subset W[G]$, since $\check{x}^G = x$ (by $G$-induction on $x \in W$). Then:

$$W[G] \models \varphi(x_1^G, \ldots, x_n^G) \longleftrightarrow \bigvee b \in G \ b \Vdash \varphi(x_1, \ldots, x_n).$$

If we suppose, moreover, that for every $b \in \mathbb{B} \setminus \{0\}$ there is a generic $G \ni b$ (e.g. if $\varphi(\mathbb{B}) \cap W$ is countable), then:

$$b \Vdash \varphi(x_1, \ldots, x_n) \longleftrightarrow (W[G] \models \varphi(x_1^G, \ldots, x_n^G) \text{ for all generic } G \ni b).$$

If $\mathbb{B}$ is complete in $\mathbf{V}$ we shall often find it useful to work in a mythical universe in which:

(*) $\mathbf{V}$ is an inner model and for every $b \in \mathbb{B} \setminus \{0\}$ there is a $G \ni b$ which is $\mathbb{B}$-generic over $\mathbf{V}$.

This is harmless, since if $\mathbb{C}$ collapsed $2^{\overline{\overline{\mathbb{B}}}}$ to $\omega$, then (*) holds of $\check{\mathbf{V}}$, $\check{\mathbb{B}}$ in $\mathbf{V}^{\mathbb{C}}$. We note that there is a $\mathring{G} \in \mathbf{V}^{\mathbb{B}}$ s.t. $\Vdash \mathring{G} \subset \check{\mathbb{B}}$ and $[[\check{b} \in \mathring{G}]] = b$ for $b \in \mathbb{B}$. ($\mathring{G}$ is in fact unique if $\mathbf{V}^{\mathbb{B}}$ is an identity model.) If then $G$ is $\mathbb{B}$-generic over $\mathbf{V}$, we have $\mathring{G}^G = G$. Thus $\Vdash \mathring{G}$ is $\check{\mathbb{B}}$-generic over $\check{\mathbf{V}}$. We call $\mathring{G}$ the *canonical* $\mathbb{B}$-*generic name*.

If our language contains predicates $\mathring{A}$ other than $0$, $\in$, we set:

$$\mathring{A}^G = \{x^G \mid [[x \in \mathring{A}]] \in G\}.$$

Since $[[x \in \check{V}]] = \bigcup_{z \in \mathbf{V}} [[x = \check{z}]]$, we get:

$$\check{\mathbf{V}}^G = \{\check{z}^G \mid z \in \mathbf{V}\} = \mathbf{V}.$$

## Sets of Conditions

By a *set of conditions* we mean $\mathbb{P} = \langle |\mathbb{P}|, \leq_{\mathbb{P}} \rangle$ s.t. $\leq = \leq_p$ is a transitive relation on $|\mathbb{P}|$. (Notationally we shall not distinguish between $\mathbb{P}$ and $|\mathbb{P}|$.) We say that two conditions $p$, $q$ are *compatible* ($p \| q$) if $\bigvee r \ r \leq p, q$. Otherwise they are *incompatible* ($p \perp q$). For each set of conditions $\mathbb{P}$ there is a *canonical complete* BA *over* $\mathbb{P}$ (BA($\mathbb{P}$)) defined as follows: For $X \subset \mathbb{P}$ set:

$$\neg X = \{q \mid \bigwedge p \in X \ p \perp q\}.$$

Then $X \subset \neg\neg X$ and $\neg\neg\neg X = \neg X$. Hence $\neg\neg$ is a hull operator on $\mathfrak{P}(\mathbb{P})$. Set $|\mathbb{B}| = \{X \subset \mathbb{P} \mid X = \neg\neg X\}$. Then BA($\mathbb{P}$) $= \langle |\mathbb{B}|, c \rangle$, where $c$ is the ordinary inclusion relation on $|\mathbb{B}|$. $\mathbb{B} = $ BA($\mathbb{P}$) is then a complete BA with the complement operation $\neg$ and intersection and union operations given by:

$$\bigcap^{\mathbb{B}} X = \bigcap X, \quad \bigcup^{\mathbb{B}} X = \neg\neg \bigcup X.$$

We say that $\Delta \subset \mathbb{P}$ is *dense in* $\mathbb{P}$ iff $\bigwedge p \in \mathbb{P} \bigvee q \subseteq p, q \in \Delta$. $\Delta$ is *predense in* $\mathbb{P}$ iff $\bigwedge p \in \mathbb{P} \bigvee q\ (q \| p$ and $q \in \Delta)$. (In other words, the closure of $\Delta$ under $\leq$ is dense in $\mathbb{P}$.)

Set: $[p] = \neg\neg\{p\}$ for $p \in \mathbb{P}$ (i.e. $[p] = \bigcap\{a \in \mathrm{BA}(\mathbb{P}) \mid a \supset b\}$). The *forcing relation* for $\mathbb{P}$ is defined by:

$$p \Vdash \varphi \longleftrightarrow_{\mathrm{Df}} [p] \subset [[\varphi]].$$

If $\mathbb{P} \in W$ and $W$ is a transitive model of ZF, we say that $G$ is $\mathbb{P}$-*generic over* $W$ iff the following hold:

- If $p, q \in G$, then $p \| q$.
- If $p \in G$ and $p \leq q$, then $q \in G$.
- If $\Delta \in W$ is dense in $\mathbb{P}$, then $G \cap \Delta \neq \emptyset$.

If $\mathbb{B} = \mathrm{BA}(\mathbb{P})_W$ is the complete BA over $\mathbb{P}$ (as defined in $W$), then it follows that $G$ is $\mathbb{P}$-generic over $W$ iff $F = F_G = \{b \in \mathbb{B} \mid b \cap G \neq \emptyset\}$ is $\mathbb{B}$-generic over $W$. Conversely, if $F$ is $\mathbb{B}$-generic, thus $G = G_F = \{p \mid [p] \in F\}$ is $\mathbb{P}$-generic.

We also note that if $\mathbb{B}$ is a complete BA, then $\langle \mathbb{B} \setminus \{0\}, \subset \rangle$ is a set of conditions, and there is an isomorphism $\sigma : \mathbb{B} \overset{\sim}{\leftrightarrow} \mathrm{BA}(\mathbb{B} \setminus \{0\})$ defined by: $\sigma(b) = \{a \mid a \supset b\}$. Moreover, $G$ is a $\mathbb{B}$-generic filter iff it is a $\mathbb{B} \setminus \{0\}$-generic set. When dealing with Boolean algebras, we shall often write: "$\Delta$ is dense in $\mathbb{B}$" to mean "$\Delta$ is dense in $\mathbb{B} \setminus \{0\}$".

### *The Two Step Iteration*

Let $\mathbb{A} \subseteq \mathbb{B}$, where $\mathbb{A}$, $\mathbb{B}$ are both complete. If (in some larger universe) $G$ is $\mathbb{A}$-generic over $\mathbf{V}$, then $G' = \{b \in \mathbb{B} \mid \bigvee a \in G\ a \subset b\}$ is a complete filter on $\mathbb{B}$ and we can form the factor algebra $\mathbb{B}/G'$ (which we shall normally denote by $\mathbb{B}/G$). It is not hard to see that $\mathbb{B}/G$ is then complete in $\mathbf{V}[G]$. By the definition of the factor algebra there is a canonical homomorphism $\sigma : \mathbb{B} \to \mathbb{B}/G$ s.t. $\sigma(b) \subset \sigma(c) \leftrightarrow \neg b \cup c \in G'$. When the context permits we shall write $b/G$ for $\sigma(b)$. We now list some basic facts about this situation.

**Fact 1** Let $\mathbb{B}_0 \subseteq \mathbb{B}_1$, $\mathbb{B}_0$ and $\mathbb{B}_1$ being complete. Let $G_0$ be $\mathbb{B}_0$-generic over $\mathbf{V}$ and let $\tilde{G}$ be $\tilde{\mathbb{B}} = \mathbb{B}_1/G$-generic over $\mathbf{V}[G]$. Set $G_1 = G_0 * \tilde{G} =_{\mathrm{Df}} \{b \in \mathbb{B}_1 \mid b/G_0 \in \tilde{G}\}$. Then $G_1$ is $\mathbb{B}_1$-generic over $\mathbf{V}$ and $\mathbf{V}[G_1] = \mathbf{V}[G_0][\tilde{G}]$.

Conversely we have:

**Fact 2** If $G_1$ is $\mathbb{B}_1$-generic over $\mathbf{V}$ and we set: $G_0 = \mathbb{B}_0 \cap G_1$, $\tilde{G} = \{b/G_0 \mid b \in G_1\}$. Then $G_0$ is $\mathbb{B}_0$-generic over $\mathbf{V}$, $\tilde{G}$ is $\mathbb{B}_1/G_0$-generic over $\mathbf{V}[G_0]$ and $G_1 = G_0 * \tilde{G}$.

**Fact 3** Let $\mathbb{A} \subseteq \mathbb{B}$ and let $h = h_{\mathbb{A},\mathbb{B}}$ as defined above. Then

$$h(b) = [[\check{b}/\mathring{G} \neq 0]]_{\mathbb{A}},$$

$\mathring{G}$ being the $\mathbb{A}$-generic name.

*Proof.* $h(b) = \bigcap\{a \in \mathbb{A} \mid a \supset b\} = \bigcap_{a \in \mathbb{A}} ([[\check{a} \supset \check{b}]] \Rightarrow a)$ where $a = [[\check{a} \in \mathring{G}]] = \bigcap_{a \in \mathbb{A}} [[\check{a} \supset \check{b} \rightarrow \check{a} \in \mathring{G}]] = [[\bigwedge a \in \check{\mathbb{A}}(a \supset \check{b} \rightarrow a \in \mathring{G})]] = [[\check{b}/\mathring{G} \neq 0]]$

QED(Fact 3)

**Fact 4** Let $\mathbb{A} \subseteq \mathbb{B}$ and $\Vdash_{\mathbb{A}} \check{b} \in \check{\mathbb{B}}/\mathring{G}$, where $\hat{b} \in \mathbf{V}^{\mathbb{A}}$. There is a unique $b \in \mathbb{B}$ s.t. $\Vdash_{\mathbb{A}} \mathring{b} = \check{b}/\mathring{G}$.

*Proof.* To see uniqueness, let $\Vdash \check{b}/\mathring{G} = \check{b}'/\mathring{G}$. Then $\Vdash \check{b} \setminus \check{b}'/\mathring{G} = 0$. Hence $h(b \setminus b') = [[\check{b} \setminus \check{b}'/\mathring{G} \neq 0]] = 0$. Hence $b \setminus b' = 0$. Hence $b \subset b'$. Similarly $b' \subset b$.
To see the existence, note that $\Delta = \{a \in \mathbb{A} \mid \bigvee b \; a \Vdash \mathring{b} = \check{b}/\mathring{G}\}$ is dense in $\mathbb{A}$. Let $X$ be a maximal antichain in $\Delta$. Let $a \Vdash \mathring{b} = \check{b}_a/\mathring{G}$ for $a \in X$. Set: $b = \bigcup_{a \in X} a \cap b_a$. Then $\Vdash \mathring{b} = \check{b}/\mathring{G}$, since if $G$ is $\mathbb{A}$-generic there is $a \in X \cap G$ by genericity. Hence $\mathring{b}^G = b_a/G = b/G$. QED(Fact 4)

Fact 2 shows that, if $\mathbb{B}_0 \subseteq \mathbb{B}_1$, then forcing with $\mathbb{B}_1$ is equivalent to a *two step iteration*: Forcing first by $\mathbb{B}_0$ to get $\mathbf{V}[G_0]$ and then by a $\tilde{\mathbb{B}} \in \mathbf{V}[G_0]$.

We now show the converse: Forcing by $\mathbb{B}_0$ and then by some $\tilde{\mathbb{B}}$ is equivalent to forcing by a single $\mathbb{B}_1$:

**Fact 5** Let $\mathbb{B}_0$ be complete and let $\Vdash_{\mathbb{B}_0} \mathring{\mathbb{B}}$ is complete. There is $\mathbb{B}_1 \supseteq \mathbb{B}_0$ s.t. $\Vdash_{\mathbb{B}_0}$ ($\mathring{\mathbb{B}}$ is isomorphic to $\check{\mathbb{B}}_1/\mathring{G}$). (Hence, whenever $G_0$ is $\mathbb{B}_0$-generic, we have $\mathbb{B}_1/G_0 \simeq \tilde{\mathbb{B}} =_{\mathrm{Df}} \mathring{\mathbb{B}}^{G_0}$.)

In order to prove this we first define:

**Definition** Let $\Vdash_{\mathbb{A}} \mathring{\mathbb{B}}$ is complete. $\mathbb{B} = \mathbb{A} * \mathring{\mathbb{B}}$ is the BA defined as follows: Assume $\mathbf{V}^{\mathbb{A}}$ to be an identity model and set:

$$|\mathbb{B}| =_{\mathrm{Df}} \{b \in \mathbf{V}^{\mathbb{A}} \mid \Vdash_{\mathbb{A}} b \in \mathring{\mathbb{B}}\}, \quad b \subset c \text{ in } \mathbb{B} \longleftrightarrow_{\mathrm{Df}} \Vdash_{\mathbb{A}} b \subset c.$$

This defines $\mathbb{B} = \langle |\mathbb{B}|, \subset \rangle$. $\mathbb{B}$ is easily seen to be a BA with the operations:

$$a \cap b = \text{ that } \ c \ \text{ s.t. } \ \Vdash_{\mathbb{A}} c = a \cap b,$$
$$a \cup b = \text{ that } \ c \ \text{ s.t. } \ \Vdash_{\mathbb{A}} c = a \cup b,$$
$$\neg b = \text{ that } \ c \ \text{ s.t. } \ \Vdash_{\mathbb{A}} c = \neg b.$$

Similarly, if $\langle b_i \mid i \in I \rangle$ is any sequence of elements of $\mathbb{B}$, there is a $\mathring{\mathbb{B}} \in \mathbf{V}^{\mathbb{A}}$ defined by:

$$\Vdash_{\mathbb{A}} \mathring{\mathbb{B}} : \check{I} \longrightarrow \mathring{\mathbb{B}}; \quad \Vdash_{\mathbb{A}} \mathring{\mathbb{B}}(\check{i}) = b_i \ \text{ for } \ i \in I.$$

We then have:

$$\bigcap_{i \in I} b_i = \text{ that } \ c \ \text{ s.t. } \ \Vdash_{\mathbb{A}} c = \bigcap_{i \in \check{I}} \mathring{\mathbb{B}}(i),$$
$$\bigcup_{i \in I} b_i = \text{ that } \ c \ \text{ s.t. } \ \Vdash_{\mathbb{A}} c = \bigcup_{i \in \check{I}} \mathring{\mathbb{B}}(i),$$

showing that $\mathbb{B}$ is complete. Now define $\sigma : \mathbb{A} \to \mathbb{B}$ by:

$$\sigma(a) = \text{ that } \ c \ \text{ s.t. } \ \Vdash_{\mathbb{A}} (a \in \mathring{G} \wedge c = 1) \vee (a \notin \mathring{G} \wedge c = 0).$$

$\sigma$ is easily shown to be a complete embedding.

Clearly, if $G$ is $\mathbb{A}$-generic, then $\sigma''G$ is $\sigma''\mathbb{A}$-generic, and $\mathbf{V}[G] = \mathbf{V}[\sigma''G]$. Set $\tilde{G} = \sigma''G$, $\tilde{\mathbb{B}} = \mathbb{B}/\tilde{G}$. We then have for $b, c \in \mathbb{B}$:

$$b/\tilde{G} \subset c/\tilde{G} \longleftrightarrow \bigvee a \in G \ \sigma(a) \subset (\neg b \cup c)$$
$$\longleftrightarrow (\neg b^G \cup c^G) = 1 \longleftarrow b^G \subset c^G \ \text{(since } \sigma(a)^G = 1 \text{ for } a \in G\text{)}.$$

Hence there is $k : \tilde{\mathbb{B}} \overset{\sim}{\leftrightarrow} \mathring{\mathbb{B}}^G$ defined by: $k(b/\tilde{G}) = b^G$. Hence:

$$\Vdash_{\mathbb{A}} (\mathring{\mathbb{B}} \ \text{ is isomorphic to } \ \check{\mathbb{B}}/G).$$

If $\mathbb{A} = \mathbb{B}_0$ and we pick $\mathbb{B}_1, \pi : \mathbb{B} \overset{\sim}{\leftrightarrow} \mathbb{B}_1$ with $\pi\sigma = \text{id}$, then $\mathbb{B}_1$ satisfies Fact 5.

QED

The algebra $\mathbb{A} * \mathring{\mathbb{B}}$ constructed above is often useful.

### *General Iterations*

It is clear from the foregoing that an $n$-step iteration – i.e. the result of $n$ successive generic extensions of $\mathbf{V}$ – can be adequately described by a sequence $\langle \mathbb{B}_i \mid i < n \rangle$ s.t. $\mathbb{B}_i \subseteq \mathbb{B}_j$ for $i \leq j < n$. The final model is the result of forcing with $\mathbb{B}_{n-1}$. What about transfinite iterations? At first glance it might seem that there is no such notion, but in fact we can define the notion by turning the previous analysis on its head. We define:

**Definition** By an *iteration* of length $\alpha > 0$ we mean a sequence $\langle \mathbb{B}_i \mid i < \alpha \rangle$ of complete BA's s.t.

- $\mathbb{B}_i \subseteq \mathbb{B}_j$ for $i \leq j < \alpha$.
- If $\lambda < \alpha$ is a limit ordinal, then $\mathbb{B}_\lambda$ is generated by $\bigcup_{i<\lambda} \mathbb{B}_i$, i.e. there is no proper $B \subset \mathbb{B}_\lambda$ s.t. $\bigcup_{i<\lambda} \mathbb{B}_i \subset B$ and $\bigcap X, \bigcup X \in B$ for all $X \subset B$.

If $G_i$ is $\mathbb{B}_{i+1}$-generic and $G_i = G \cap \mathbb{B}_i$, then $\mathbf{V}[G] = \mathbf{V}[G_i][\tilde{G}_i]$ where $\tilde{G}_i = \{b/G_i \mid b \in G\}$ is $\tilde{\mathbb{B}}_i = \mathbb{B}_{i+1}/G_i$-generic. If $G$ is $\lambda$-generic for a limit $\lambda$, then $\mathbf{V}[G]$ can be regarded as a "limit" of successive $\tilde{\mathbb{B}}_i$-generic extensions, where $G_i = G \cap \mathbb{B}_i$, $\tilde{\mathbb{B}}_i = \mathbb{B}_{i+1}/G_i$ for $i < \lambda$.

In practice, we usually at the $i$-th stage pick a $\overset{\circ}{\mathbb{B}}_i$ s.t. $\Vdash_{\mathbb{B}_i}$ ($\overset{\circ}{\mathbb{B}}_i$ is a complete BA), and arrange that:

$$\Vdash_{\mathbb{B}_i} (\check{\mathbb{B}}_i/\overset{\circ}{G} \text{ is isomorphic to } \overset{\circ}{\mathbb{B}}).$$

If the construction of the $\mathbb{B}_i$'s is sufficiently canonical, then the iteration is completely characterized by the sequence of $\overset{\circ}{\mathbb{B}}_i$'s. However, our definition of "iteration" gives us great leeway in choosing $\mathbb{B}_\lambda$ for limit $\lambda < \alpha$. We shall make use of that freedom in these notes. Traditionally, however, a handful of standard limiting procedures has been used. The *direct limit* takes $\mathbb{B}_\lambda$ as the minimal completion of the Boolean algebra $\bigcup_{i<\lambda} \mathbb{B}_i$. It is characterized up to isomorphism by the property that $\bigcup_{i<\lambda} \mathbb{B}_i \setminus \{0\}$ lies dense in $\mathbb{B}_\lambda$. (If $\mathbb{B}^* = \mathrm{BA}(\bigcup_{i<\lambda} \mathbb{B}_i \setminus \{0\})$, there is then a unique isomorphism of $\mathbb{B}_\lambda$ onto $\mathbb{B}^*$ taking $b$ to $[b]$ for $b \in \bigcup_{i<\lambda} \mathbb{B}_i \setminus \{0\}$.) Another frequently used variant is the *inverse limit*, which can be defined as follows: By a *thread* in $\langle \mathbb{B}_i \mid i < \lambda \rangle$ we mean a $b = \langle b_i \mid i < \lambda \rangle$ s.t. $b_j \in \mathbb{B}_j \setminus \{0\}$ and $h_{\mathbb{B}_i \mathbb{B}_j}(b_j) = b_i$ for $i \leq j < \lambda$. We call $\mathbb{B}_\lambda$ an *inverse limit* of $\langle \mathbb{B}_i \mid i < \lambda \rangle$ iff

- If $b$ is a thread, then $b^* = \bigcap_{i<\lambda} b_i \neq 0$ in $\mathbb{B}_\lambda$.
- The set of such $b^*$ is dense in $\mathbb{B}_\lambda$.

$\mathbb{B}_\lambda$ is then characterized up to isomorphism by these conditions. (If $T$ is the set of all threads, we can define a partial ordering of $T$ by: $b \leq c$ iff $\bigwedge i < \lambda\, b_i \subset c_i$.) If we then set: $\mathbb{B}^* = \mathrm{BA}(T)$, there is a unique isomorphism of $\mathbb{B}_\lambda$ onto $\mathbb{B}^*$ taking $b^*$ to $[b]$ for each thread $b$.)

By the *support* of a thread we mean the set of $j < \lambda$ s.t. $b_i \neq b_j$ for all $i < j$. The *countable support* (CS) limit is defined like the inverse limit using only those threads which have a countable support. A *CS iteration* is one in which $\mathbb{B}_\lambda$ is a CS limit for all limit $\lambda < \alpha$. (This is equivalent to taking the inverse limit at $\lambda$ of cofinality $\omega$ and otherwise the direct limit.) Countable support iterations tend to work well if no cardinal has its cofinality changed to $\omega$ in the course of the iteration. Otherwise – e.g. if we are trying to iterate Namba forcing – we can use the *revised countable support* (RCS) *iteration*, which was invented by Shelah. The present definition is due to Donder: By an RCS *thread* we mean a thread $b$ s.t. *either* there is $i < \lambda$ s.t. $b_i \Vdash_{\mathbb{B}_i} \mathrm{cf}(\check{\lambda}) = \omega$ *or* the support of $b$ is bounded in $\lambda$. The RCS limit is then defined like the inverse limit, using only RCS threads. An RCS iteration is one which uses the RCS limit at all limit points.

**Note** Almost all iterations which have been employed to date make use of sublimits of the inverse limit – i.e. $\{b^* \mid b \text{ is a thread } \wedge b^* \neq 0\}$ is dense in $\mathbb{B}_\lambda$ for all limit $\lambda$. This means that $(\prod\limits_{i<\lambda} \overline{\overline{\mathbb{B}}}_i)^+$ remains regular. In these notes, however, we shall see that it is sometimes necessary to employ larger limits which do not have this consequence.

In dealing with iterations we shall employ the following conventions: If $\mathbb{B} = \langle \mathbb{B}_i \mid i < \alpha \rangle$ is an iteration we assume the $\mathbf{V}^{\mathbb{B}_i}$ to be so constructed that $\mathbf{V}^{\mathbb{B}_i} \subseteq \mathbf{V}^{\mathbb{B}_j}$ (in the sense of our earlier definition). In particular $[[\varphi(\vec{x})]]_{\mathbb{B}_i} = [[\varphi(\vec{x}]]_{\mathbb{B}_j}$ for $x_1, \ldots, x_n \in \mathbf{V}^{\mathbb{B}_i}$, $i \leq j$, when $\varphi$ is a $\Sigma_0$ formula. We shall also often simplify the notation by using the indices $i < \alpha$ as in: $h_{ij}$ for $h_{\mathbb{B}_i\mathbb{B}_j}$, $\Vdash_i$ for $\Vdash_{\mathbb{B}_i}$, $[[\varphi]]_i$ for $[[\varphi]]_{\mathbb{B}_j}$. If $i_0 < \alpha$ and $G$ is $\mathbb{B}_{i_0}$-generic, we set: $\mathbb{B}/G = \langle \mathbb{B}_{i_0+j}/G \mid j < \alpha - i_0 \rangle$. We can assume the factor algebras to be so defined that $\mathbb{B}_{i_0+h}/G \subseteq \mathbb{B}_{i_0+j}/G$ for $h \leq j < \alpha - i_0$. ($\tilde{\mathbb{B}} = \bigcup\limits_{i<\alpha} \mathbb{B}_i$ is a BA. Hence we can form $\tilde{\mathbb{B}}/G$ and identify $\mathbb{B}_{i_0+j}/G$ with $\{b/G \mid b \in \mathbb{B}_{i_0+j}\}$.) It then follows easily that $\mathbb{B}/G$ is an iteration in $\mathbf{V}[G]$.

## 1. Admissible Sets

### 1.1. *Introduction*

Let $H = H_\omega$ be the collection of hereditarily finite sets. We use the usual Levy hierarchy of set theoretic formulae:

$\Pi_0 = \Sigma_0 =$ the set of all formulae in which all quantifiers are bounded.

$\Sigma_{n+1} =$ the set of all formulae $\bigvee x\, \varphi$ where $\varphi$ is $\Pi_n$.

$\Pi_{n+1} =$ the set of all formulae $\bigwedge x\, \varphi$ where $\varphi$ is $\Sigma_n$.

The use of $H$ offers an elegant way to develop ordinary recursion theory. Call a relation $R \subset H^n$ *r.e.* (or "$H$-r.e.") iff $R$ is $\Sigma_1$-definable over $H$. We call $R$ *recursive* (or *$H$-recursive*) iff it is $\Delta_1$-definable (i.e. $R$ and its complement $\neg R$ are $\Sigma_1$-definable). Then $R \subset \omega^n$ is rec (r.e.) in the usual sense iff it is the restriction of an $H$-rec. ($H$-r.e.) relation to $\omega$. Moreover, there is an $H$-recursive function $\pi : \omega \leftrightarrow H$ s.t. $R \subset H^n$ is $H$-recursive iff $\{\langle x_1, \ldots, x_n\rangle \mid R(\pi(x_1), \ldots, \pi(x_n))\}$ is recursive. (Hence $\{\langle x, y\rangle \mid \pi(x) \in \pi(y)\}$ is recursive.)

⋆ ⋆ ⋆ ⋆ ⋆

This suggests a way of relativizing the concepts of recursion theory to transfinite domains: Let $N = \langle |N|, \in, A_1, A_2, \ldots\rangle$ be a transitive structure (with finitely or infinitely many predicates). We define:

$$R \subset N^n \text{ is } N\text{-r.e. } (N\text{-rec.}) \text{ iff } R \text{ is } \underline{\Sigma}_1 \ (\underline{\Delta}_1) \text{ definable over } N.$$

Since $N$ may contain infinite sets, we must also relativize the notion "finite":

$$u \text{ is } N\text{-}\textit{finite} \text{ iff } u \in N.$$

There are, however, certain basic properties which we expect any recursion theory to possess. In particular:

- If $A$ is recursive and $u$ finite, then $A \cap u$ is finite.
- If $u$ is finite and $F : u \to N$ is recursive, then $F''u$ is finite.

The transitive structures $N = \langle |N|, \in, A_1, A_2, \ldots\rangle$ which yield a satisfactory recursion theory are called *admissible*. They were characterized by Kripke and Platek as those which satisfy the following axioms:

(1) $\emptyset$, $\{x, y\}$, $\bigcup x$ are sets.
(2) The $\Sigma_0$-axiom of subsets (Aussonderung)
$x \cap \{z \mid \varphi(z)\}$ is a set, where $\varphi$ is any $\Sigma_0$ formula.
(3) The $\Sigma_0$-axiom of collection
$\bigwedge x \bigvee y\, \varphi(x, y) \to \bigwedge u \bigvee v \bigwedge x \in u \bigvee y \in v\, \varphi(x, y)$ where $\varphi$ is any $\Sigma_0$ formula.

**Note** Applying (3) to: $x \in u \to \varphi(x, y)$, we get:

$$\bigwedge x \in u \bigvee y\, \varphi(x, y) \longrightarrow \bigvee v \bigwedge x \in u \bigvee y \in v\, \varphi(x, y).$$

**Note** *Kripke-Platek set theory* (KP) consists of the above axioms together with the axiom of extensionality and the full axiom of foundation (i.e. for all formulae, not just $\Sigma_0$ ones). These latter axioms of course hold trivially in transitive domains. KPC (KP with choice) is KP augmented by: Every set is enumerable by an ordinal.

We now show that admissible structures satisfy the criteria stated above.

**Lemma 1** *Let $u \in M$. Let $A$ be $\underline{\Delta}_1(M)$. Then $A \cap u \in M$.*

*Proof.* Let $Ax \leftrightarrow \bigvee y\, A_0yx$, $\neg Ax \leftrightarrow \bigvee y\, A_1yx$, where $A_0$, $A_1$ are $\underline{\Sigma}_0$. Then $\bigwedge x \bigvee y(A_0yx \vee A_1yx)$. Hence there is $v \in M$ s.t. $\bigwedge x \in u \bigvee y \in v(A_0yx \vee A_1yx)$. Hence $u \cap A = u \cap \{x \mid \bigvee y \in v\, A_0yx\} \in M$. QED(Lemma 1)

Before verifying the second criterion we prove:

**Lemma 2** *$M$ satisfies:*

$$\bigwedge x \in u \bigvee y_1 \dots y_n\, \varphi(x, \vec{y}) \longrightarrow \bigvee v \bigwedge x \in u \bigvee y_1 \dots y_n \in v\, \varphi(x, \vec{y})$$

*for $\Sigma_0$ formulas $\varphi$.*

*Proof.* Assume $\bigwedge x \in u \bigvee y_1 \dots y_n\, \varphi(x, \vec{y})$. Then

$$\bigwedge x \bigvee w(\underbrace{x \in u \to \bigvee y_1 \dots y_n \in w\, \varphi(x, \vec{y})}_{\Sigma_0}).$$

Hence there is $v' \in M$ s.t.
$\bigwedge x \in u \bigvee w \in v' \bigvee y_1 \dots y_n \in w\, \varphi(x, \vec{y})$. Take $v = \bigcup v'$. QED(Lemma 2)

Finally we get:

**Lemma 3** *Let $u \in M$, $u \subset \mathrm{dom}(F)$, where $F$ is $\underline{\Sigma}_1(M)$. Then $F''u \in M$.*

*Proof.* Let $y = F(x) \leftrightarrow \bigvee z\, F'zyx$, where $F'$ is $\underline{\Sigma}_0(M)$. Since $\bigwedge x \in u \bigvee y\; y = F(x)$, there is $v$ s.t. $\bigwedge x \in u \bigvee y, z \in v\, F'zyx$. Hence $F''u = v \cap \{y \mid \bigvee x \in u \bigvee z \in v\, F'zyx\}$. QED(Lemma 3)

By similarly straightforward proofs we get:

**Lemma 4** *If $Ry\vec{x}$ is $\Sigma_1$, so is $\bigvee y\, Ry\vec{x}$.*

**Lemma 5** *If $Ry\vec{x}$ is $\Sigma_1$, so is $\bigwedge y \in u\, Ry\vec{x}$ (since $\bigwedge y \in u \bigvee z\, \varphi(y, z) \leftrightarrow \bigvee v \underbrace{\bigwedge y \in v \bigvee z \in v\, \varphi(y, z)}_{\Sigma_0}$).*

**Lemma 6** *If $R, Q \subset M^n$ are $\Sigma_1$, then so are $R \cup Q$, $R \cap Q$.*

**Lemma 7** *If $R'(y_1, \dots, y_n)$ is $\Sigma_1$ and $f(x_1, \dots, x_m)$ is a $\Sigma_1$ function for $i = 1, \dots, n$, then $R(f_1(\vec{x}), \dots, f_n(\vec{x}))$ is $\Sigma_1$.*

*Proof.* $R(\vec{f}(\vec{x})) \leftrightarrow \bigvee y_1 \dots y_n (\bigwedge_{i=1}^{n} y_i = f_i(\vec{x}) \wedge R(\vec{y}))$. QED(Lemma 7)

**Note** The boldface versions of Lemmas 4–7 follow immediately.

**Corollary 8** *If the functions $f(z_1,\dots,z_n)$, $g_i(\vec{x})$ $(i=1,\dots,n)$ are $\Sigma_1$ in a parameter $p$, then so is $h(\vec{x}) \simeq f(g_1(\vec{x}),\dots,g_n(\vec{x}))$.*

**Lemma 9** *The following functions are $\Delta_1$: $\bigcup x$, $x \cup y$, $x \cap y$, $x \setminus y$ (set difference), $\{x_1,\dots,x_n\}$, $\langle x_1,\dots,x_n\rangle$, $\mathrm{dom}(x)$, $\mathrm{rng}(x)$, $x''y$, $x \upharpoonright y$, $x^{-1}$, $x \times y$, $(x)^n_i$, where: $(\langle z_0,\dots,z_{n-1}\rangle)_i = z_i$; $(u)^n_i = \emptyset$ otherwise;*

$$x[z] = \begin{cases} x(z) & \text{if } x \text{ is a function and } z \in \mathrm{dom}(x), \\ \emptyset & \text{if not.} \end{cases}$$

**Note** As a corollary of Lemma 3 we have: If $f$ is $\underline{\Sigma}_1$, $u \in M$, $u \subset \mathrm{dom}(f)$. Then $f \upharpoonright u \in M$, since $f \upharpoonright u = g''u$, where $g(x) \simeq \langle f(x), x\rangle$.

**Lemma 10** *If $f : M^{n+1} \to M$ is $\Sigma_1$ in the parameter $p$, then so are:*

$$F(u,\vec{x}) = \{f(z,\vec{x}) \mid z \in u\}, \quad F'(u,\vec{x}) = \langle f(z,\vec{x}) \mid z \in u\rangle.$$

*Proof.* $y = F(u,\vec{x}) \leftrightarrow \bigwedge z \in y \bigvee v \in u\ z = f(y,\vec{z}) \wedge \bigwedge v \in u \bigvee z \in y\ z = f(y,\vec{x})$.
But $F'(u,\vec{x}) = \{f'(z,\vec{x}) \mid z \in u\}$, where $f'(y,\vec{x}) = \langle f(y,\vec{x}),\vec{x}\rangle$.

QED(Lemma 10)

(**Note** The proof of Lemma 10 shows that, even if $f$ is not defined everywhere, $F$ is $\Sigma_1$ in $p$, where:

$$F(u,\vec{x}) \simeq \{f(y,\vec{x}) \mid y \in u\},$$

where this equation means that $F(u,\vec{x})$ is defined and has the displayed value iff $f(y,\vec{x})$ is defined for all $y \in u$. Similarly for $F'$.)

**Lemma 11** (Set Recursion Theorem) *Let $G$ be an $n+2$-ary $\Sigma_1$ function in the parameter $p$. Then there is $F$ which is also $\Sigma_1$ in $p$ s.t.*

$$F(y,\vec{x}) \simeq G(y,\vec{x},\langle F(z,\vec{x}) \mid z \in y\rangle)$$

*(where this equation means that $F$ is defined with the displayed value iff $F(z,\vec{x})$ is defined for all $z \in y$ and $G$ is defined at $\langle y,\vec{x},\langle F(z,\vec{x}) \mid z \in y\rangle\rangle$.)*

*Proof.* Set $u = F(y,\vec{x}) \leftrightarrow \bigvee f(\varphi(f,\vec{x}) \wedge \langle u,y\rangle \in f)$, where

$$\begin{aligned}\varphi(f,\vec{x}) \longleftrightarrow (f \text{ is a function } \wedge \textstyle\bigcup \mathrm{dom}(f) \subset \mathrm{dom}(f) \\ \wedge \textstyle\bigwedge y \in \mathrm{dom}(f)\ f(y) = G(y,\vec{x},f \upharpoonright y)).\end{aligned}$$

The equation is verified by $\in$-induction on $y$. QED(Lemma 11)

**Corollary 12** *TC, rn are $\Delta_1$ functions, where*

$$\text{TC}(x) = \text{ the transitive closure of } x = x \cup \bigcup_{z \in x} \text{TC}(z),$$

$$\text{rn}(x) = \text{ the rank of } \text{lub}\{\text{rn}(z) \mid z \in x\}.$$

**Lemma 13** *$\omega$, $On \cap M$ are $\Sigma_0$ classes.*

*Proof.* $x \in On \leftrightarrow (\bigcup x \subset x \wedge \bigwedge z, w \in x(z \in w \vee w \in z))$, $x \in \omega \leftrightarrow (x \in On \wedge \neg\text{Lim}(x) \wedge \bigwedge y \in x \neg\text{Lim}(y))$, where $\text{Lim}(x) \leftrightarrow (x \neq 0 \wedge x \in On \wedge x = \bigcup x)$.

**Corollary 14** *The ordinal functions $\alpha + 1, \alpha + \beta, \alpha \cdot \beta, \alpha^\beta, \ldots$ are $\Delta_1$.*

An even more useful version of Lemma 11 is

**Lemma 15** *Let $G$ be as in Lemma 11. Let $h : M \to M$ be $\Sigma_1$ in $p$ s.t. $\{\langle x, z\rangle \mid x \in h(z)\}$ is well founded. There is $F$ which is $\Sigma_1$ in $p$ s.t.,*

$$F(y, \vec{x}) \simeq G(y, \vec{x}, \langle F(z, \vec{x}) \mid z \in h(y)\rangle).$$

The proof is just as before. We also note:

**Lemma 16.1** *Let $u \in H_\omega$. Then the class $u$ and the constant function $f(x) = u$ are $\Sigma_0$.*

*Proof.* $\in$-induction on $u$: $x \in u \leftrightarrow \bigvee_{z \in u} x = z$, $x = u \leftrightarrow (\bigwedge z \in x\ z \in u \wedge \bigwedge_{z \in u} z \in x)$.

QED

**Lemma 16.2** *If $\omega \in M$, then the constant function $x = \omega$ is $\Sigma_0$.*

*Proof.* $x = \omega \leftrightarrow (\bigwedge z \in x\ z \in \omega \wedge \emptyset \in x \wedge \bigwedge z \in x\ z \cup \{z\} \in x)$.

**Lemma 16.3** *If $\omega \in M$, the constant for $x = H_\omega$ is $\Sigma_1$ (hence $\Delta_1$).*

*Proof.* $x = H_\omega \leftrightarrow (\bigwedge z \in x \bigvee u \bigvee f \bigvee n \in \omega(\bigcup n \subset u \wedge x \subset u \wedge f : n \leftrightarrow x)) \wedge \emptyset \in x \wedge \bigwedge z, w \in x(\{z, w\}, z \cup w \in x)$.

**Lemma 17** *Fin, $\mathfrak{P}_\omega(x)$ are $\Delta_1$, where* Fin $= \{x \in M \mid \overline{\overline{x}} < \omega\}$, $\mathfrak{P}_\omega(x) = \text{Fin} \cap \mathfrak{P}(x)$.

*Proof.* $x \in \text{Fin} \leftrightarrow \bigvee n \in \omega \bigvee f \text{ fin} \leftrightarrow x$,
$x \notin \text{Fin} \leftrightarrow \bigvee y(y = \omega \wedge \bigwedge n \in y \bigvee f \bigvee n \subset x \text{ fin} \leftrightarrow n)$,
$y = \mathfrak{P}_\omega(x) \leftrightarrow \bigwedge u \in y(u \in \text{Fin} \wedge u \subset x) \wedge \bigwedge z \in x(\{z\} \in y \wedge \bigwedge u, v \in y\ u \cup v \in y)$.

QED

The constructible hierarchy relative to a class $A$ is defined by:

$$L_0[A] = \emptyset; \quad L_{\nu+1}[A]d = \mathrm{Def}(\langle L_\nu[A], A \cap L_\nu[A]\rangle)$$
$$L_\lambda[A] = \bigcup_{\nu<\lambda} L_\nu[A] \quad \text{for limit } \lambda,$$

where $\mathrm{Def}(\mathfrak{A})$ is the set of $B \subset \mathfrak{A}$ which are $\mathfrak{A}$-definable in parameters from $\mathfrak{A}$. We also define $L_\nu = L_\nu[\emptyset]$.

The constructible hierarchy *over a set $u$* is defined by:

$$L_0(u) = \mathrm{TC}(\{u\}), \quad L_{\nu+1}(u) = \mathrm{Def}(L_\nu(u)),$$
$$L_\lambda(u) = \bigcup_{\nu<\lambda} L_\nu(u) \quad \text{for limit } \lambda.$$

It is easily seen that:

**Lemma 18** *If $A \subset M$ is $\Delta_1(M)$ in $p$, then $\langle L_\nu[A] \mid \nu \in M\rangle$ is $\Delta_1(M)$ in $p$.*

- *If $u \in M$, then $\langle L_\nu(u) \mid \nu \in M\rangle$ is $\Delta_1(M)$ in $u$.*

*By set recursion we can also define a sequence $\langle <^A_\nu \mid \nu < \infty\rangle$ s.t.*

- *$<^A_\nu$ well orders $L_\nu[A]$.*
- *$<^A_\mu$ end extends $<^A_\nu$ for $\nu \leq \mu$.*

Then:

**Lemma 19** *If $A \in M$ is $\Delta_1(M)$ in $p$, then $\langle <^A_\nu \mid \nu \in M\rangle$ is $\Delta_1(M)$ in $p$.*

**Definition** $L^A_\nu = \langle L_\nu[A], A \cap L_\nu[A]\rangle$.

$$\langle L^A_\nu, B_1, B_2, \ldots\rangle = \langle L_\nu[A], A \cap L_\nu[A], B_1, B_2, \ldots\rangle.$$

It follows easily that:

**Lemma 20** *Let $M = \langle L^A_\alpha, B_1, \ldots\rangle$ be admissible. Then $<_M =_{\mathrm{Df}} \bigcup_{\nu<\alpha} <^A$ is a $\Delta_1(M)$ well ordering of $M$. Moreover, there is a $\Delta_1(M)$ map $h : M \to M$ s.t. $h(x) = \{z \mid z <_M x\}$.*

Using this, it follows easily that every $\Sigma_1(M)$ relation is uniformizable by a $\Sigma_1(M)$ function.

Thus the KP axioms give us a "reasonable" recursion theory. They do not suffice, however, to get $\Sigma_1$-uniformization. In fact, since we have not posited the axiom of choice, we do not even have $N$-finite uniformization. However, the admissible structures dealt with in these notes will almost always satisfy $\Sigma_1$-uniformization. This can happen in different ways. If $N =$

$L_\tau^A =_{\text{Df}} \langle L_\tau[A], A\rangle$, there is a well ordering $<$ of $N$ s.t. the function $h(x) = \{z \mid z < x\}$ is $\Sigma_1$. We can then uniformize $R(y, \vec{x})$ as follows: Let $R(y, \vec{x}) \leftrightarrow \bigvee z\, R'(y, z, \vec{x})$, where $R'$ is $\Sigma_0$. $R$ is then uniformized by:

$$\bigvee z (R'(y, z, \vec{x}) \wedge \bigwedge \langle y', z'\rangle \in h(\langle y, z\rangle) \neg R(u', z'\vec{x})).$$

The same holds for $N = L_\tau(a)$ where $a$ is a transitive set with a well ordering constructible from $a$ below $\tau$. If $N$ is a ZFC$^-$ model with a definable well ordering $<$, then every definable relation has a definable uniformization. If $N^* = \langle N, A_1, A_2, \ldots\rangle$ is the result of adding all $N$-definable predicates to $N$, then the $\Sigma_1(N^*)$ relations are exactly the $N$-definable relations, so uniformization holds trivially.

## 1.2. *Ill founded ZF$^-$ models*

We now prove a lemma about arbitrary (possibly ill founded) models of ZF$^-$ (where the language of ZF$^-$ may contain predicates other than '$\in$'). Let $\mathfrak{A} = \langle A, \in_{\mathfrak{A}}, B_1, B_2, \ldots\rangle$ be such a model. For $X \subset A$ we of course write $\mathfrak{A}|X = \langle X, \in_{\mathfrak{A}} \cap X^2, \ldots\rangle$. By the *well founded core* of $\mathfrak{A}$ we mean the set of all $x \in A$ s.t. $\in_{\mathfrak{A}} \cap\, C(x)^2$ is well founded, where $C(x)$ is the closure of $\{x\}$ under $\in_{\mathfrak{A}}$. Let $\text{wfc}(\mathfrak{A})$ denote the restriction of $\mathfrak{A}$ to its well founded core. Then $\text{wfc}(\mathfrak{A})$ is a well founded structure satisfying the axiom of extensionality, and is, therefore, isomorphic to a transitive structure. Hence there is $\mathfrak{A}'$ s.t. $\mathfrak{A}'$ is isomorphic to $\mathfrak{A}$ and $\text{wfc}(\mathfrak{A}')$ is transitive. We say that a model $\mathfrak{A}$ of ZF$^-$ is *solid* iff $\text{wfc}(\mathfrak{A})$ is transitive and $\in_{\text{wfc}(\mathfrak{A})} = \in \cap\, \text{wfc}(\mathfrak{A})^2$. Thus every consistent set of sentences in ZF$^-$ has a solid model. Note that if $\mathfrak{A}$ is solid, then $\omega \subset \text{wfc}(\mathfrak{A})$. By $\Sigma_0$-absoluteness we of course have:

(1) $\quad \text{wfc}(\mathfrak{A}) \models \varphi(\vec{x}) \longleftrightarrow \mathfrak{A} \models \varphi(\vec{x})$

if $x_1, \ldots, x_n \in \text{wfc}(\mathfrak{A})$ and $\varphi$ is a $\Sigma_0$-formula. By $\in$-induction on $x \in \text{wfc}(\mathfrak{A})$ it follows that the rank function is absolute:

(2) $\quad \text{rn}(x) = \text{rn}^{\mathfrak{A}}(x) \quad \text{for} \quad x \in \text{wfc}(\mathfrak{A})$.

Using this we prove:

**Lemma 21** *Let $\mathfrak{A}$ be a solid model of ZF$^-$. Then* $\text{wfc}(\mathfrak{A})$ *is admissible.*

*Proof.* Let $\varphi$ be $\Sigma_0$ and let

(3) $\quad \text{wfc}(\mathfrak{A})) \models \bigwedge x \bigvee y\, \varphi(x, y, \vec{z})$

where $z_1, \ldots, z_n \in \text{wfc}(\mathfrak{A})$. Let $u \in \text{wfc}(\mathfrak{A})$. By (3) and $\Sigma_0$ absoluteness:

(4) $\quad \mathfrak{A} \models \bigwedge x \in u \bigvee y\, \varphi(x, y, \vec{z})$.

Since $\mathfrak{A}$ is a ZFC$^-$ model, there must then be $v \in \mathfrak{A}$ of minimal $\mathfrak{A}$-rank $\text{rn}^{\mathfrak{A}}(v)$ s.t.

(5) $\quad \mathfrak{A} \models \bigwedge x \in u \bigvee y \in v\, \varphi(x, y, \vec{z})$.

It suffices to note that $\text{rn}^{\mathfrak{A}}(v) \in \text{wfc}(\mathfrak{A})$, hence $\text{rn}^{\mathfrak{A}}(v) = \text{rn}(v)$ and $v \in \text{wfc}(\mathfrak{A})$. (Otherwise there is $r \in \mathfrak{A}$ s.t. $\mathfrak{A} \models r < \text{rn}(v)$ and there is $v' \in \mathfrak{A}$ s.t. $\mathfrak{A} \models v' = \{x \in v \mid \text{rn}(x) < r\}$. Hence $v'$ satisfies (5) and $\text{rn}^{\mathfrak{A}}(v') < \text{rn}^{\mathfrak{A}}(v)$. Contradiction!) By $\Sigma_0$ absoluteness, then:

(6) $\quad \text{wfc}(\mathfrak{A}) \models \bigwedge x \in u \bigvee y \in v\, \varphi(x, y, \vec{z})$.

QED (Lemma 21)

As immediate corollaries we have:

**Corollary 21.1** *Let $\delta = On \cap \text{wfc}(\mathfrak{A})$. Then $L_\delta(a)$ is admissible for $a \in \text{wfc}(\mathfrak{A})$.*

**Corollary 21.2** *$L_\delta^A = \langle L_\delta[A], A \cap L_\delta[A]\rangle$ admissible whenever $A$ is $\mathfrak{A}$-definable.*
(*Proof.* We may suppose w.l.o.g. that $A$ is one of the predicates of $\mathfrak{A}$.)

**Note** In Lemma 21 we can replace ZF$^-$ by KP. In this form it is known as *Ville's Lemma.* However, a form of Lemma 21 was first employed in our paper [NA] with Harvey Friedman. If memory serves us, the idea was due to Friedman.

### 1.3. *Primitive recursive set functions*

A function $f : \mathbf{V} \to \mathbf{V}$ is called *primitive recursive* (pr) iff it is generated by successive applications of the following schemata:

(i) $f(\vec{x}) = x_i$ (here $\vec{x}$ is $x_1, \ldots, x_n$)
(ii) $f(\vec{x}) = \{x_i, x_j\}$
(iii) $f(\vec{x}) = x_i \setminus x_j$
(iv) $f(\vec{x}) = g(h_1(\vec{x}), \ldots, h_m(\vec{x}))$
(v) $f(y, \vec{x}) = \bigcup_{z \in y} g(z, \vec{x})$
(vi) $f(y, \vec{x}) = g(y, \vec{x}, \langle f(z, \vec{x}) \mid z \in y\rangle)$

We call $A \subset \mathbf{V}^n$ a pr *relation* iff its characteristic function is a pr function. (*However*, a function can be a pr relation without being a pr function.) pr functions are ubiquitous. It is easily seen for instance that the functions listed in Lemma 9 are pr. Lemmas 4–7 hold with '$\Sigma_1$' replaced by 'pr'. The functions $\text{TC}(x)$, $\text{rn}(x)$ are easily seen to be pr. We call $f$ :

$On^n \to \mathbf{V}$ a pr function if it is the restriction of a pr function to $On$. The functions $\alpha + 1, \alpha + \beta, \alpha \cdot \beta, \alpha^\beta, \ldots$ etc. are then pr.

Since the pr functions are proper classes, the above discussion is carried out in second order set theory. However, all that needs to be said about pr functions can, in fact, be adequately expressed in ZFC. To do this we talk about pr *definitions*:

By a pr definition we mean a finite list of schemata of the form (i)–(vi) s.t.

- the function variable on the left side does not occur in a previous equation in the list.
- every function variable on the right side occurs previously on the left side.

Clearly, every pr definition $s$ defines a pr function $F_s$. Moreover, for each $s$, $F_s$ has a *canonical* $\Sigma_1$ *definition* $\varphi_s(y, x_1, \ldots, x_n)$. (Indeed, the relation $\{\langle x, s\rangle \mid x \in F_s\}$ is $\Sigma_1$.) The canonical definition has some remarkable absoluteness properties. If $u$ is transitive, let $F_s^u$ be the function obtained by relativizing the canonical definition to $u$. Hence $F_s^u \subset F_s$ is a partial map on $u$. Then:

- If $u$ is pr closed, then $F_s^u = F_s \cap u$.
- If $\alpha$ is closed under the functions $\nu + 1, \nu \cdot \tau, \nu^\tau, \ldots$ etc., then $L_\alpha[A]$ is pr closed for every $A \subset \mathbf{V}$.

These facts are provable in ZFC$^-$. The proofs can be found in [AS] or [PR] As corollaries we get:

(1) *Let* $\mathbf{V}[G]$ *be a generic extension of* $\mathbf{V}$. *Then* $\mathbf{V} \cap F_s^{\mathbf{V}[G]} = F_s^{\mathbf{V}}$.
(2) *Let* $\mathfrak{A}$ *be a solid model of ZFC*$^-$. *Let* $A = \text{wfc}(\mathfrak{A})$. *Then*

$$F_s^{\mathfrak{A}} \cap A = F_s^A = F_s.$$

*Proof.* We prove (2). Clearly $F_s^A = F_s$, since $A$ being admissible, is pr closed. But each $x \in A$ is an element of a transitive pr closed $u \in A$, since $A$ is admissible. Hence $y = F_s^{\mathfrak{A}}(x) \leftrightarrow y = F_s^u(x) \leftrightarrow y = F_s^A(x)$. QED

## 2. Barwise Theory

Jon Barwise worked out the syntax and model theory of certain infinitary (but $M$-finitary) languages on countable admissible structures $M$. In so doing, he created a powerful and flexible tool for set theorists, which enables us to construct transitive structures using elementary model theory. In this section we give an introduction to Barwise' work, whose potential for set theory has, we feel, been unduly neglected.

Let $M$ be admissible. Barwise develops a first order theory in which arbitrary $M$-finite conjunctions and disjunctions are allowed. The predicates, however, have only a (genuinely) finite number of argument places and there are no infinite strings of quantifiers. If we wish to make use of the notion of $M$-finiteness, we must "arithmetize" the language – i.e. identify its symbols with objects in $M$. A typical arithmetization is:

**Predicates:** $P^n_x = \langle 0, \langle n, x \rangle \rangle$ $(x \in M, 1 \leq n < \omega)$
$(P^n_x =$ the $x$-th $n$-place predicate$)$

**Constants:** $c_x = \langle 1, x \rangle$ $(x \in M)$

**Variables:** $v_x = \langle 2, x \rangle$ $(x \in M)$

**Note** The set of variables must be $M$-infinite, since otherwise a single formula could exhaust all the variables. We let $P^2_0$ be the identity predicate $(\dot{=})$ and also reserve $P^2_1$ as the $\in$-predicate $(\dot{\in})$, which will be a part of most interesting languages.

By a *primitive formula* we mean $Pt_1 \ldots t_n = \langle 3, \langle P, t_1, \ldots, t_n \rangle \rangle$, where $P$ is an $n$-place predicate and $t_1, \ldots, t_n$ are variables and constants. We then define:

$$\neg\varphi = \langle 4, \varphi \rangle, \quad (\varphi \vee \psi) = \langle 5, \langle \varphi, \psi \rangle \rangle, \quad (\varphi \wedge \psi) = \langle 6, \langle \varphi, \psi \rangle \rangle,$$
$$(\varphi \to \psi) = \langle 7, \langle \varphi, \psi \rangle \rangle, \quad (\varphi \leftrightarrow \psi) = \langle 8, \langle \varphi, \psi \rangle \rangle, \quad \bigwedge v\, \varphi = \langle 9, \langle v, \varphi \rangle \rangle,$$
$$\bigvee v\, \varphi = \langle 10, \langle v, \varphi \rangle \rangle, \quad \text{and: } \mathbb{V} f = \langle 11, f \rangle, \quad \mathbb{A} f = \langle 12, f \rangle.$$

The set Fml of 1-st order $M$-*formulas* is the smallest set $X$ which contains all primitive formulae, is closed under $\neg$, $\vee$, $\wedge$, $\to$, $\leftrightarrow$, and s.t.

- If $v$ is a variable and $\varphi \in X$, then $\bigwedge v\, \varphi, \bigvee v\, \varphi \in X$.
- If $f = \langle \varphi_i \mid i \in I \rangle \in M$ and $\varphi_i \in X$ for $i \in I$, then $\mathbb{V}_{i \in I} \varphi_i =_{\text{Df}} \mathbb{V} f$ and $\mathbb{A}_{i \in I} \varphi_i =_{\text{Df}} \mathbb{A} f$ are in $I$.

Then the usual syntactical notions are $\Delta_1$, including: Fml, Cnst (set of constants), Vbl (set of variables), Sent (set of all sentences), $\mathrm{Fr}(\varphi) =$ the set of free variables in $\varphi$, and: $\varphi({}^{v_1, \ldots, v_n}/_{t_1, \ldots, t_n}) \simeq$ the result of replacing all free occurences of the vbl $v_i$ by $t_i$ (where $t_i \in \mathrm{Vbl} \cup \mathrm{Const}$), as long as this can be done without any new occurence of a variable $t_i$ being bound; otherwise undefined.

That Vbl, Const are $\Delta_1$ (in fact $\Sigma_0$) is immediate. The characteristic function $\mathcal{X}$ of Fml is definable by a recursion of the form:

$$\mathcal{X}(x) = G(x, \langle \mathcal{X}(z) \mid z \in \mathrm{TC}(x) \rangle).$$

Similarly for the functions $\mathrm{Fr}(\varphi)$ and $\varphi(\vec{v}/\vec{t})$. Then Sent $= \{\varphi \mid \mathrm{Fr}(\varphi) = \emptyset\}$.

**Note** We of course employ the usual notation, writing $\varphi(t_1, \ldots, t_n)$ for $\varphi(v_1, \ldots, v_n / t_1, \ldots, t_n)$, where the sequence $v_1, \ldots, v_n$ is taken as known.

*M-finite predicate logic* has as axioms all instances of the usual predicate logical axiom schemata together with:

$$\bigwedge_{i \in u} \varphi_i \longrightarrow \varphi_j, \qquad \varphi_j \longrightarrow \bigvee_{i \in u} \varphi_i \quad \text{for } j \in u \in M.$$

The *rules of inference* are:

$$\frac{\varphi, \varphi \to \psi}{\psi} \quad \text{(modus ponens)},$$

$$\frac{\varphi \to \psi}{\varphi \to \bigwedge x\, \psi}, \quad \frac{\psi \to \varphi}{\bigvee x\, \psi \to \varphi} \quad \text{for } x \notin \mathrm{Fr}(\varphi),$$

$$\frac{\varphi \to \psi_i \ (i \in u)}{\varphi \to \bigwedge_{i \in u} \varphi_i}, \quad \frac{\psi_i \to \varphi \ (i \in u)}{\bigvee_{i \in u} \psi_i \to \varphi}.$$

We say that $\varphi$ is *provable* from a set of statements $A$ if $\varphi$ is in the smallest set which contains $A$ and the axioms and is closed under the rules of inference. We write $A \vdash \varphi$ to mean that $\varphi$ is provable from $A$. (Note: By the last rule, $\bigvee \emptyset \to \varphi$ for every $\varphi$, hence $\vdash \neg \bigvee \emptyset$. Similarly $\vdash \bigwedge \emptyset$.)

A formula is provable if and only if it has a proof. Because we have not assumed choice to hold in our admissible structure $M$, we must use a somewhat unorthodox concept of proof, however.

**Definition** By a *proof from $A$* we mean a sequence $\langle p_i \mid i < \alpha \rangle$ s.t. $\alpha \in On$ and for each $i < \alpha$, if $\psi \in p_i$, then either $\psi \in A$ or $\psi$ is an axiom or $\psi$ follows from $\bigcup_{h<i} p_h$ by a single application of one of the rules.

$p = \langle p_i \mid i < \alpha \rangle$ is a *proof of* $\varphi$ iff $\varphi \in \bigcup_{i<\alpha} p_i$.

If $A$ is $\Sigma_1(M)$ in a parameter $q$ it follows easily that $\{p \in M \mid p$ is a proof from $A\}$ is $\Sigma_1(M)$ in the same parameter. It is also easily seen that $A \vdash \varphi$ iff there exists a proof of $\varphi$ from $A$. A more interesting conclusion is:

**Lemma 1** *Let $A$ be $\underline{\Sigma}_1(M)$. Then $A \vdash \varphi$ iff there is an $M$-finite proof of $\varphi$ from $A$.*

*Proof*. $(\leftarrow)$ is trivial. We prove $(\rightarrow)$.
Let $X =$ the set of $\varphi$ s.t. there exists a $p \in M$ which proves $\varphi$ from $A$.

**Claim** $\{\varphi \mid A \vdash \varphi\} \subset X$.

*Proof*. We know that $A \subset X$ and all axioms lie in $X$. Hence it suffices to show that $X$ is closed under the rules of proof. This must be demonstrated rule by rule. As an example we show:

**Claim** Let $\varphi \rightarrow \psi_i \in X$ for $i \in u$, where $u \in M$. Then $\varphi \rightarrow \bigwedge\limits_{i \in u} \psi_i \in X$.

*Proof*. Let $P(p, \psi)$ mean: $p$ is a proof of $\psi$ from $A$. Then $P$ is $\underline{\Sigma}_1(M)$. By our assumption:

(1) $\quad \bigwedge i \in u \ \bigvee p \, P(p, \varphi \rightarrow \psi_i)$.

Now let $P(p, \psi) \leftrightarrow \bigvee z \, P'(z, p, \psi)$, where $P'$ is $\Sigma_0$. We then have:

(2) $\quad \bigwedge i \in u \ \bigvee z \ \bigvee p \, P'(z, p, \varphi \rightarrow \psi_i)$

whence follows easily that there is $v \in M$ with:

(3) $\quad \bigwedge i \in u \ \bigvee z \in v \ \bigvee p \in v \, P'(z, p, \varphi \rightarrow \psi_i)$.

Set $w = \{p \in v \mid \bigvee i \in u \ \bigvee z \in v \ P'(z, p, \psi)\}$. Then

(4) $\quad \bigwedge i \in u \ \bigvee p \in w \ P(p, \varphi \rightarrow \psi_i)$ and $w$ consists of proofs from $A$.

Let $\alpha \in M$, $\alpha \geq \mathrm{dom}(p)$ for all $p \in w$. Define a proof $p^*$ of length $\alpha + 1$ by:

$$p^*(i) = \begin{cases} \bigcup\{p_i \mid p \in w \wedge i \in \mathrm{dom}(p)\} & \text{for } i < \alpha, \\ \{\varphi \rightarrow \bigwedge\limits_{i \in u} \psi_i\} & \text{for } i = \alpha. \end{cases}$$

Then $p^* \in M$ proves $\varphi \rightarrow \bigwedge\limits_{i \in u} \psi_i$ from $A$. QED(Lemma 1)

From this we get the *M-finiteness lemma*:

**Lemma 2** *Let $A$ be $\underline{\Sigma}_1(M)$. Then $A \vdash \varphi$ iff there is $u \in M$ s.t. $u \subset A$ and $u \vdash \varphi$.*

*Proof*. $(\leftarrow)$ is trivial. We prove $(\rightarrow)$.
Let $p \in M$ be a proof of $\varphi$ from $A$. Let $u =$ the set of $\psi$ s.t. for some $i \in \mathrm{dom}(p)$, $\psi \in p_i$, but $\psi$ is not an axiom and does not follow from $\bigcup\limits_{h<i} p_h$ by a single application of a rule. Then $u \in M$, $u \subset A$, and $p$ is a proof from $u$. Hence $u \vdash \varphi$. QED(Lemma 2)

Another consequence of Lemma 1 is

**Lemma 3** *Let $A$ be $\Sigma_1(M)$ in $q$. Then $\{\varphi \mid A \vdash \varphi\}$ is $\Sigma_1(M)$ in the same parameter $q$ (uniformly in the $\Sigma_1$ definition of $A$ from $q$).*

*Proof.* $\{\varphi \mid A \vdash \varphi\} = \{\varphi \mid \bigvee p \in M \;\; p \text{ proves } \varphi \text{ from } A\}$. QED

**Corollary 4** *Let $A$ be $\Sigma_1(M)$ in $q$. Then "$A$ is consistent" is $\Pi_1(M)$ in the same parameter $q$ (uniformly in the $\Sigma_1$ definition of $A$ from $q$).*

Note that, since $u \in M$ is uniformly $\Sigma_1(M)$ in itself, we have:

**Corollary 5** $\{\langle u, \varphi\rangle \mid u \in M \wedge u \vdash \varphi\}$ *is* $\Sigma_1(M)$.

Similarly:

**Corollary 6** $\{u \in M \mid u \textit{ is consistent}\}$ *is* $\Pi_1(M)$.

**Note** Call a proof $p$ *strict* iff $\overline{\overline{p}}_i = 1$ for $i \in \text{dom}(p)$. This corresponds to the more usual notion of proof. If $M$ satisfies the axiom of choice in the form: Every set is enumerable by an ordinal, then Lemma 1 holds with "strict proof" in place of "proof". We leave this to the reader.

### *Languages*

We will normally not employ all of the predicates and constants in our $M$-finitary first order logic, but cut down to a smaller set of symbols which we intend to interpret in a model. Thus we define a *language* to be a set $\mathcal{L}$ of predicates and constants. By a *model* of $\mathcal{L}$ we mean a structure

$$\mathfrak{A} = \langle |\mathfrak{A}|, \langle t^{\mathfrak{A}} \mid t \in \mathcal{L}\rangle\rangle$$

s.t. $|\mathfrak{A}| \neq \emptyset$, $P^{\mathfrak{A}} \subset |\mathfrak{A}|^n$ whenever $P$ is an $n$-place predicate, and $c^{\mathfrak{A}} \in |\mathfrak{A}|$ whenever $|\mathfrak{A}|$ is a constant. By a *variable assignment* we mean a map $f : \text{Vbl} \to \mathfrak{A}$ (Vbl being the set of all variables). The *satisfaction relation* $\mathfrak{A} \models \varphi[f]$ is defined in the usual way, where $\mathfrak{A} \models \varphi[f]$ means that the formula $\varphi$ becomes true in $\mathfrak{A}$ if the free variables in $\varphi$ are interpreted by $f$. We leave the definition to the reader, remarking only that:

$$\mathfrak{A} \models \bigwedge\!\!\!\bigwedge_{i \in u} \varphi_i[f] \quad \text{iff} \quad \bigwedge i \in u \;\; \mathfrak{A} \models \varphi_i[f],$$

$$\mathfrak{A} \models \bigvee\!\!\!\bigvee_{i \in u} \varphi_i[f] \quad \text{iff} \quad \bigvee i \in u \;\; \mathfrak{A} \models \varphi_i[f].$$

We adopt the usual conventions of model theory, writing $\mathfrak{A} = \langle |\mathfrak{A}|, t_1^{\mathfrak{A}}, \ldots\rangle$ if we think of the predicates and constants of $\mathcal{L}$ as being arranged in a fixed sequence $t_1, t_2, \ldots$ Similarly, if $\varphi = \varphi(v_1, \ldots, v_n)$ is a formula in which at most the variables $v_1, \ldots, v_n$ occur free, we write: $\mathfrak{A} \models \varphi[x_1, \ldots, x_n]$ for: $\mathfrak{A} \models \varphi[f]$ where $f(v_i) = x_i$ $(i = 1, \ldots, n)$. If $\varphi$ is a statement, we write: $\mathfrak{A} \models \varphi$. If $A$ is a set of statements we write: $\mathfrak{A} \models A$ to mean: $\mathfrak{A} \models \varphi$ for all $\varphi \in A$.

The *correctness theorem* says that if $A$ is a set of $\mathcal{L}$-statements and $\mathfrak{A} \models A$, then $A$ is consistent. (We leave this to the reader.)

*Barwise' Completeness Theorem* says that the converse holds if our admissible structure $M$ is countable:

**Theorem 7** *Let $M$ be a countable admissible structure. Let $A$ be a set of statements in the $M$-language $\mathcal{L}$. If $A$ is consistent in $M$-finite predicate logic, then $A$ has a model $\mathfrak{A}$.*

*Proof* (sketch). We make use of the following theorem of Rasiowa and Sikorski: Let $\mathbb{B}$ be a Boolean algebra. Let $X_i \subset \mathbb{B}$ $(i < \omega)$ s.t. the Boolean union $\bigcup X_i = b_i$ exists in the sense of $\mathbb{B}$. Then $\mathbb{B}$ has an ultrafilter $U$ s.t.

$$b_i \in U \longleftrightarrow X_i \cap U \neq \emptyset \quad \text{for } i < \omega.$$

(*Proof*. Successively choose $c_i$ $(i < \omega)$ by $c_0 = 1$, $c_{i+1} = c_i \cap b \neq 0$, where $b \in X_i \cup \{\neg b_i\}$. Let $\overline{U} = \{a \in \mathbb{B} \mid V_i\, c_i \subset a\}$. Then $\overline{U}$ is a filter and extends to an ultrafilter on $\mathbb{B}$.)
Extend the language $\mathcal{L}$ by adding an $M$-infinite set $C$ of new constants. Call the extended language $\mathcal{L}^*$ and set:

$$[\varphi] = \{\psi \mid A \vdash \psi \leftrightarrow \varphi\}$$

for $\mathcal{L}^*$-statements $\varphi$. Then

$$\mathbb{B} = \{[\varphi] \mid \varphi \in \mathrm{St}_{\mathcal{L}^*}\}$$

in the Lindenbaum algebra of $\mathcal{L}^*$ with the operations:

$$[\varphi] \cup [\psi] = [\varphi \vee \psi], \quad [\varphi] \cap [\psi] = [\varphi \wedge \psi], \quad \neg[\varphi] = [\neg\varphi],$$
$$\bigcup_{i \in u} [\varphi_i] = \Big[\bigvee_{i \in u} \varphi_i\Big] \quad (u \in M), \quad \bigcap_{i \in u} [\varphi_i] = \Big[\bigwedge_{i \in u} \varphi_i\Big] \quad (u \in M),$$
$$\bigcup_{c \in C} [\varphi(c)] = [\bigvee v\, \varphi(v)], \quad \bigcap_{c \in C} [\varphi(c)] = [\bigwedge v\, \varphi(v)].$$

The last two equations hold because the constants in $C$, which do not occur in the axioms $A$, behave like free variables. By Rasiowa and Sikorski there is then an ultrafilter $U$ on $\mathbb{B}$ which respects the above operations. We define a model $\mathfrak{A} = \langle |\mathfrak{A}|, \langle t^{\mathfrak{A}} \mid t \in \mathcal{L} \rangle \rangle$ as follows: For $c \in C$ set $[c] = \{c' \in C \mid [c = c'] \in U\}$. If $P \in \mathcal{L}$ is an $n$-place predicate, set:

$$P^{\mathfrak{A}}([c_1], \ldots, [c_n]) \longleftrightarrow [Pc_1 \ldots c_n] \in U.$$

If $t \in \mathcal{L}$ is a constant set:

$$t^{\mathfrak{A}} = [c], \text{ where } c \in C, \quad [t = c] \in U.$$

A straightforward induction then shows:

$$\mathfrak{A} \models \varphi[[c_1], \ldots, [c_n]] \longleftrightarrow [\varphi(c_1, \ldots, c_n)] \in U$$

for formulae $\varphi = \varphi(v_1, \ldots, v_n)$ with at most the free variables $v_1, \ldots, v_n$. In particular $\mathfrak{A} \models \varphi \leftrightarrow [\varphi] \in U$ for $\mathcal{L}^*$-statements $\varphi$. Hence $\mathfrak{A} \models A$, since $[\varphi] = 1$ for all $\varphi \in A$. QED(Theorem 7)

Combining the completeness theorem with the $M$-finiteness lemma, we get the well known *Barwise compactness theorem*:

**Corollary 8** *Let $M$ be countable. Let $\mathcal{L}$ be $\underline{\Delta}_1$ and $A$ be $\underline{\Sigma}_1$. If every $M$-finite subset of $A$ has a model, then so does $A$.*

By a *theory* or *axiomatized language* we mean a pair $\mathcal{L} = \langle \mathcal{L}_0, A \rangle$ s.t. $\mathcal{L}_0$ is a language and $A$ a set of $\mathcal{L}_0$-statements. We say that $\mathfrak{A}$ *models* $\mathcal{L}$ iff $\mathfrak{A}$ is a model of $\mathcal{L}_0$ and $\mathfrak{A} \models A$. We also write: $\mathcal{L} \vdash \varphi$ for $(A \vdash \varphi \wedge \varphi \in \mathrm{Fml}_{\mathcal{L}_0})$. We say that $\mathcal{L} = \langle \mathcal{L}_0, A \rangle$ is $\Sigma_1(M)$ (in the parameter $p$) iff $\mathcal{L}_0$ is $\Delta_1(M)$ (in $p$) and $A$ is $\Sigma_1(M)$ (in $p$). Similarly for: $\mathcal{L}$ is $\Delta_1(M)$ (in $p$).

*****

We now consider the class of axiomatized languages containing a fixed predicate $\dot{\in}$, the special constants $\underline{x}$ ($x \in M$) (We can set e.g. $\underline{x} = \langle 1, \langle 0, x \rangle \rangle$.) and the *basic axioms*

- Extensionality
- $\bigwedge v (v \,\dot{\in}\, \underline{x} \leftrightarrow \bigvee\!\!\!\bigvee_{z \in x} v = \underline{z}) \quad (x \in M)$

(Further predicates, constants, and axioms are allowed, of course.) We call any such theory an "$\in$-theory". Then:

**Lemma 9** *Let $\mathfrak{A}$ be a solid model of the $\in$-theory $\mathcal{L}$. Then $\underline{x}^{\mathfrak{A}} = x \in \mathrm{wfc}(\mathfrak{A})$ for $x \in M$.*

*Proof*. $\in$-induction on $x$.

**Definition** Let $\mathcal{L}$ be an $\in$-theory. $\mathrm{ZF}^-_{\mathcal{L}}$ is the set of (really) finite $\mathcal{L}$-statements which are axioms of $\mathcal{L}$. (Similarly for $\mathrm{ZFC}^-_{\mathcal{L}}$.)

We write $\mathcal{L} \vdash \mathrm{ZF}^-$ for $\mathcal{L} \vdash \mathrm{ZF}^-_{\mathcal{L}}$. (Similarly for $\mathcal{L} \vdash \mathrm{ZFC}^-$.)

⋆⋆⋆⋆⋆

$\in$-theories are a useful tool in set theory. We now bring some typical applications. We recall that an ordinal $\alpha$ is called *admissible* if $L_\alpha$ is admissible and *admissible in* $a \subset \alpha$ if $L^a_\alpha = \langle L_\alpha[a], a\rangle$ is admissible.

**Lemma 10** *Let $\alpha > \omega$ be a countable admissible ordinal. There is $a \subset \omega$ s.t. $\alpha$ is the least ordinal admissible in $a$.*

This follows straightforwardly from:

**Lemma 11** *Let $M$ be a countable admissible structure. Let $\mathcal{L}$ be a consistent $\Sigma_1(M)$ $\in$-theory s.t. $\mathcal{L} \vdash \mathrm{ZF}^-$. Then $\mathcal{L}$ has a solid model $\mathfrak{A}$ s.t. $On \cap \mathrm{wfc}(\mathfrak{A}) = On \cap M$.*

We first show that Lemma 11 implies Lemma 10, and then prove Lemma 11. Take $M = L_\alpha$, where $\alpha$ is as in Lemma 10. Let $\mathcal{L}$ be the $M$-theory with:
**Predicate:** $\dot{\in}$
**Constants:** $\underline{x}$ $(x \in M)$, $\mathring{a}$
**Axioms:** Basic axioms + $\mathrm{ZF}^-$, and $\underline{\beta}$ is not admissible in $\mathring{a}$ $(\beta < \alpha)$.

Then $\mathcal{L}$ is consistent, since $\langle H_{\omega_1}, \in, a\rangle$ is a model, where $a$ is any $a \subset \omega$ which codes a well ordering of type $\geq \alpha$ (and $\underline{x}$ is interpretedly $x$ for $x \in M$). Now let $\mathfrak{A}$ be a solid model of $\mathcal{L}$ s.t. $On \cap \mathrm{wfc}(\mathfrak{A}) = \alpha$. Then $\mathrm{wfc}(\mathfrak{A})$ is admissible by Section 1, Lemma 21. Hence so is $L^a_\alpha$, where $a = \mathring{a}^{\mathfrak{A}}$. But $\beta$ is not admissible in $a$ for $\omega < \beta < \alpha$, since "$\beta$ is admissible in $a$" is $\Sigma_1(L^a_\alpha)$; hence the same $\Sigma_1$ statement would hold of $\beta$ in $\mathfrak{A}$. Contradiction!

QED(Lemma 10)

**Note** Pursuing this method a bit further we can prove: Let $\omega < \alpha_0 < \ldots < \alpha_{n-1}$ be a sequence of countable admissible ordinals. There is $a \subset \omega$ s.t. $\alpha_i =$ the $i$-th $\alpha > \omega$ which is admissible in $a$ $(i < n)$. A similar theorem holds for countable infinite sequences, but the proof requires forcing and is much more complex. It is given in §5 and §6 [AS].

We now prove Lemma 11 by modifying the proof of the completeness theorem. Let $\Gamma(v)$ be the set of formulae $v \in On$, $v > \underline{\beta}$ $(\beta \in M)$. Add an $M$-infinite (but $\Delta_1(M)$) set $E$ of new constants to $\mathcal{L}$. Let $\mathcal{L}'$ be $\mathcal{L}$ with the new constants and the new axioms $\Gamma(e)$ $(e \in E)$. Then $\mathcal{L}'$ is consistent, since any $M$-finite subset of the axioms can be modeled by interpreting the new constants as ordinals in $\mathrm{wfc}(\mathfrak{A})$, $\mathfrak{A}$ being any solid model of $\mathcal{L}$. As in the proof of completeness we then add a new class $C$ of constants which is not $M$-finite. We assume, however, that $C$ is $\Delta_1(M)$. We add no further axioms, so the elements of $C$ behave like free variables. The so extended language $\mathcal{L}''$ is clearly $\underline{\Sigma}_1(M)$. Now set:

$$\Delta(v) = \{v \notin On\} \cup \bigcup_{\beta \in M} \{v \leq \underline{\beta}\} \cup \bigcup_{e \in E} \{e < v\}.$$

**Claim** Let $c \in C$. Then $\bigcup\{[\varphi] \mid \varphi \in \Delta(c)\} = 1$ in the Lindenbaum algebra of $\mathcal{L}''$.

*Proof*. Suppose not. Set $\Delta' = \{\neg\varphi \mid \varphi \in \Delta(c\}$. Then there is an $\mathcal{L}''$ statement $\psi$ s.t. $A \cup \{\psi\}$ is consistent, where $\mathcal{L}'' = \langle \mathcal{L}_0'', A\rangle$ and $A \cup \{\psi\} \vdash \Delta'$. Pick an $e \in E$ which does not occur in $\psi$. Let $A^*$ be the result of omitting the axioms $\Gamma(e)$ from $A$. Then $A^* \cup \{\psi\} \cup \Gamma(e) \vdash c \leq e$. By the $M$-finiteness lemma there is $\beta \in M$ s.t. $A^* \cup \{\psi\} \cup \{\underline{\beta} \leq e\} \vdash c \leq e$. But $e$ behaves here like a free variable, so $A^* \cup \{\psi\} \vdash c \leq \underline{\beta}$. But $A \supset A^*$ and $A \cup \{\psi\} \vdash \underline{\beta} < c$. Thus $A \cup \{\psi\}$ is inconsistent. Contradiction! QED(Claim)

Now let $U$ be an ultrafilter on the Lindenbaum algebra of $\mathcal{L}''$ which respects both the operations listed in the proof of the completeness theorem and the unions $\bigcup\{[\varphi] \mid \varphi \in \Delta(c)\}$ for $c \in C$. Let $X = \{\varphi \mid [\varphi] \in U\}$. Then as before, $\mathcal{L}''$ has a model $\mathfrak{A}$, all of whose elements have the form $c^{\mathfrak{A}}$ for a $c \in C$ and such that $\mathfrak{A} \models \varphi \leftrightarrow \varphi \in X$ for $\mathcal{L}''$-statements $\varphi$. We assume w.l.o.g. that $\mathfrak{A}$ is solid. It suffices to show that $Y = \{x \in \mathfrak{A} \mid x > \underline{\nu} \text{ in } \mathfrak{A} \text{ for all } v \in m\}$ has no minimal element in $\mathfrak{A}$. Let $x \in Y$, $x = c^{\mathfrak{A}}$. Then $\mathfrak{A} \models e < c$ for some $e \in E$. But $e^{\mathfrak{A}} \in Y$. QED(Lemma 11)

Another – very typical – application is:

**Lemma 12** *Let $W$ be an inner model of ZFC. Suppose that, in $W$, $U$ is a normal measure on $\kappa$. Let $\tau > \kappa$ be regular in $W$ and set $M = \langle H_\tau^W, U\rangle$. Assume that $M$ is countable in* $\mathbf{V}$. *Then for any $\alpha \leq \kappa$ there is $\overline{M} = \langle \overline{H}, \overline{U}\rangle$ s.t. $\overline{U}$ is a normal measure in $\overline{M}$ and $\overline{M}$ iterates to $M$ in exactly $\alpha$ many steps. (Hence $\overline{M}$ is iterable, since $M$ is.)*

*Proof*. The case $\alpha = 0$ is trivial, so assume $\alpha > 0$. Let $\delta$ be least s.t. $L_\delta(M)$ is admissible. Then $N = L_\delta(M)$ is countable. Let $\mathcal{L}$ be the $\in$-theory on $N$ with:

**Predicate:** $\dot{\in}$

**Constants:** $\underline{x}$ $(x \in N)$, $\mathring{M}$

**Axioms:** The basic axioms; ZFC$^-$; $\mathring{M} = \langle \mathring{H}, \mathring{U}\rangle$ is a transitive ZFC$^-$ model; $\mathring{M}$ iterates to $\underline{M}$ in $\underline{\alpha}$ many steps.

It suffices to prove:

**Claim** $\mathcal{L}$ is consistent.

We first show that the claim implies the theorem. Let $\mathfrak{A}$ be a solid model of $\mathcal{L}$. Then $N \subset \text{wfc}(\mathfrak{A})$. Hence $M, \overline{M} \in \text{wfc}(\mathfrak{A})$, where $\overline{M} = \mathring{M}^{\mathfrak{A}}$. There is $\langle \overline{M}_i \mid i < \alpha \rangle$ which, in $\mathfrak{A}$, is an iteration from $\overline{M}$ to $M$. But then $\langle \overline{M}_i \mid i < \alpha \rangle \in \text{wfc}(\mathfrak{A})$ really is an iteration by absoluteness. QED

We now prove the claim.

*Case 1* $\alpha < \kappa$.
Iterate $\langle W, U \rangle$ $\alpha$ many times, getting $\langle W_i, U_i \rangle$ $(i \leq \alpha)$ with iteration maps $\pi_{ij} : \langle W_i, U_i \rangle \prec \langle W_j, U_j \rangle$. Set $M_i = \pi_{0i}(M)$. Then $\langle M_i, U_i \rangle$ $(i \leq \alpha)$ is the iteration of $\langle M, U \rangle$ with maps $\pi'_{ij} = \pi_{ij} \upharpoonright M_i$. It suffices to show that $\mathcal{L}_\alpha = \pi_{0,\alpha}(\mathcal{L})$ is consistent. This is clear, however, since $\langle H_{\tau^+}, M \rangle$ models $\mathcal{L}_\alpha$ (with $M$ interpreting the constant $\mathring{M}_\alpha = \pi_{0,\alpha}(\mathring{M})$). QED(Case 1)

*Case 2* $\alpha = \kappa$.
This time we iterate $\langle W, U \rangle$ $\beta$ many times where $\pi_{0\beta}(\kappa) = \beta$ and $\beta \leq \kappa^+$. $\langle H_{\tau^+}, M \rangle$ again models $\mathcal{L}_\beta$. QED(Lemma 12)

Barwise theory is useful in situations where one is given a transitive structure $Q$ and wishes to find a transitive structure $\overline{Q}$ with similar properties inside an inner model. Another tool used in such situations is Schoenfield's lemma, which, however requires coding $\overline{Q}$ by a real. Unsurprisingly, Schoenfield's lemma can itself be derived from Barwise theory. We first note the well known fact that every $\Sigma^1_2$ condition on a real is equivalent to a $\Sigma_1(H_{\omega_1})$ condition, and conversely. Thus it suffices to show:

**Lemma 13** *Let $H_{\omega_1} \models \varphi[a]$, $a \subset \omega$, where $\varphi$ is $\Sigma_1$. Then*

$$H_{\omega_1}^{L[a]} \models \varphi[a].$$

*Proof*. Let $\varphi = \bigvee z\, \psi$, where $\psi$ is $\Sigma_0$. Let $H_{\omega_1} \models \psi[z, a]$, where $\text{rn}(z) < \alpha$ and $\alpha$ is admissible in $a$. Let $\mathcal{L}$ be the language on $L_\alpha(a)$ with:
**Predicate:** $\dot{\in}$
**Constants:** $\mathring{z}$, $\underline{x}$ $(x \in L_\alpha(a))$
**Axioms:** Basic axioms, ZFC$^-$, $\psi(\mathring{z}, \underline{a})$.

Then $\mathcal{L}$ is consistent since $\langle H_{\omega_1}, z \rangle$ is a model. Applying Löwenheim-Skolem in $L(a)$, we find a countable $\overline{\alpha}$ and a map $\pi : L_{\overline{\alpha}}(a) \prec L_\alpha(a)$. Let w.l.o.g. $\pi(\mathring{z}) = \mathring{z}$ and let $\overline{\mathcal{L}}$ be defined over $L_{\overline{\alpha}}(a)$ like $\mathcal{L}$ over $L_\alpha(a)$. Then

$\overline{\mathcal{L}}$ is consistent and has a solid model $\mathfrak{A}$ in $L(a)$. Let $z = \mathring{z}^{\mathfrak{A}}$. Then $z \in L(a)$ and $H_{\omega_1} \models \psi[z, a]$ in $L(a)$.

QED(Lemma 13)

## 3. Subcomplete Forcing

### 3.1. *Introduction*

In §10 of [PF] Shelah defines the notion of *complete forcing*:

**Definition** Let $\mathbb{B}$ be a complete BA. $\mathbb{B}$ is a *complete forcing* iff for sufficiently large $\theta$ we have: Let $\mathbb{B} \in H_\theta$. Let $\sigma : \overline{H} \prec H$, where $\overline{H}$ is countable and transitive. Let $\sigma(\overline{\mathbb{B}}) = \mathbb{B}$. If $\overline{G}$ is $\overline{\mathbb{B}}$-generic over $\overline{H}$, then there is $b \in \mathbb{B}$ which forces that, that whenever $G \ni b$ is $\mathbb{B}$-generic, then $\sigma''\overline{G} \subset G$.

**Note** If $\overline{G}$, $G$, $\overline{H}$, $H$, $\sigma$ are as above, then $\sigma$ extends uniquely to a $\sigma^*$ s.t. $\sigma^* : \overline{H}[\overline{G}] \prec H[G]$ and $\sigma^*(\overline{G}) = G$.

*Proof*. To see uniqueness, note that each $x \in \overline{H}[\overline{G}]$ has the form $x = t^{\overline{G}}$ where $t \in \overline{H}$ is a $\overline{\mathbb{B}}$-name. Thus $\sigma^*(x) = \sigma(t)^G$. To see existence, note that:

$$\overline{H}[\overline{G}] \models \varphi(t_1^{\overline{G}}, \ldots, t_n^{\overline{G}}) \longleftrightarrow \bigvee b \in \overline{G}\; b \Vdash^{\overline{H}}_{\overline{\mathbb{B}}} \varphi(t_1, \ldots, t_n)$$
$$\longrightarrow \bigvee b \in G \Vdash^{H}_{\mathbb{B}} \varphi(\sigma(t_1), \ldots, \sigma(t_n)) \longrightarrow H[G] \models \varphi(\sigma(t_1)^G, \ldots, \sigma(t_n)^G).$$

Hence there is $\sigma^* : \overline{H}[\overline{G}] \prec H[G]$ defined by: $\sigma^*(t^{\overline{G}}) = \sigma(t)^G$. But then $\sigma^* \supset \sigma$ since

$$\sigma^*(x) = \sigma^*(\check{x}^{\overline{G}}) = \sigma(\check{x})^G = \sigma(x)$$

for $x \in \overline{H}$. Letting $\dot{\overline{G}}$ be the $\overline{\mathbb{B}}$-generic name and $\dot{G}$ the $\mathbb{B}$-generic name we then have:

$$\sigma^*(\overline{G}) = \sigma^*(\dot{\overline{G}}^{\overline{G}}) = \dot{G}^G = G.$$

QED

**Lemma 1.1** *Let $\mathbb{B}$ be a complete forcing. Let $G$ be $\mathbb{B}$-generic. Then* $\mathbf{V}[G]$ *has no new countable sets of ordinals.*

*Proof*. Let $\Vdash f : \check{\omega} \to On$.
**Claim** $f^G \in \mathbf{V}$.
Suppose not. Then $b \Vdash f \notin \check{\mathbf{V}}$ for some $b$. Let $\theta$ be big enough and let $\sigma : \overline{H} \prec H_\theta$ s.t. $\sigma(\overline{f}, \overline{b}, \overline{B}) = f, b, \mathbb{B}$, where $\overline{H}$ is countable and transitive. Let $\overline{G} \ni \overline{b}$ be $\overline{\mathbb{B}}$-generic over $\overline{H}$. Let $G$ be $\mathbb{B}$-generic s.t. $\sigma''\overline{G} \subset G$. Let $\sigma^*$ be the above mentioned extension of $\sigma$. Then $\sigma^*(\overline{f}^{\overline{G}}) = f^G$. But clearly $\sigma^*(\overline{f}^{\overline{G}}) = \sigma''\overline{f}^{G} \in \mathbf{V}$, where $b = \sigma(\overline{b}) \in G$. Contradiction! QED(Lemma 1)

We note without proof that

**Lemma 1.2** *If $\mathbb{B}$ is the result of a countable support iteration of complete forcings, then $\mathbb{B}$ is complete.*

**Remark** In fact, the notion of complete forcing reduces to that of an $\omega$-closed set of conditions. ($\mathbb{P}$ is called $\omega$-closed iff whenever $\langle p_i \mid i < \omega \rangle$ is a sequence with $p_i \leq p_j$ for all $j \leq i$, then there is $q$ with $q \leq p_i$ for all $i$.) It is shown in [FA] that:

**Lemma 1.3** *$\mathbb{B}$ is a complete forcing iff it is isomorphic to* $\mathrm{BA}(\mathbb{P})$ *for some $\omega$-closed set of conditions $\mathbb{P}$.*

The properties of $\omega$-closed forcing are well known and Lemmas 1.1, 1.2 follow easily from Lemma 1.3.

The knowledgable reader will recognize the complete forcings as being a subclass of Shelah's *proper forcings*. Proper forcings satisfy Lemma 1.2 but not Lemma 1.1. In fact, many proper forcings add new reals. However, a proper forcing can never change the cofinality of an uncountable regular cardinal to $\omega$. Thus, the notion is useless in dealing e.g. with Namba forcing. What we want is a class of forcings which do not add new reals but do permit new sets of ordinals – even to the extent of changing cofinalities. We of course want these forcings to be iterable – i.e. some reasonable analogue of Lemma 1.2 should hold. The proof of Lemma 1.1 gives us a clue as to how such a class might be defined: The proof depends strongly on the fact that $\sigma''\overline{G} \subset G$ for a $\sigma \in \mathbf{V}$. Instead, we might require that, if $\overline{H}$, $\sigma$, $\theta$, $\mathbb{B}$, $\overline{G}$ are as in the definition of "completed forcing", then there is $b \in \mathbb{B}$ which forces that, if $G \ni b$ is $\mathbb{B}$-generic, there is $\sigma' \in \mathbf{V}[G]$ s.t. $\sigma' : \overline{H} \prec H_\theta$, $\sigma'(\overline{\mathbb{B}}) = \mathbb{B}$ and $\sigma'\,''\overline{G} \subseteq G$. We can even require $b$ to force $\sigma'(\overline{s}) = \sigma(\overline{s})$ for an arbitrarily chosen $\overline{s} \in \overline{H}$. If we now try to carry out the proof of Lemma 1 with a $\sigma' : \overline{H} \prec H_\theta$ s.t. $\sigma'(\overline{f}, \overline{b}, \overline{\mathbb{B}}) = f, b, \mathbb{B}$, in place of $\sigma$, we can conclude only that $f^G = \sigma'\,''\overline{f}^{\overline{G}}$. Since we do not know that $\sigma' \in \mathbf{V}$, we cannot conclude that $f^G \in \mathbf{V}$. However, if we assume $\Vdash f : \omega \to \omega$, then $f^G = \sigma'\,''\overline{f}^{\overline{G}}$, where $\overline{f}^{\overline{G}} \in \mathbf{V}$ and $\overline{f}^{\overline{G}} \subset \omega^2$. Since $\sigma' \upharpoonright \omega = \mathrm{id}$, we can then conclude that $f^G \in \mathbf{V}$.

Thus such forcings will add no reals, but may permit us to add new countable sets of ordinals.

In order to carry out this program we must address several difficulties, the first being this: Suppose that $H_\theta$ has definable Skolem functions. (This is certainly the case if $\mathbf{V} = L$.) We could then form $\sigma : \overline{H} \prec H_\theta$ s.t. $\sigma(\overline{b}, \overline{f}, \overline{\mathbb{B}}) = b, f, \mathbb{B}$ simply by transitivizing the Skolem closure of $\{b, f, \mathbb{B}\}$.

But then $\sigma$ is the *only possible* elementary map to $H_\theta$ with $\sigma(\overline{b}, \overline{f}, \overline{\mathbb{B}}) = b, f, \mathbb{B}$. Thus we perforce have: $\sigma' = \sigma$. In order to avoid this we must place a stronger condition on $\overline{H}$ which implies the possibility of many maps to the top. We shall define such a condition for the case that $\overline{H}$ is a ZFC$^-$-model.

**Definition** Let $N$ be transitive. $N$ is *full* iff $\omega \in N$ and there is $\gamma$ s.t. $L_\gamma(N)$ models ZFC$^-$ and $N$ is *regular* in $L_\gamma(N)$ – i.e. if $f : x \to N$, $x \in N$, $f \in L_\gamma(N)$, then $\text{rng}(f) \in N$.

It follows that $N$ itself is a ZFC$^-$ model. In fact, regularity in $L_\gamma(N)$ is equivalent to saying that $N$ models 2nd order ZFC$^-$ in $L_\gamma(N)$.

If $\overline{N}$ is full and $\sigma : \overline{N} \prec N$, then there will, indeed, be many different maps $\sigma' : \overline{N} \prec N$. Since fullness is a property of ZFC$^-$ models, however, we shall have to reformulate Shelah's definition so that we do not work directly with $H_\theta$ but rather with ZFC$^-$ models containing $H_\theta$. It also turns out that, in order to prove iterability, we must apparently impose a stronger similarity between $\sigma'$ and $\sigma$ than we have hitherto stated. In order to formulate this we define:

**Definition** Let $\mathbb{B}$ be a complete BA.

$\delta(\mathbb{B}) =$ the smallest cardinality of a set which lies dense in $\mathbb{B} \setminus \{0\}$.

**Note** If we were working with sets $\mathbb{P}$ of conditions rather than complete BA's, we would normally choose $\mathbb{P}$ to have cardinality $\delta(BA(\mathbb{P}))$. Hence the above definition would be superfluous and we would work with $\overline{\overline{\mathbb{P}}}$ instead.

**Definition** Let $N = L_\tau^A =_{\text{Df}} \langle L_\tau[A], \in, A \cap L_\tau[A] \rangle$ be a ZFC$^-$ model. Let $X \cup \{\delta\} \subset N$.

$$C_\delta^N(X) =_{\text{Df}} \text{ the smallest } Y \prec N \text{ s.t. } X \cup \{\delta\} \subset Y.$$

We are now ready to define:

**Definition** Let $\mathbb{B}$ be a complete BA. $\mathbb{B}$ is a *subcomplete forcing* iff for sufficiently large cardinals $\theta$ we have: $\mathbb{B} \in H_\theta$ and for any ZFC$^-$ model $N = L_\tau^A$ s.t. $\theta < \tau$ and $H_\theta \subset N$ we have: Let $\delta : \overline{N} \prec N$ where $\overline{N}$ is countable and full. Let $\sigma(\overline{\theta}, \overline{s}, \overline{\mathbb{B}}) = \theta, s, \mathbb{B}$ where $\overline{s} \in \overline{N}$. Let $\overline{G}$ be $\overline{\mathbb{B}}$-generic over $\overline{N}$. Then there is $b \in \mathbb{B} \setminus \{0\}$ s.t. whenever $G \ni b$ is $\mathbb{B}$-generic over $\mathbf{V}$, there is $\sigma' \in \mathbf{V}[G]$ s.t.

(a) $\sigma' : \overline{N} \prec N$,
(b) $\sigma'(\overline{\theta}, \overline{s}, \overline{\mathbb{B}}) = \theta, s, \mathbb{B}$,
(c) $C_\delta^N(\text{rng}(\sigma')) = C_\delta^N(\text{rng}(\sigma))$ where $\delta = \delta(\mathbb{B})$,

(d) $\sigma' {}''\overline{G} \subset G$.

(Hence $\sigma'$ extends uniquely to a $\sigma^* : \overline{N}[\overline{G}] \prec N[G]$ s.t. $\sigma^*(\overline{G}) = G$.)

**Note** We define $N[G]$ in such a way that $A$ is still a predicate. Thus $N = L_\tau^A$ is $N[G]$-definable.

**Note** This is expressible in $\mathbf{V}$, since the last part can be expressed as:

$$\bigvee b \in \mathbb{B} \ \ b \Vdash \varphi(\check{\mathbb{B}}, \check{\overline{N}}, \check{N}, \mathring{\sigma}, \overline{G}, \mathring{G}),$$

$\mathring{G}$ being the generic name.

**Note** If we omitted (c) from the definition of subcompleteness, the resulting class of forcings would still satisfy Lemma 1.2 for countable support iterations of length $\leq \omega_2$. Since such forcings might change the cofinality of $\omega_2$ to $\omega$, we would thereafter have to use the revised countable support (RCS) iteration. (We will also have to make some further assumptions on the component forcings $\mathbb{B}_i$ of the iteration $\langle \mathbb{B}_i \mid i < \alpha \rangle$.) (c) appears to be needed to get past regular limits points $\lambda$ of the iteration s.t. $\lambda > \delta(\mathbb{B}_i)$ for $i < \lambda$.

**Definition** $\theta$ *verifies* the subcompleteness of $\mathbb{B}$ iff $\theta$ is as in the definition of subcompleteness.

In the following discussion write 'ver$(\mathbb{B}, \theta)$' to mean '$\theta$ verifies the subcompleteness of $\mathbb{B}$'. Now let $\mathbb{B} \in H_\theta$ and let $\theta' > \overline{\overline{H}}_\theta$ be a cardinal. A Löwenheim-Skolem argument that, in order to determine whether ver$(\mathbb{B}, \theta)$, we need only consider $N = L_\tau^A$ s.t. $N \in H_{\theta'}$. By the well known fact: $H_{\theta'}[G] = (H_{\theta'})^{\mathbf{V}[G]}$ for $\mathbb{B}$-generic $G$, where $\mathbb{B} \in H_{\theta'}$, we see that, in fact, the definition of ver$(\mathbb{B}, \theta)$ relativizes to $H_{\theta'}$ – i.e.

**Lemma 2.1** *Let $\mathbb{B} \in H_\theta$. Let $\theta' > \overline{\overline{H}}_\theta$ be a cardinal. The statement* ver$(\mathbb{B}, \theta)$ *is absolute in $H_{\theta'}$.*

This holds in particular for $\theta' = (\overline{\overline{H}}_\theta)^+$. But then the elements of $H_{\theta'}$ can be coded by subsets of $H_\theta$ and we get:

**Lemma 2.2** *Let $\theta > \omega_1$ be a cardinal.* $\{\mathbb{B} \mid \text{ver}(\mathbb{B}, \theta)\}$ *is uniformly 2nd order definable over $H_\theta$.*

Hence:

**Corollary 2.3** *Let $W$ be an inner model s.t. $\mathfrak{P}(H_\theta) \subset W$. Then* ver$(\mathbb{B}, \theta)$ *is absolute in $W$.*

Finally, we note:

**Lemma 2.4** *Let $\theta$ verify the subcompleteness of $\mathbb{B}$. Then $\mathbb{B}$ is subcomplete.*

(Thus "sufficiently large $\theta$" can be replaced by "some $\theta$" in the definition of 'subcomplete'.)

*Proof of Lemma* 2.4. It suffices to show:

**Claim** Let $\mathbb{B} \in H_\theta$. Let $\theta' > \overline{\overline{H}}_\theta$ be a cardinal. Then ver$(\mathbb{B}, \theta')$.

*Proof*. We can assume w.l.o.g. that $\theta$ is least with ver$(\mathbb{B}, \theta)$. Then by Lemma 2.1:

(1) $\quad H_{\theta'} \models \theta$ is least s.t. ver$(\mathbb{B}, \theta)$.

Now let $N = L_\tau^A$ s.t. $\theta' < \tau$ and $H_{\theta'} \subset N$. Then $\theta < \tau$ and $H_\theta \subset N$. Let $\sigma : \overline{N} \prec N$ where $\overline{N}$ is countable and full. Let $\sigma(\overline{\mathbb{B}}, \overline{\theta}', \overline{s}) = \mathbb{B}, \theta', s$. By (1) there is $\overline{\theta}$ s.t. $\sigma(\overline{\theta}) = \theta$. By ver$(\mathbb{B}, \theta)$ there is then a $b \in \mathbb{B}$ with the desired property.

QED(Lemma 2.4)

When actually verifying the subcompleteness of a specific $\mathbb{B}$ we often find it convenient to employ an additional parameter. Thus we define:

**Definition** $\langle \theta, p \rangle$ verifies the subcompleteness of $\mathbb{B}$ (ver$(\mathbb{B}, \theta, p)$) iff $p, \mathbb{B} \in H_\theta$ and for any ZFC$^-$ model $N = L_\tau^A$ with $\theta < \tau$ and $H_\theta \subset N$ we have: Let $\sigma : \overline{N} \prec N$ where $\overline{N}$ is countable and full. Let $\sigma(\overline{p}, \overline{\theta}, \overline{s}, \overline{\mathbb{B}}) = p, \theta, s, \mathbb{B}$. Let $\overline{G}$ be $\overline{\mathbb{B}}$-generic over $\overline{N}$. Then the previous conclusion holds.

The natural analogues of Lemma 2.1 – Corollary 2.3 follow as before. But then we can repeat the proof of Lemma 2.4 to get:

**Lemma 2.5** *Let $\langle \theta, p \rangle$ verify the subcompleteness of $\mathbb{B}$. Then $\mathbb{B}$ is subcomplete.*

This will often be tacitly used in verifications of subcompleteness.

### 3.2. *Liftups*

In order to better elucidate the concept of fullness, we make a digression on the topic of cofinal embeddings.

**Definition** Let $\overline{\mathfrak{A}}$, $\mathfrak{A}$ be models which satisfy the extensionality axiom. Let $\pi : \overline{\mathfrak{A}} \to \mathfrak{A}$ be a structure preserving map. We call $\pi$ *cofinal* (in symbols: $\pi : \overline{\mathfrak{A}} \to \mathfrak{A}$ cofinally) iff for all $x \in \mathfrak{A}$ there is $u \in \overline{\mathfrak{A}}$ s.t. $x \in_{\mathfrak{A}} \pi(u)$.

**Note** In this definition we did not require $\overline{\mathfrak{A}}$, $\mathfrak{A}$ to be transitive or even well founded. Most of our applications will be to transitive models, but we must occasionally deal with ill founded structures. We shall, however, normally assume such structures to be *solid* in the sense of Section 1. (I.e. the well founded core of $\mathfrak{A}$ (wfc($\mathfrak{A}$)) is transitive and $\in^{\mathfrak{A}} \cap \text{wfc}(\mathfrak{A})^2 = \in \cap \text{wfc}(\mathfrak{A})^2$.)

**Definition** Let $\tau$ be a cardinal in $\mathfrak{A}$. $H_\tau^{\mathfrak{A}}$ = the set of $x$ s.t. $\mathfrak{A} \models x \in H_\tau$.

**Note** Even if $\mathfrak{A}$ were a transitive ZFC$^-$ model, we would *not* necessarily have: $H_\tau^{\mathfrak{A}} \in \mathfrak{A}$.

**Definition** Let $\tau \in \overline{\mathfrak{A}}$ be a cardinal in $\overline{\mathfrak{A}}$. We call $\pi : \overline{\mathfrak{A}} \to \mathfrak{A}$ $\tau$-*cofinal* iff for all $x \in \mathfrak{A}$ there is $u \in \overline{\mathfrak{A}}$ s.t. $\overline{\overline{u}} < \tau$ in $\overline{\mathfrak{A}}$ and $x \in_{\mathfrak{A}} \pi(u)$.

We shall generally work with elementary embeddings but must sometimes consider a finer degree of preservation:

**Definition** $\pi : \overline{\mathfrak{A}} \to \mathfrak{A}$ is $\Sigma_n$-*preserving* ($\pi : \overline{\mathfrak{A}} \to_{\Sigma_n} \mathfrak{A}$) iff for all $\Sigma_n$-formulae $\varphi$ and all $x_1, \ldots, x_n \in \overline{\mathfrak{A}}$:

$$\overline{\mathfrak{A}} \models \varphi[x_1, \ldots, x_n] \longleftrightarrow \mathfrak{A} \models \varphi[\pi(x_1), \ldots, \pi(x_n)].$$

**Definition** Let $\overline{\mathfrak{A}}$ be a solid model of ZFC$^-$. Let $\tau \in \text{wfc}(\overline{\mathfrak{A}})$ be an uncountable cardinal in $\overline{\mathfrak{A}}$. Set $\overline{H} = H_\tau^{\overline{\mathfrak{A}}}$. (Hence $\overline{H} \subset \text{wfc}(\overline{\mathfrak{A}})$.) Let $\overline{\pi} : \overline{H} \to_{\Sigma_0} H$ cofinally, where $H$ is transitive. Then by a *liftup* of $\langle \overline{\mathfrak{A}}, \overline{\pi} \rangle$ we mean a pair $\langle \mathfrak{A}, \pi \rangle$ s.t $\pi \supset \overline{\pi}$, $H \subset \text{wfc}(\mathfrak{A})$, and $\pi : \overline{\mathfrak{A}} \to_{\Sigma_0} \mathfrak{A}$ $\tau$-cofinally, where $\mathfrak{A}$ is solid.
(We also say: $\pi : \overline{\mathfrak{A}} \to \mathfrak{A}$ is a *liftup* of $\overline{\mathfrak{A}}$ by $\overline{\pi} : \overline{H} \to H$.)

**Lemma 3.1** *Let $\overline{\mathfrak{A}}$, $\tau$, $\overline{H}$, $H$, $\overline{\pi}$ be as in the above definition. The liftup $\langle \mathfrak{A}, \pi \rangle$ of $\langle \overline{\mathfrak{A}}, \overline{\pi} \rangle$ (if it exists) is determined up to isomorphism (i.e. if $\langle \mathfrak{A}', \pi' \rangle$ is another liftup, there is $\sigma : \mathfrak{A} \stackrel{\sim}{\leftrightarrow} \mathfrak{A}'$ with $\sigma\pi = \pi'$).*

*Proof*. Set $\Delta$ = the set of $f \in \overline{\mathfrak{A}}$ s.t. $\overline{\mathfrak{A}} \models (f$ is a function $\wedge \text{dom}(f) \in H_\tau)$. For each $f \in \Delta$ let $d(f)$ = that $u \in \overline{H}$ s.t. $u = \text{dom}(f)$ in $\overline{\mathfrak{A}}$. Set:

$$\Gamma = \{\langle f, x \rangle \mid f \in \Delta \wedge x \in \overline{\pi}(d(f))\}.$$

It is easily seen by $\tau$-cofinality that each $a \in \mathfrak{A}$ has the form: $a = \pi(f)(x)$ in $\mathfrak{A}$, where $\langle f, x \rangle \in \Gamma$. The same holds for $\langle \mathfrak{A}', \pi' \rangle$ if $\langle \mathfrak{A}', \pi' \rangle$ is another liftup. But:

$$\begin{aligned} \pi(f)(x) \in \pi(g)(y) \text{ in } \mathfrak{A} &\longleftrightarrow \langle x, y \rangle \in \overline{\pi}(\{\langle z, w \rangle \mid f(z) \in g(w) \text{ in } \overline{\mathfrak{A}}\}) \\ &\longleftrightarrow \pi'(f)(x) \in \pi'(g)(y) \text{ in } \mathfrak{A}\,! \end{aligned}$$

Similarly:

$$\pi(f)(x) = \pi(g)(y) \text{ in } \mathfrak{A} \longleftrightarrow \pi'(f)(x) = \pi'(g)(y) \text{ in } \mathfrak{A}'.$$

Hence there is $\sigma : \mathfrak{A} \widetilde{\leftrightarrow} \mathfrak{A}'$ defined by $\sigma(\pi(f)(x)_{\mathfrak{A}}) = \pi'(f)(x)_{\mathfrak{A}}$ for $\langle f, x\rangle \in \Gamma$. But for any $a \in \overline{\mathfrak{A}}$, we have: $\overline{\mathfrak{A}} \models a = k_a(0)$, where $k_a = \{\langle a, 0\rangle\}$ in $\overline{\mathfrak{A}}$. Thus $\pi(a) = \pi(k_a)(0)$ in $\mathfrak{A}$, where $\langle k_a, 0\rangle \in \Gamma$. Hence $\sigma(\pi(a)) = \pi'(k_a)(0) = \pi'(a)$. QED(Lemma 3.1)

Since the identity is the only isomorphism of a transitive structure onto a transitive structure, we have:

**Corollary 3.2** *Let $\langle \mathfrak{A}, \pi\rangle$ be the liftup $\langle \overline{\mathfrak{A}}, \overline{\pi}\rangle$, where $\mathfrak{A}$, $\overline{\mathfrak{A}}$ are transitive. Then $\langle \mathfrak{A}, \pi\rangle$ is the unique liftup.*

*Proof*. Let $\langle \mathfrak{A}', \pi'\rangle$ be a liftup. Let $\sigma : \mathfrak{A} \widetilde{\leftrightarrow} \mathfrak{A}'$ s.t. $\pi' = \sigma\pi$. Then $\mathfrak{A}'$ is well founded, hence transitive, by solidity. Hence $\sigma = \text{id}$ and $\pi' = \pi$, $\mathfrak{A}' = \mathfrak{A}$.

QED(Corollary 3.2)

A transitive liftup does not always exist, even when $\overline{\mathfrak{A}}$ is transitive. However, a straightforward modification of the ultrapower construction does give us:

**Lemma 3.3** *Let $\overline{\mathfrak{A}}$ be a solid model of ZFC$^-$. Let $\tau > \omega$, $\tau \in \text{wfc}(\overline{\mathfrak{A}})$ be a cardinal in $\overline{\mathfrak{A}}$ and set: $\overline{H} = H_\tau^{\overline{\mathfrak{A}}}$. Let $\overline{\pi} : \overline{H} \to_{\Sigma_0} H$ cofinally, where $H$ is transitive. Then $\langle \overline{\mathfrak{A}}, \overline{\pi}\rangle$ has a liftup $\langle \mathfrak{A}, \pi\rangle$.*

*Proof*. Define $\Delta$, $\Gamma$ as above. Let $\mathfrak{A} = \langle |\mathfrak{A}|, \in_{\mathfrak{A}}, A_1^{\mathfrak{A}}, \ldots, A_n^{\mathfrak{A}}\rangle$. Define an equality model $\Gamma^* = \langle \Gamma, =^*, \in^*, A_1^*, \ldots, A_n^*\rangle$ by:

$$\langle f, x\rangle =^* \langle g, y\rangle \longleftrightarrow \langle x, y\rangle \in \overline{\pi}(\{\langle z, w\rangle \mid f(z) \in g(w) \text{ in } \overline{\mathfrak{A}}\})$$

$$\langle f, x\rangle \in^* \langle g, y\rangle \longleftrightarrow \langle x, y\rangle \in \overline{\pi}(\{\langle z, w\rangle \mid f(z) \in g(w) \text{ in } \overline{\mathfrak{A}}\})$$

$$\langle f, x\rangle \in A_i^* \longleftrightarrow x \in \overline{\pi}(\{z \mid f(z) \in A_i \text{ in } \overline{\mathfrak{A}}\}).$$

A straightforward modification of the usual proof gives us *Łos' Theorem* for $\Gamma^*$:

(1)

$$\Gamma^* \models \varphi[\langle f_1, x_1\rangle, \ldots, \langle f_n, x_n\rangle] \longleftrightarrow \langle x_1, \ldots, x_n\rangle \in \overline{\pi}(\{\langle \vec{z}\rangle \mid \overline{\mathfrak{A}} \models \varphi[f_1(z_1), \ldots, f_n(z_n)]\}).$$

This is proven by induction on $\varphi$. The case that $\varphi$ is a primitive formula is immediate. We display the induction step for $\varphi = \varphi(v_1 \ldots, v_n) = \bigvee v_0\, \psi(v_0, \ldots, v_n)$.

($\rightarrow$) Let $\Gamma^* \models \varphi[\langle f_1, x_1\rangle, \ldots, \langle f_n, x_n\rangle]$. Then $\Gamma^* \models \psi[\langle f_0, x_0\rangle, \ldots, \langle f_n, x_n\rangle]$

for some $\langle f_0, x_0\rangle \in \Gamma$. Hence

$$\langle x_0,\ldots,x_n\rangle \in \overline{\pi}(\{\langle\vec{z}\rangle \mid \overline{\mathfrak{A}} \models \psi[f_0(z_0),\ldots,f_n(z_n)\})$$
$$\bigcap$$
$$\overline{\pi}(d(f_0) \times \{\langle\vec{z}\rangle \mid \overline{\mathfrak{A}} \models \varphi[f_1(z_1),\ldots,f_n(z_n)]\})$$
$$\longrightarrow \langle x_1,\ldots,x_n\rangle \in \overline{\pi}(\{\langle\vec{z}\rangle \mid \overline{\mathfrak{A}} \models \varphi[f_1(z_1),\ldots,f_n(z_n)]\}).$$

($\leftarrow$) Set $u = \{\langle\vec{z}\rangle \mid \overline{\mathfrak{A}} \models \varphi[f_1(z_1),\ldots,f_n(z_n)]\}$. Then $u \in \overline{H}$ and $\langle\vec{x}\rangle \in \pi(u)$. In $\overline{\mathfrak{A}}$ we have $\bigwedge\vec{z} \bigvee y(y, f_1(z_1),\ldots,f_n(z_n))$. Hence, by $\mathrm{ZFC}^-$, there is $f_0 \in \overline{\mathfrak{A}}$ s.t.

$$\bigwedge\vec{z}\ \psi(f_0(\vec{z}), f_1(z_1),\ldots,f_n(z_n)) \quad \text{in } \overline{\mathfrak{A}}.$$

But then $\langle f_0, \langle\vec{z}\rangle\rangle \in \Gamma$ and

$$\langle\langle\vec{x}\rangle, x_1,\ldots,x_n\rangle \in \overline{\pi}(\{\langle\vec{z}\rangle \mid \overline{\mathfrak{A}} \models \psi[z_0,\ldots,z_n]\}).$$

Hence $\Gamma^* \models \psi[\langle f_0, \langle\vec{x}\rangle\rangle, \langle f_1, x_1\rangle,\ldots,\langle f_n, x_n\rangle]$. QED(1)

Now let $\Gamma' = \langle|\Gamma'|, \in', A'_1,\ldots,A'_n\rangle$ be the result of factoring $\Gamma^*$ by $=^*$, the elements being the $=^*$-equivalence classes $x'$ of $x \in \Gamma$. Since $\Gamma'$ satisfies extensionality, there is an isomorphism $\sigma : \Gamma' \overset{\sim}{\leftrightarrow} \mathfrak{A}$, where $\mathfrak{A}$ is solid. Set: $[f,x] = \sigma(\langle f,x\rangle')$, where $\langle f,x\rangle \in \Gamma$. Then $\mathfrak{A} \models \mathrm{ZFC}^-$ by (1). We now define $\pi : \overline{\mathfrak{A}} \prec \mathfrak{A}$ by:

**Definition** For $a \in \overline{\mathfrak{A}}$ let $k = \{\langle a, 0\rangle\}$ in $\overline{\mathfrak{A}}$. Set: $\pi(a) =_{\mathrm{Df}} [k, 0]$. Then:

(2) $\quad \pi : \overline{\mathfrak{A}} \prec \mathfrak{A}$.

*Proof.*

$$\begin{aligned}\overline{\mathfrak{A}} \models \varphi[a_1,\ldots,a_n] &\longleftrightarrow \langle 0-0\rangle \in \{\langle\vec{z}\rangle \mid \overline{\mathfrak{A}} \models \varphi[k_{a_1}(z_1),\ldots,k_{a_n}(z_n)]\}\\ &\longleftrightarrow \langle 0-0\rangle \in \pi(\{\langle\vec{z}\rangle \mid \overline{\mathfrak{A}} \models \varphi[k_{a_1}(z_1),\ldots,k_{a_n}(z_n)]\})\\ &\longleftrightarrow \mathfrak{A} \models \varphi[\pi(a_1),\ldots,\pi(a_n)]\end{aligned}$$

by (1). QED(2)

Now set:

**Definition** $\Delta^0 =$ the set of functions $f \in \overline{H}$.

$$\Gamma^0 = \text{ the set of } \langle f,x\rangle \text{ s.t. } f \in \Delta^0 \text{ and } x \in \overline{\pi}(\mathrm{dom}(f)).$$

Since $\overline{\pi} : \overline{H} \to H$ cofinally, $H$ is the set of $\overline{\pi}(f)(x)$ s.t. $\langle f,x\rangle \in \Gamma^0$. Now set:

**Definition** $\tilde{H} = \{[f,x] \mid \langle f,x\rangle \in \Gamma^0\}$.

(3) $\tilde{H}$ is "$\mathfrak{A}$-transitive" – i.e if $a \in_{\mathfrak{A}} b \in \tilde{H}$, then $a \in \tilde{H}$.

*Proof*. Let $a = [f,x]$, $b = [g,y]$, where $\langle g,y\rangle \in \Gamma^0$ and $\langle f,x\rangle \in \Gamma$. Set: $u = \{z \in d(f) \mid f(z) \in \overline{H}\}$: Then $\langle f,x\rangle \in^* \langle g,y\rangle$ implies $\langle f,x\rangle =^* \langle f \restriction u, x\rangle$, where $\langle f \restriction u, x\rangle \in \Gamma^0$. QED(3)

But for $\langle f,x\rangle, \langle g,y\rangle \in \Gamma^0$ we have:

$$[f,x] \in [g,y] \text{ in } \mathfrak{A} \longleftrightarrow \langle x,y\rangle \in \overline{\pi}(\{\langle z,w\rangle \mid f(z) \in g(w)\}) \\ \longleftrightarrow \overline{\pi}(f)(x) \in \overline{\pi}(g)(y).$$

Similarly: $[f,x] = [g,y] \leftrightarrow \overline{\pi}(f)(x) = \pi(g)(y)$. Hence there is an isomorphism $\sigma : \langle \tilde{H}, \in_{\mathfrak{A}}\rangle \overset{\sim}{\leftrightarrow} \langle H, \in\rangle$ defined by: $\sigma([f,x]) = \overline{\pi}(f)(x)$ for $\langle f,x\rangle \in \Gamma^0$. Hence $\langle \tilde{H}, \in_{\mathfrak{A}}\rangle$ is well founded. Since $\tilde{H}$ is $\mathfrak{A}$-transitive it follows that $\tilde{H} \subset \text{wfc}(\mathfrak{A})$; hence $\in_{\mathfrak{A}} \cap \tilde{H}^2 = \in \wedge H^2$ by solidity. Hence $\tilde{H}$ is transitive. Thus $\sigma = \text{id}$ and

(4) $\tilde{H} = H \subset \text{wfc}(\mathfrak{A})$ and $[f,x] = \overline{\pi}(f)(x)$ for $\langle f,x\rangle \in \Gamma^0$.

But then:

(5) $[f,x] = \pi(f)(x)$ in $\mathfrak{A}$ for all $\langle f,x\rangle \in \Gamma$.

*Proof*. $x \in \pi(d(f))$, where

$$d(f) = \{x \mid f(x) = f(x)\} = \{x \mid f(x) = (k_f(0))(\text{id} \restriction d(f))(x) \text{ in } \overline{\mathfrak{A}}\}$$

where $k_f = \{\langle f,0\rangle\}$ in $\overline{\mathfrak{A}}$. Hence

$$\langle x,0,x\rangle \in \pi(\{\langle z,y,w\rangle \mid f(z) = k_f(y)(\text{id} \restriction d(f))(z) \text{ in } \overline{\mathfrak{A}}\}.$$

Thus $[f,x] = [k_f,0]([(\text{id} \restriction d(f)),x])$ in $\mathfrak{A}$, where: $[k_f,0] = \pi(f)$ and $[\text{id} \restriction d(f),x] = \overline{\pi}(\text{id} \restriction d(f))(x) = x$ by (4). QED(5)

(6) $\pi \restriction \overline{H} = \overline{\pi}$, since for $a \in \overline{H}$ we have $\pi(a) = [k_a,0] = \overline{\pi}(k_a)(0) = k_{\overline{\pi}(a)}(0) = \overline{\pi}(a)$ by (4).

Finally, since every $a \in \mathfrak{A}$ has the form $\pi(f)(x)$ for an $x \in H$, it follows that $a \in \pi(\text{rng}(f))$ in $\mathfrak{A}$, where $\overline{\overline{\text{rng}(f)}} < \tau$ in $\overline{\mathfrak{A}}$. Thus

(7) $\pi : \overline{\mathfrak{A}} \prec \mathfrak{A}$ $\tau$-cofinally. QED(Lemma 3.3)

The above proof yields more than we have stated. For instance:

**Lemma 3.4** *Let $\pi : \overline{N} \to_{\Sigma_0} N$ confinally, where $\overline{N}$ is a ZFC$^-$ model and $N$ is transitive. Then $\pi : \overline{N} \prec N$. (Hence $N$ is a ZFC$^-$ model.)*

*Proof*. Repeat the above proof with $\tau = On \cap \overline{N}$ (hence $\overline{H} = \overline{N}$). All steps go through and we get $\mathfrak{A} = \tilde{H} = N$. QED(Lemma 3.4)

**Lemma 3.5** *Let $\overline{\mathfrak{A}}$, $\mathfrak{A}$, $\overline{H}$, $H$, $\tau$, $\pi$ be as in Lemma 3.3. Set $\tilde{\tau} = On \cap H$. Then $\tilde{\tau} \in \mathrm{wfc}(\mathfrak{A})$ and $H = H^{\mathfrak{A}}_{\tilde{\tau}}$.*

*Proof*. By the definition of wfc($\mathfrak{A}$) we have:

(*) If $x \in \mathfrak{A}$ and $y \in \mathrm{wfc}(\mathfrak{A})$ whenever $y \in_{\mathfrak{A}} x$, then $x \in \mathrm{wfc}(\mathfrak{A})$.

We consider two cases:

**Case 1** $\tau$ is regular in $\overline{\mathfrak{A}}$.

**Claim** $H = H^{\mathfrak{A}}_{\pi(\tau)}$ (hence $\pi(\tau) = \tilde{\tau} \in \mathrm{wfc}(\mathfrak{A})$).

*Proof*. ($\subset$) is trivial. We prove ($\supset$).
Let $x \in H_{\pi(\tau)}$ in $\mathfrak{A}$. We claim that $x \in H$. Let $x \in \pi(u)$ in $\mathfrak{A}$, where $u \in \overline{\mathfrak{A}}$, $\overline{\overline{u}} < \tau$ in $\overline{\mathfrak{A}}$. Let $v = u \cap H_\tau$ in $\overline{\mathfrak{A}}$. Then $v \in \overline{H} = H^{\overline{\mathfrak{A}}}_\tau$ by regularity of $\tau$. But then $x \in \pi(v) \in H$. Hence $x \in H$. QED(Case 1)

**Case 2** Case 1 fails.

Let $\kappa = \mathrm{cf}(\tau)$ in $\overline{\mathfrak{A}}$. Then $\kappa \in \overline{H}$. Let $f : \kappa \to \tau$ in $\overline{\mathfrak{A}}$ be normal and cofinal in $\tau$. Then $f \in \mathrm{wfc}(\overline{\mathfrak{A}})$ by (*). Let $\tilde{\kappa} = \sup \pi''\kappa$. Then $\tilde{\kappa} \leq \pi(\kappa) \in H$. Hence $\tilde{\kappa} \in H$. Let $g = \pi(f) \restriction \tilde{\kappa}$ in $\mathfrak{A}$. It follows easily by (*) that $g \in \mathrm{wfc}(\mathfrak{A})$. Thus $\tilde{\tau} = \sup g''\tilde{\kappa} \in \mathrm{wfc}(\mathfrak{A})$.

**Claim** $H = H^{\mathfrak{A}}_{\tilde{\tau}}$

($\subset$) Let $x \in H$. Then $x \in \pi(u)$ where $u \in \overline{H}$. Hence $x \in \pi(u) \in H_{\tilde{\tau}}$. Hence $x \in H_{\tilde{\tau}}$.

($\supset$) Let $x \in H^{\mathfrak{A}}_{\tilde{\tau}}$. Then $x \in H^{\mathfrak{A}}_{\pi(\nu)}$ for a $\nu < \tau$ which is regular in $\overline{\mathfrak{A}}$, since $\tilde{\tau} = \sup \pi''\tau$ and $\tilde{\tau}$ is a limit cardinal in $\mathfrak{A}$. Let $x \in \pi(u)$ in $\mathfrak{A}$, where $u \in \overline{\mathfrak{A}}$, $\overline{\overline{u}} < \tau$ in $\overline{\mathfrak{A}}$. We can choose $\nu$ large enough that $\overline{\overline{u}} < \nu$ in $\overline{\mathfrak{A}}$. Let $v = u \cap H_\nu$ in $\overline{\mathfrak{A}}$. Then $v \in H_\nu \subset \overline{H}$ and $x \in \pi(v) \in H$. QED(Lemma 3.5)

An immediate corollary of the proof is:

**Corollary 3.6** *If $\tau$ is regular or $\mathrm{cf}(\tau) = \omega$ in $\overline{\mathfrak{A}}$. Then $\tilde{\tau} = \pi(\tau)$ and $H = H^{\mathfrak{A}}_{\pi(\tau)}$.*

Note that if $\overline{N}$, $N$ are transitive ZFC$^-$ models, $\tau \in \overline{N}$ is a cardinal in $\overline{N}$ and $\pi : \overline{N} \prec N$ $\tau$-cofinally, then $\pi$ is $\kappa$ cofinal for every $\kappa \geq \tau$ which is a cardinal in $\overline{N}$. Hence, by Corollary 3.6 we conclude:

**Corollary 3.7** *Let $\pi : \overline{N} \to_{\Sigma_0} N$ $\tau$-cofinally, where $\overline{N}$, $N$ are transitive, $\tau \in \overline{N}$ is a cardinal in $\overline{N}$, and $\overline{N} \models$ ZFC$^-$. Let $\kappa \geq \tau$ be regular in $\overline{N}$ or*

$\mathrm{cf}(\kappa) = \omega$ *in* $\overline{N}$. *Then* $\pi(\kappa) = \sup \pi''\kappa$ *and* $H^N_{\pi(\kappa)} = \bigcup_{u \in H^{\overline{N}}_\kappa} \pi(u)$.

⋆ ⋆ ⋆ ⋆ ⋆

We are now ready to develop the concept of fullness further. We first generalize it as follows:

**Definition** Let $N$ be a transitive $\mathrm{ZFC}^-$ model. $N$ is *almost full* iff $\omega \in N$ and there is a solid $\mathfrak{A}$ s.t.

- $\mathfrak{A} \models \mathrm{ZFC}^-$,
- $N \in \mathrm{wfc}(\mathfrak{A})$,
- $N$ is regular in $\mathfrak{A}$ – i.e. if $f : x \in N$, $x \in N$, and $f \in \mathfrak{A}$, then $\mathrm{rng}(f) \in N$.

The last condition can be alternatively expressed by: $|N| = H^{\mathfrak{A}}_\tau$, where $\tau = On \cap N$.

**Definition** $\mathfrak{A}$ *verifies* the almost fullness of $N$ iff the above holds.

Clearly every full structure is almost full. By Lemmas 3.3 and 3.5 we then have:

**Lemma 4.1** *Let* $\overline{N}$ *be almost full. Let* $\overline{\pi} : \overline{N} \to_{\Sigma_0} N$ *cofinally, where* $N$ *is transitive. Then* $N$ *is almost full. (In fact, if* $\overline{\mathfrak{A}}$ *verifies the almost fullness of* $\overline{N}$ *and* $\langle \mathfrak{A}, \pi \rangle$ *is a liftup of* $\langle \overline{\mathfrak{A}}, \overline{\pi} \rangle$, *then* $\mathfrak{A}$ *verifies the almost fullness of* $N$.)

**Definition** Let $N$ be a transitive $\mathrm{ZFC}^-$ model. $\delta_N =$ the least $\delta$ s.t. $L_\delta(N)$ is admissible.

By Section 1 Corollary 21.1 we then have:

**Lemma 4.2** *If* $\mathfrak{A}$ *verifies the almost fullness of* $N$, *then* $L_{\delta_N}(N) \subset \mathrm{wfc}(\mathfrak{A})$.

Combining this with Lemma 4.1 we get a conclusion that is rich in consequences:

**Lemma 4.3** *Let* $\pi : \overline{N} \to_{\Sigma_0} N$ *cofinally where* $\overline{N}$ *is almost full and* $N$ *is transitive. Let* $\varphi$ *be a* $\Pi_1$ *condition. Let* $a_1, \ldots, a_n \in \overline{N}$. *Then*

$$L_{\delta_{\overline{N}}}(\overline{N}) \models \varphi[\overline{N}, \vec{a}\,] \longrightarrow L_{\delta_N}(N) \models \varphi[N, \pi(\vec{a}\,)].$$

*Proof*. Let $\overline{\mathfrak{A}}$ verify the almost fullness of $\overline{N}$ and let $\langle \mathfrak{A}, \tilde{\pi} \rangle$ be a liftup of $\langle \overline{\mathfrak{A}}, \overline{\pi} \rangle$. We assume:

$$L_{\delta_N}(N) \models \psi[N, \pi(\vec{a}\,)],$$

where $\psi$ is a $\Sigma_1$ condition, and prove:

**Claim** $L_{\delta_{\overline{N}}}(\overline{N}) \models \psi[\overline{N}, \vec{a}]$.
Set: $\nu =$ the least ordinal s.t. $L_\nu(N) \models \psi[N, \pi(\vec{a})]$. Then $\nu < \delta_N$. Noting that $\mathfrak{A} \models \psi[N, \pi(\vec{a})]$, we see that $\nu$ is $\mathfrak{A}$-definable, hence has a preimage $\overline{\nu}$ under $\tilde{\pi}$

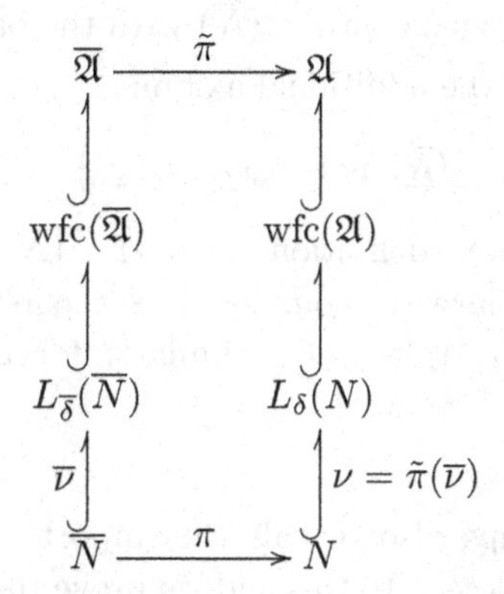


Since $\nu \in \text{wfc}(\mathfrak{A})$, we conclude that $\overline{\nu} \in \text{wfc}(\overline{\mathfrak{A}})$. Hence $L_{\overline{\nu}}(\overline{N}) \models \psi[\overline{N}, \vec{a}]$. But $L_\eta(N)$ is not admissible for any $\eta \leq \nu$. Hence $L_\eta(\overline{N})$ is not admissible for any $\eta \leq \overline{\nu}$. Hence $\overline{\nu} < \delta_{\overline{N}}$ and the conclusion follows. QED(Lemma 4.3)

We now combine this with Barwise' theory. Recall that by a *theory* or *axiomatized language* on an admissible structure $M$ we mean a pair $\langle \mathcal{L}_0, A \rangle$ where $\mathcal{L}_0$ is a language (i.e. a set of predicates and constants) in $M$-finitary predicate logic, and $A$ is a set of axioms in $\mathcal{L}_0$.

We defined $\mathcal{L} = \langle \mathcal{L}_0, A \rangle$ to be $\Sigma_1(M)$ in parameters $p_1, \ldots, p_n \in M$ iff $\mathcal{L}_0$ is $\Delta_1(M)$ in $\vec{p}$ and $A$ is $\Sigma_1(M)$ in $\vec{p}$.

By Section 2 Corollary 4 we get:

**Lemma 4.4** *Let $M$ be admissible. Let $\mathcal{L} = \langle \mathcal{L}_0, A \rangle$ be a theory on $M$ which is $\Sigma_1(M)$ in parameters $p_1, \ldots, p_n \in M$. The statement: '$\mathcal{L}$ is consistent' is then $\Pi_1(M)$ in $\vec{p}$ (uniformly in the $\Sigma_1$ definition of $A$ from $\vec{p}$).*

Hence

**Lemma 4.5** *Let $\pi : \overline{N} \to_{\Sigma_0} N$ cofinally, where $\overline{N}$ is almost full. Let $\overline{\mathcal{L}}$ be an infinitary theory on $L_{\delta_{\overline{N}}}(\overline{N})$ which is $\Sigma_1$ in parameters $\overline{N}, p_1, \ldots, p_n \in \overline{N}$. Let the theory $\mathcal{L}$ on $L_{\delta_N}(N)$ be $\Sigma_1$ in $N$, $\pi(\vec{p})$ by the same definition. If $\overline{\mathcal{L}}$ is consistent, so is $\mathcal{L}$.*

A typical application is:

**Corollary 4.6** *Let $\pi : \overline{N} \to_{\Sigma_0} N$ cofinally, where $\overline{N}$ is almost full. Let $\varphi(v_1, \ldots, v_n)$ be a first order (finite) formula in the $\overline{N}$-language with one* *additional predicate* $\mathring{A}$. *Let* $\text{card}(\overline{N}) = \overline{\tau}$, $\text{card}(N) = \tau$. *Let* $x_1, \ldots, x_n \in \overline{N}$. *If* $\text{coll}(\omega, \overline{\tau})$ *forces* $\bigvee A\langle \overline{N}, A\rangle \models \varphi[\vec{x}]$. *Then* $\text{coll}(\omega, \tau)$ *forces* $\bigvee A\langle N, A\rangle \models \varphi[\pi(\vec{x})]$, ($\text{coll}(\omega, \tau)$ *being the usual conditions for collapsing* $\tau$ *to* $\omega$).

*Proof*. Let $\overline{\mathcal{L}}$ be the language on $L_{\delta_{\overline{N}}}(\overline{N})$ with the basic axioms. The additional constant $\mathring{a}$, and the additional axiom:

$$\langle \underline{\overline{N}}, \mathring{a}\rangle \models \varphi[\underline{x}_1, \ldots, \underline{x}_n].$$

Let $\mathcal{L}$ have the same definition over $L_{\delta_N}(N)$ in the parameters $\pi(x_1), \ldots, \pi(x_n)$. By Barwise' completeness theorem, $\overline{\mathcal{L}}$ is consistent iff $\text{coll}(\omega, \overline{\tau})$ forces $\bigvee A\langle \overline{N}, A\rangle \models \varphi[\vec{x}]$. Similarly for $\mathcal{L}$, $N$, $\pi(\vec{x})$. The conclusion then follows by Lemma 4.5.

QED(Corollary 4.6)

The theory of liftups also reveals the import of condition (c) in the definition of "subcomplete". To this end we prove the *interpolation lemma*:

**Lemma 5.1** *Let* $\pi : \overline{N} \prec N$ *where* $\overline{N}$ *is a transitive ZFC*$^-$ *model and* $N$ *is transitive. Let* $\tau$ *be a cardinal in* $\overline{N}$. *Set:* $\overline{H} = H_\tau^{\overline{N}}$ *and* $\tilde{H} = \bigcup\{\pi(u) \mid u \in \overline{N} \text{ and } \overline{\overline{u}} < \tau \text{ in } \overline{N}\}$. *Then:*

*(a) The transitive liftup* $\langle \tilde{N}, \tilde{\pi}\rangle$ *of* $\langle \overline{N}, \pi \upharpoonright \overline{H}\rangle$ *exists.*
*(b) There is* $\sigma : \tilde{N} \prec N$ *s.t.* $\sigma\tilde{\pi} = \pi$ *and* $\sigma \upharpoonright \tilde{H} = \text{id}$.
*(c)* $\sigma$ *is the unique* $\sigma' : \tilde{N} \to_{\Sigma_0} N$ *s.t.* $\sigma'\tilde{\pi} = \pi$ *and* $\sigma' \upharpoonright \tilde{\tau} = \text{id}$, *where* $\tilde{\tau} = On \cap \tilde{H}$.

*Proof*. Let $\langle \mathfrak{A}, \tilde{\pi}\rangle$ be a liftup of $\langle \overline{N}, \pi \upharpoonright \overline{H}\rangle$. Letting $\Gamma$ be as in the proof of Lemma 3.3 we see that each $y \in \mathfrak{A}$ has the form $\tilde{\pi}(f)(x)$ in $\mathfrak{A}$ for some $\langle f, x\rangle \in \Gamma$. Moreover:

$$\begin{aligned} &\mathfrak{A} \models \varphi[\tilde{\pi}(f_1)(x_1), \ldots, \tilde{\pi}(f_n)(x_n)] \\ &\qquad \longleftrightarrow \langle x_1, \ldots, x_n\rangle \in \pi(\{\langle\vec{z}\rangle \mid \overline{N} \models \varphi[f_1(z_1), \ldots, f_n(z_n)]\}) \\ &\qquad \longleftrightarrow N \models \varphi[\pi(f_1)(x_1), \ldots, \pi(f_n)(x_n)]. \end{aligned}$$

Hence there is $\sigma : \mathfrak{A} \prec N$ defined by: $\sigma(\tilde{\pi}(f)(x)) = \pi(f)(x)$ for $\langle f, x\rangle \in \Gamma$. Thus $\mathfrak{A}$ is well founded, hence transitive by solidity. This proves (a), (b). We now prove (c). Let $\sigma'$ be as in (c). Since $\pi : \overline{N} \prec \tilde{N}$ $\tau$-cofinally, it follows that any $y \in \tilde{N}$ has the form $\pi(f)(\nu)$ for an $\langle f, \nu\rangle \in \Gamma$ s.t. $\text{dom}(f) \subset \tau$. Hence $\sigma'(y) = \pi(f)(\nu) = \sigma(y)$.

QED(Lemma 5.1)

Just as in the proof of Lemma 3.4 we can repeat this using $\tau = On \cap N$, getting:

**Lemma 5.2** *Let $\pi : \overline{N} \prec N$ where $\overline{N}$, $N$ are transitive ZFC$^-$ models. Set: $\tilde{N} = \bigcup_{u \in \overline{N}} \pi(u)$. (Hence $\pi : \overline{N} \prec \tilde{N}$ cofinally.) Then $\tilde{N} \prec N$.*

We now utilize this to examine the meaning of (c) in the definition of "subcomplete".

**Lemma 5.3** *Let $\sigma : \overline{N} \prec N$ where $\overline{N} = L_\alpha^A$ is a ZFC$^-$ model and $N$ is transitive. Let $\sigma(\overline{\delta}) = \delta$, where $\delta$ is a cardinal in $N$. Set $C = C_\delta^N(\mathrm{rng}(\sigma))$, $\overline{H} = (H_{\delta^+})^{\overline{N}}$, $\tilde{H} = \bigcup_{u \in \overline{H}} \sigma(u)$. Let $\langle \tilde{N}, \tilde{\sigma} \rangle$ be the liftup of $\langle \overline{N}, \sigma \restriction \overline{H} \rangle$ and let $k = \tilde{N} \prec N$ s.t. $k\tilde{\sigma} = \sigma$ and $k \restriction \tilde{H} = \mathrm{id}$. Then $C = \mathrm{rng}(k)$.*

*Proof*. ($\subset$) $\mathrm{rng}(\sigma) \subset \mathrm{rng}(k)$ and $\sigma \subset \mathrm{rng}(k)$.
($\supset$) Let $x \in \mathrm{rng}(k)$, $x = k(\tilde{x})$ where $\tilde{x} \in \tilde{\sigma}(u)$, $u \in \overline{N}$, $\overline{\overline{u}} < \overline{\delta}^+$ in $\overline{N}$. Let $f \in N$, $f : \overline{\delta} \xrightarrow{\text{onto}} u$. Then $x = k\tilde{\sigma}(f)(\nu) = \sigma(f)(\nu)$ for a $\nu < \delta$. Hence $x \in C$.

QED(Lemma 5.3)

Stating this differently, we can recover $\tilde{N}$, $k$ from $C$ by the definition: $k : \tilde{N} \stackrel{\sim}{\leftrightarrow} C$, where $\tilde{N}$ is transitive. We can then recover $\tilde{\sigma}$ from $C$ by $\tilde{\sigma} = k^{-1} \cdot \sigma$. If we now have another $\sigma' : \overline{N} \prec N$ s.t. $\sigma'(\overline{\delta}) = \delta$ and $C = C_\delta^N(\mathrm{rng}(\sigma'))$, then $\langle \tilde{N}, \tilde{\sigma}' \rangle$ is the liftup of $\langle \overline{N}, \sigma' \restriction \overline{H} \rangle$, where $\tilde{\sigma}' = k^{-1}\sigma'$. Thus $\sigma = k\tilde{\sigma}$, $\sigma' = k\sigma'$ where $\tilde{\sigma}$, $\tilde{\sigma}'$ are determined entirely by $\sigma \restriction \overline{H}$, $\sigma' \restriction \overline{H}$, respectively. Hence

**Corollary 5.4** *Let $\sigma$, $\sigma'$ be as above. Let $\overline{\tau} \in \overline{N}$ be regular in $\overline{N}$ s.t. $\overline{\tau} > \overline{\delta}$ and $\sigma(\tau) = \sigma'(\tau)$. Then $\sup \sigma''\tau = \sup \sigma'{}''\tau$.*

*Proof*. Let $k(\tilde{\tau}) = \sigma(\tau) = \sigma'(\tau)$. Then $\tilde{\tau} = \sup \tilde{\sigma}''\tau = \sup \tilde{\sigma}'{}''\tau$, since $\tilde{\sigma}$, $\tilde{\sigma}'$ are $\tau$-cofinal and $\tau$ is regular in $\overline{N}$. But then: $\sup \sigma''\tau = \sup \sigma'{}''\tau = \sup k''\tilde{\tau}$.

QED(Corollary 5.4)

A similar argument yields:

**Corollary 5.5** *Let $\tau = On \cap \overline{N}$, where $\sigma$, $\sigma'$ are as above. Then $\sup \sigma''\tau = \sup \sigma'{}''\tau = \sup k''\tilde{\tau}$, where $\tilde{\tau} = On \cap \tilde{N}$.*

Our original version of (c) was weaker, and can be stated as:

(c$'$) Let $\overline{s} = \langle \overline{s}_0, \overline{\lambda}_1, \ldots, \overline{\lambda}_n \rangle$ and $s = \langle s_0, \lambda_1, \ldots, \lambda_n \rangle$ where $\lambda_i > \delta$ is regular in $N$. Let $\overline{\lambda}_0 = On \cap \overline{N}$. Then $\sup \sigma''\overline{\lambda}_i = \sup \sigma'{}''\overline{\lambda}_i$ for $i = 0, \ldots, n$.

This is, of course, an immediate consequence of the above two corollaries. The weaker definition of 'subcomplete' should not be forgotten, since we might someday encounter a forcing which satisfies the weaker version but not the stronger one. That has not happened to date, however, and in fact our original verifications of (c′) turned essentially on first verifying (c).

Before leaving the topic of $\tau$-cofinal embeddings, we mention that these concepts can be applied to structures that are not $\mathrm{ZFC}^-$ models. For our purposes it will suffice to deal with the class of *smooth* models:

**Definition** Let $N$ be a transitive model. $N$ is *smooth* iff either $N \models \mathrm{ZFC}^-$ or else there is a sequence $\langle\langle N_i, \alpha_i\rangle \mid i < \lambda\rangle$ of limit length s.t. $N = \bigcup_i N_i$ and $N_j \models \mathrm{ZFC}^-$, $N_j$ is transitive, and $N_i \in N_j$ s.t. $\alpha_i$ is regular in $N_j$ and $N_i = H_{\alpha_i}^{N_j}$ for $i < j < \lambda$.

Then:

**Lemma 5.6** *If $\overline{N}$ is smooth, $N$ transitive, and $\pi : \overline{N} \to_{\Sigma_0} N$ cofinally, then $N$ is smooth.*

*Proof*. If $\overline{N} \models \mathrm{ZFC}^-$, this is immediate from the foregoing. Otherwise there is a sequence $\langle\langle \overline{N}_i, \overline{\alpha}_i\rangle$ which verifies the smoothness of $\overline{N}$. Set $N_i = \pi(\overline{N}_i)$, $\alpha_i = \pi(\overline{\alpha}_i)$. Then $\langle\langle N_i, \alpha_i\rangle \mid i < \lambda\rangle$ verifies the smoothness of $N$. QED(Lemma 5.6)

**Note** It does *not* follow that $\pi : \overline{N} \prec N$.

The concepts "$\tau$-cofinal" and "liftup" are defined as before, and it follows as before that if $\overline{N}$ is smooth, $\tau$ is a cardinal in $\overline{N}$ and $\overline{\pi} : H_\tau^{\overline{N}} \to_{\Sigma_0} H$ cofinally, then $\langle \overline{N}, \overline{\pi}\rangle$ has at most one transitive liftup.

**Lemma 5.7** *Let $\pi : \overline{N} \to_{\Sigma_0} N$ $\tau$-cofinally, where $\overline{N}$ is smooth. Let $\kappa \in \overline{N}$ be regular in $\overline{N}$, where $\kappa > \tau$. Let $\overline{H} = H_\kappa^{\overline{N}}$, $H = H_{\pi(\kappa)}^N$. Then $\pi \upharpoonright \overline{H} : \overline{H} \to_{\Sigma_0} H$ $\tau$-cofinally.*

*Proof*. Exactly as in Case 1 of Lemma 3.5.

**Lemma 5.8** *Let $\langle\langle \overline{N}_i, \overline{\alpha}_i\rangle \mid i < \lambda\rangle$ verify the smoothness of $\overline{N}$. Let $\tau \in \overline{N}$ be a cardinal. Let $\overline{\pi} : H_\tau^{\overline{N}} \to_{\Sigma_0} H$ cofinally. The transitive liftup of $\langle \overline{N}, \overline{\pi}\rangle$ exists iff for each $i$ s.t. $\tau < \alpha_i$ the transitive liftup of $\langle \overline{N}_i, \overline{\pi}\rangle$ exists.*

*Proof*. ($\to$) Let $\langle N, \pi\rangle$ be the liftup of $\langle \overline{N}, \overline{\pi}\rangle$. Set: $\alpha_i = \pi(\overline{\alpha}_i)$, $N_i = \pi(\overline{N}_i)$. Then $\langle N_i, \pi \upharpoonright \overline{N}_i\rangle$ is the liftup of $\langle \overline{N}_i, \overline{\pi}\rangle$.
($\leftarrow$) Let $\langle N_i, \pi_i\rangle$ be the liftup of $\langle \overline{N}_i, \overline{\pi}\rangle$ for $\tau < \overline{\alpha}_i$. By Lemma 5.7 we

have: $\pi_j \upharpoonright \overline{N}_i : \overline{N}_i \to N_i$ $\tau$-cofinally. Hence $\pi_j \upharpoonright \overline{N}_i = \pi_i$ and we can set: $\pi = \bigcup_i \pi_i$. $\pi : \overline{N} \to_{\Sigma_0} N$ is then $\tau$-cofinal and $\overline{\pi} = \pi \upharpoonright \overline{H}$. QED(Lemma 5.8)

**Lemma 5.9** *Let $\overline{N}$ be smooth and $\pi : \overline{N} \to_{\Sigma_0} N$, where $N$ is transitive. Let $\tau$ be a cardinal in $\overline{N}$. Set $\overline{H} = H_\tau^{\overline{N}}$. Then:*

*(a) The transitive liftup $\langle \tilde{N}, \tilde{\pi} \rangle$ of $\langle \overline{N}, \pi \upharpoonright \overline{H} \rangle$ exists.*
*(b) There is $\sigma : \tilde{N} \to_{\Sigma_0} N$ s.t. $\sigma\tilde{\pi} = \pi$ and $\sigma \upharpoonright \tilde{H} = \text{id}$, where $\tilde{H} = \bigcup_{u \in \overline{H}} \pi(u)$.*
*(c) $\sigma$ is the unique $\sigma : \tilde{N} \to_{\Sigma_0} N$ s.t. $\sigma\tilde{\pi} = \pi$ and $\sigma \upharpoonright \tilde{\tau} = \text{id}$, where $\tilde{\tau} = On \cap \tilde{H}$.*

*Proof*.
**Case 1** $\overline{N} \models \text{ZFC}^-$.
Set: $N' = \bigcup_{u \in \overline{N}} \pi(u)$. Then $\pi : \overline{N} \to_{\Sigma_0} N'$ cofinally. Hence $\pi : \overline{N} \prec N' \subset N$ and we apply our previous lemmas.
**Case 2** Case 1 fails.
Let $\langle \langle \overline{N}_i, \overline{\alpha}_i \rangle \mid i < \lambda \rangle$ verify the smoothness of $\overline{N}$. Assume w.l.o.g. that $\tau \in \overline{N}_0$. (Hence $\overline{H} = H_\tau^{\overline{N}_i}$ for all $i < \lambda$.) (a) follows by Lemma 5.8. Moreover $\langle \overline{N}_i, \tilde{N}_i \rangle$ is the liftup of $\langle \overline{N}, \pi \upharpoonright \overline{H} \rangle$ by Lemma 5.7, where $\tilde{N}_i = \pi(\overline{N}_i)$. Let $\sigma_i : \tilde{N}_i \to_{\Sigma_0} \pi(\overline{N}_i)$ be defined by $\sigma_i\tilde{\pi}_i = \pi \upharpoonright \overline{N}_i$, $\sigma_i \upharpoonright \tilde{H} = \text{id}$. Set $\sigma = \bigcup_i \sigma_i$.

Then $\sigma : \tilde{N} \to_{\Sigma_0} N$ and $\sigma\tilde{\pi} = \pi$, $\sigma \upharpoonright \tilde{H} = \text{id}$. This proves (b). But $\sigma_i$ is unique s.t. $\sigma_i : \tilde{N}_i \to_{\Sigma_0} \pi(\overline{N}_i)$, $\sigma_i\tilde{\pi} = \pi \upharpoonright \overline{N}_i$ and $\sigma_i \upharpoonright \tilde{\tau} = \text{id}$. Hence $\sigma \upharpoonright \tilde{N}_i = \sigma_i$ for $i < \lambda$ if $\sigma$ is as in (c). This proves (c). QED(Lemma 5.9)

### 3.3. *Examples*

We are now ready to prove that some specific forcings are subcomplete. Since these forcings will be presented as sets of conditions rather than Boolean algebras, we set:

**Definition** Let $\mathbb{P}$ be a set of conditions.

$$\mathbf{V}^{\mathbb{P}} =_{\text{Df}} \mathbf{V}^{\text{BA}(\mathbb{P})}, \quad \delta(\mathbb{P}) =_{\text{Df}} \delta(\text{BA}(\mathbb{P}))$$

where $\text{BA}(\mathbb{P})$ is the canonical Boolean algebra over $\mathbb{P}$ as defined in Section 0.

We may refer to the elements of $\mathbf{V}^{\mathbb{P}}$ as '$\mathbb{P}$-names'. We note:

**Fact 1** Let $N = L_\tau^A$ be a $\text{ZFC}^-$ model. Let $\text{BA}(\mathbb{P}) \in H_\theta \subset N$. Let $\delta \subset C \prec N$, where $\text{BA}(\mathbb{P}) \in C$ and $\delta = \delta(\mathbb{P})$. Then for each $p \in \mathbb{P}$ there is

$q \in C \cap \mathbb{P}$ s.t. $[q] \subset [p]$. (Hence every set predense in $C \cap \mathbb{P}$ is predense in $\mathbb{P}$.)

*Proof*. Let $\mathbb{B} = \text{BA}(\mathbb{P})$. By definition there are $f, \Delta \in H_\theta$ s.t. $\Delta$ is dense in $\mathbb{B}$ and $f : \delta \leftrightarrow \Delta$. Hence there are such $f, \Delta \in C$. But $\Delta \subset C$, since $\delta \subset C$. Let $p \in \mathbb{P}$. There is $b \in \Delta$ s.t. $b \subset [p]$. Hence there is $q \in C \cap \mathbb{P}$ s.t. $[q] \subset b$, since $C \prec N$. Hence $[q] \subset [p]$. QED(Fact 1)

Our first example is Prikry forcing.

**Lemma 6.1** *Prikry forcing is subcomplete.*

*Proof*. Let $U$ be a normal ultrafilter on a measurable cardinal $\kappa$. We define the Prikry forcing determined by $U$ to be the set $\mathbb{P} = \mathbb{P}_U$ consisting of all pairs $\langle s, X \rangle$ s.t. $X \in U$ and $s \subset \kappa$ is finite. The extension relation $\leq_{\mathbb{P}}$ is defined by:

$$\langle s, X \rangle \leq \langle t, Y \rangle \quad \text{iff} \quad X \subset Y, s \supset t, \text{ and } \quad t = \text{lub}(t) \cap s.$$

$\mathbb{P}$ does not collapse cardinals or add new bounded subsets of $\kappa$ If $G$ is $\mathbb{P}$-generic, the $\mathbb{P}$-*sequence* added by $G$ is

$$S = S_G = \bigcup \{s \mid \bigvee X \langle s, X \rangle \in G\}.$$

Then $S$ is unbounded in $\kappa$ and has order type $\omega$. $G$ is, in turn, definable from $S$ by:

$$G = G_S = \{\langle s, X \rangle \in \mathbb{P} \mid s = S \cap \text{lub}(s) \wedge S \setminus s \subset X\}.$$

**Definition** We call $S \subset \kappa$ a $\mathbb{P}$-*sequence* (or *Prikry sequence*) iff $S = S_G$ for some $\mathbb{P}$-generic $G$.

The following characterization of Prikry sequences is well known:

**Fact 2** $S$ is a Prikry sequence iff $S \subset \kappa$ has order type $\omega$ and is almost contained in every $X \in U$ (i.e. $\bigvee \nu < \kappa \; S \setminus \nu \subset X$).

We now prove that $\mathbb{P}$ is subcomplete. To this end we let $\theta > 2^{2^\kappa}$ and let $N = L_\tau^A$ be a ZFC$^-$ model s.t. $\tau > \theta$ and $H_\theta \subset N$. Furthermore we assume that $\sigma : \overline{N} \prec N$ where $\overline{N}$ is countable and full. We also suppose that

$$\sigma(\overline{\theta}, \overline{U}, \overline{\mathbb{P}}, \overline{s}) = \theta, U, \mathbb{P}, s.$$

Hence $\sigma(\overline{\mathbb{B}}) = \mathbb{B}$, where $\mathbb{B} = \text{BA}(\mathbb{P})$ and $\overline{B} = \text{BA}(\overline{\mathbb{P}})$ in $\overline{N}$. We must show:

**Main Claim** There is $p \in \mathbb{P}$ s.t. whenever $\mathbb{G} \ni p$ is $\mathbb{P}$-generic. Then there is $\sigma' \in \mathbf{V}[G]$ s.t.

(a) $\sigma' : \overline{N} \prec N$,
(b) $\sigma'(\overline{\theta}, \overline{U}, \overline{\mathbb{P}}, \overline{s}) = \theta, U, \mathbb{P}, s$,
(c) $C^N_\delta(\operatorname{rng}\sigma') = C^N_\delta(\operatorname{rng}\sigma)$, where $\delta = \delta(\mathbb{P})$.
(d) $\sigma'{}''\overline{G} \subset G$.

Note that if we set: $S = S_G$ and $\overline{S} = S_{\overline{G}}$ in $\overline{N}[\overline{G}]$, then (d) becomes equivalent to:

(d′) $\sigma'{}''\overline{S} = S$.

Let $C = C^N_\delta(\operatorname{rng}\sigma)$. Using Fact 1 we get:

(1) Let $X \in U$. Then there is $Y \in C \cap U$ s.t. $Y \subset X$.

*Proof*. Suppose not. Then for each $\nu < \kappa$ the set $\Delta_\nu$ is dense in $\mathbb{P} \cap C$ where

$$\Delta_\nu = \{\langle s, Y\rangle \in \mathbb{P} \cap C \mid s \setminus \nu \not\subset X\}.$$

Hence $\Delta_\nu$ is predense in $\mathbb{P}$ by Fact 1. Let $G$ be $\mathbb{P}$-generic. Then $G \cap \Delta_\nu \neq \emptyset$ for $\nu < \kappa$. Hence $S_G$ is not almost contained in $X$. Contradiction! by Fact 2.

QED(1)

Hence:

(2) $S$ is a $\mathbb{P}$-generic sequence iff $S$ has order type $\omega$ and is almost contained in every $X \in C \cap U$.
(3) $\delta \geq \kappa$,

since otherwise $\overline{\overline{C}} < \kappa$ and $C \cap U$ would have a minimal element $Y = \bigcap(C \cap U)$.

**Definition** We define $N_0$, $k_0$, $\sigma_0$ by: $k_0 : N_0 \stackrel{\sim}{\leftrightarrow} C$, where $N_0$ is transitive $\sigma_0 = k_0^{-1} \circ \sigma$. We also set: $\Theta_0, \mathbb{P}_0, U_0, s_0 = \sigma_0(\overline{\theta}, \overline{\mathbb{P}}, \overline{U}, \overline{s})$.

By Section 3.2, however, we have an alternative characterization:

(4) Let $\sigma_0(\overline{\delta}) = \delta$, $\overline{\nu} = \overline{\delta}^{+\overline{N}}$, $\overline{H} = H^{\overline{N}}_{\overline{\nu}}$. Then $\langle N_0, \sigma_0\rangle$ is the liftup of $\langle \overline{N}, \sigma \upharpoonright \overline{H}\rangle$. Moreover $k_0$ is defined by the condition:

$$k_0 : N_0 \prec N, \quad k_0\sigma_0 = \sigma, \quad k_0 \upharpoonright \nu_0 = \mathrm{id},$$

where $\nu_0 = \sup \sigma''\overline{\nu}$.

Since $\overline{\nu}$ is regular in $\overline{N}$, we conclude:

(5) $\sigma_0$ is a $\overline{\nu}$-cofinal map and $\sigma_0(\overline{\nu}) = \nu_0$.

**Definition** $\alpha_0 = \delta_{N_0} =$ the least $\alpha$ s.t. $L_\alpha(N_0)$ is admissible.

Our Main Claim will reduce to the assertion that a certain language $\mathcal{L}_0$ on $L_{\alpha_0}(N_0)$ is consistent. We define:

**Definition** $\mathcal{L}_0$ is the language on $L_{\alpha_0}(N_0)$ with:

**Predicate:** $\in$

**Constants:** $\mathring{S}, \mathring{\sigma}, \underline{x}$ $(x \in L_{\alpha_0}(N_0))$

**Axioms:**
- Basic axioms and ZFC$^-$
- $\mathring{S}$ is $\underline{\mathbb{P}}_0$-sequence over $N_0$
- $\mathring{\sigma} : \underline{\overline{N}} \prec \underline{N}_0$ $\underline{\overline{\kappa}}$-cofinally, where $\sigma(\overline{\kappa}) = \kappa$
- $\mathring{\sigma}(\underline{\overline{\theta}}, \underline{\overline{\mathbb{P}}}, \underline{\overline{U}}, \underline{\overline{s}}) = \underline{\theta}_0, \underline{\mathbb{P}}_0, \underline{U}_0, \underline{s}_0$
- $\mathring{\sigma}''\underline{\overline{S}} = \mathring{S}$.

We first show that $\mathcal{L}_0$ is consistent. To this end we define:

**Definition** $\langle N_1, \sigma_1 \rangle =$ the liftup of $\langle \overline{N}, \sigma \upharpoonright H^{\overline{N}}_{\overline{\kappa}} \rangle$.
$k_1 =$ the unique $k : N_1 \prec N_0$ s.t. $k\sigma_1 = \sigma_0$ and $k \upharpoonright \kappa_1 = \text{id}$, where $\kappa_1 = \sup \sigma''\overline{\kappa}$.
$\theta_1, \mathbb{P}_1, U_1, s_1 = \sigma_1(\overline{\theta}, \overline{\mathbb{P}}, \overline{U}, \overline{s})$, $S_1 = \sigma_1{}''\overline{S}$.

Note that $\kappa_1 = \sigma_1(\overline{\kappa})$, since $\sigma_1$ is $\overline{\kappa}$-cofinal into $N_1$ and $\overline{\kappa}$ is regular in $\overline{N}$. Then:

(6) (a) $S_1$ is a $\mathbb{P}_1$-sequence over $N_1$,
(b) $\sigma_1 : \overline{N} \prec N_1$ $\overline{\kappa}$-cofinally,
(c) $\sigma_1(\overline{\theta}, \overline{\mathbb{P}}, \overline{U}, \overline{s}) = \theta_1, \mathbb{P}_1, U_1, s_1$,
(d) $\sigma_1{}''\overline{S} = S_1$.

*Proof*. (b)–(d) are immediate. (a) follows by:

**Claim** Let $X \in U_1$. Then $S_1$ is almost contained in $X$.

*Proof*. Let $X \in \sigma_1(w)$, where $\overline{\overline{w}} < \overline{\kappa}$ in $\overline{N}$. Then $\overline{Y} = \bigcap(\overline{U} \cap w)$ is almost contained in every $z \in \overline{U} \cap w$ and $\overline{Y} \in \overline{U}$. Hence $Y = \sigma_1(Y)$ is almost contained in every $Y \in U_1 \cap \sigma_1(w)$. In particular, $Y$ is almost contained in $X$. But $S_1$ is almost contained in $Y$. QED(6)

Now let:

$$\alpha_1 = \delta_{N_1} = \text{ the least } \alpha \text{ s.t. } L_\alpha(N_1) \text{ is admissible.}$$

Let $\mathcal{L}_1$ be the language $L_{\alpha_1}(N_1)$ which is defined as $\mathcal{L}_0$ was defined on $L_{\alpha_0}(N_0)$, substituting $\theta_1$, $\mathbb{P}_1$, $U_1$, $s_1$, $\kappa_1$ for $\theta_0$, $\mathbb{P}_0$, $U_0$, $s_0$, $\kappa_0$. Then

(7) $\mathcal{L}_1$ is consistent.

*Proof*. $\langle H_\kappa, S_1, \sigma_1 \rangle$ models $\mathcal{L}_1$ by (6). QED(7)

Note, however, that:

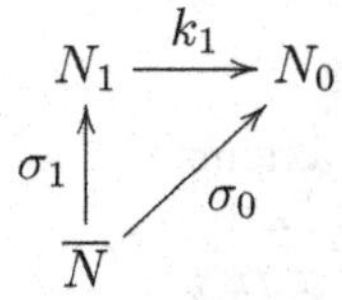

$$\begin{array}{ccc} N_1 & \xrightarrow{k_1} & N_0 \\ \sigma_1 \uparrow & \nearrow \sigma_0 & \\ \overline{N} & & \end{array}$$

where all maps are cofinal and all models are almost full.

Then $\mathcal{L}_0$ is $\Sigma_1(L_{\alpha_0}[N_0])$ in $N_0$ and the parameters:

$$\kappa,\ \mathbb{P}_0,\ \overline{\kappa},\ \overline{N},\ \overline{\theta},\ \overline{\mathbb{P}},\ \overline{U},\ \overline{s},\ \theta_0,\ \mathbb{P}_0,\ U_0,\ s_0.$$

But $\mathcal{L}_1$ is $\Sigma_1(L_{\alpha_1}[N_1])$ in $N_1$ and the $k_1$-preimages of these parameters by the same $\Sigma_1$-formula. Since $N_1$ is almost full and $k_1 : N_1 \prec N_0$ cofinally, we conclude by Lemma 4.5:

(8) $\mathcal{L}_0$ is consistent.

From this we now derive the Main Claim: Work in a generic extension $\mathbf{V}[F]$ of $\mathbf{V}$ in which $L_{\alpha_0}[N]$ is countable. Then $\mathcal{L}_0$ has a solid model

$$\mathfrak{A} = \langle |\mathfrak{A}|, \mathring{S}^{\mathfrak{A}}, \mathring{\sigma}^{\mathfrak{A}} \rangle.$$

Set $S = \mathring{S}^{\mathfrak{A}}$, $\sigma' = k_0 \circ \mathring{\sigma}^{\mathfrak{A}}$. Then

(8) (a) $\sigma' : \overline{N} \prec N$,
 (b) $S$ is $\mathbb{P}$-generic over $\mathbf{V}$,
 (c) $\sigma'(\overline{\theta}, \overline{\mathbb{P}}, \overline{U}, \overline{s}) = \theta, \mathbb{P}, U, s$,
 (d) $C_\delta^N(\operatorname{rng} \sigma') = C$,
 (e) $S = \sigma'{}''\overline{S}$.

*Proof*. (a), (c) are immediate. To see (d) note that $N_0 = C_\delta^{N_0}(\operatorname{rng}\mathring{\sigma}^{\mathfrak{A}})$, since $\mathring{\sigma}^{\mathfrak{A}}$ is $\overline{\kappa}$-cofinal and $\delta \geq \kappa = \mathring{\sigma}^{\mathfrak{A}}(\overline{\kappa})$. Hence:

$$C = k_0{}''N_0 = C_\delta^N(\operatorname{rng} k_0 \circ \mathring{\sigma}^{\mathfrak{A}}).$$

Since $k_0 \restriction (\kappa + 1) = \mathrm{id}$ we have $U \cap N_0 = U \cap C$. Hence (b) follows by (2). (e) follows by $\sigma' \restriction \overline{\kappa} = \mathring{\sigma}^{\mathfrak{A}} \restriction \overline{\kappa}$. QED(9)

We have almost proven the Main Claim, the only problem being that $\sigma'$ is not necessarily an element of $\mathbf{V}[S]$. We now show:

(10) There is $\sigma' \in \mathbf{V}[S]$ satisfying (9).

*Proof*. Work in $\mathbf{V}[S]$. Let $\mu$ be regular in $\mathbf{V}[S]$ s.t. $N \in H_\mu$. Set: $M = \langle H_\mu, N, S, \theta, \mathbb{P}, U, s, \sigma \rangle$. We define a language $\mathcal{L}_2$ on the admissible structure $M$ as follows:

**Definition** $\mathcal{L}_2$ is the language on $M$ with

**Predicate:** $\in$

**Constants:** $\mathring{\sigma}$, $\underline{x}$, $(x \in M)$

**Axioms:**
- ZFC$^-$ and basic axioms
- $\mathring{\sigma} : \underline{\overline{N}} \prec \underline{N}$
- $\mathring{\sigma}(\underline{\overline{\theta}}, \underline{\overline{\mathbb{P}}}, \underline{\overline{U}}, \underline{\overline{s}}) = \underline{\theta}, \underline{\mathbb{P}}, \underline{U}, \underline{s}$
- $C^{\underline{N}}_{\underline{\delta}}(\mathrm{rng}\, \mathring{\sigma}) = \underline{C}$
- $\underline{S} = \mathring{\sigma}''\underline{\overline{S}}$

$\mathcal{L}_2$ is clearly consistent, since $\langle M, \sigma' \rangle$ is a model of $\mathcal{L}_2$ in $\mathbf{V}[F]$, where $\sigma'$ is defined as above.

Now let $\pi : \tilde{M} \prec M$, where $\tilde{M}$ is countable and transitive. Let $\tilde{\mathcal{L}}_2$ be the language on $\tilde{M}$ with the same $\Sigma_1$ definition, replacing all parameters by their preimages under $\pi$. Then $\tilde{\mathcal{L}}_2$ is consistent. Since $\tilde{M} \in H^{\mathbf{V}[S]}_{\omega_1} = H^{\mathbf{V}}_{\omega_1}$, it follows that $\tilde{\mathcal{L}}_2$ has a solid model $\tilde{\mathfrak{A}}$ in $\mathbf{V}$. Let $\tilde{\sigma} = \mathring{\sigma}^{\tilde{\mathfrak{A}}}$ and set: $\sigma' = \pi \circ \tilde{\sigma}$. The verification of (9) is then straightforward. QED(10)

But, since $S$ is a Prikry sequence, there must be $p \in G_S$ which forces the existence of such a $\sigma'$. This proves the Main Claim. QED(Lemma 6.1)

**Lemma 6.2** *Assume CH. Then Namba forcing is subcomplete.*

*Proof*. We first define Namba forcing. The set $\omega_2^{<\omega}$ of monotone finite sequences in $\omega_2$ is a tree ordered by inclusion. The set $\mathbb{N}$ of *Namba conditions* is the collection of all subtrees $T \neq \emptyset$ of $\omega_2^{<\omega}$ s.t. $T$ is downward closed in $\omega_2^{<\omega}$ and for each $s \in T$ the set $\{t \mid s \leq_T t\}$ has cardinality $\omega_2$. The extension relation $\leq_{\mathbb{N}}$ is defined by:

$$T \leq T' \longleftrightarrow_{\mathrm{Df}} T \subset T'.$$

If $G$ is $\mathbb{N}$-generic, then $S = \bigcup \bigcap G$ is a cofinal map of $\omega$ into $\omega_2^{\mathbf{V}}$. We rite $S = S_G$ and call any such $S$ a *Namba sequence*. $G$ is then recoverable from $S$ by:

$$G = G_S = \{T \in \mathbb{N} \mid \bigwedge n < \omega \;\; S \upharpoonright n \in T\}.$$

It is known that, if CH holds, then Namba forcing adds no reals.
We shall also make use of the following fact, which is proven in the Appendix to [DSF]:

**Fact** Let $S$ be a Namba sequence. Let $S' \in \mathbf{V}[S]$ be a cofinal $\omega$-sequence in $\omega_2^{\mathbf{V}}$. Then $S'$ is a Namba sequence and $\mathbf{V}[S'] = \mathbf{V}[S]$.

Note that $\delta(\mathbb{N}) \geq \omega_2$, since otherwise $\omega_2$ would not be collapsed.

We now turn to the proof. Let $\theta > 2^{2^{\omega_2}}$. Let $N = L_\tau^A$ be a ZFC$^-$ model s.t. $\tau > \theta$ and $H_\theta \subset N$. Let $\sigma : \overline{N} \prec N$ where $\overline{N}$ is countable and full. Let $\sigma(\overline{\theta}, \overline{\mathbb{N}}, \overline{s}) = \theta, \mathbb{N}, s$. Let $\overline{G}$ be $\overline{\mathbb{N}}$-generic over $\overline{N}$. It suffices to show:

**Main Claim** There is $p \in \mathbb{N}$ s.t. whenever $G \ni p$ is $\mathbb{N}$-generic, then there is $\sigma' \in \mathbf{V}[G]$ with:

(a) $\sigma' : \overline{N} \prec N$,
(b) $\sigma'(\overline{\theta}, \overline{\mathbb{N}}, \overline{s}) = \theta, \mathbb{N}, s$,
(c) $C_\delta^N(\operatorname{rng}\sigma') = C_\delta^N(\operatorname{rng}\sigma)$ where $\delta = \delta(\mathbb{N})$,
(d) $\sigma'{}''\overline{G} \subset G$.

**Note** We shall actually prove a stronger form of (c):

$$C_{\omega_2}^N(\operatorname{rng}\sigma') = C_{\omega_2}^N(\operatorname{rng}\sigma).$$

**Note** (d) can equivalently be replaced by:

$$\sigma'{}''\overline{S} = S, \text{ where } \overline{S} = S_{\overline{G}}, \quad S = S_G.$$

**Definition** Set $C = C_{\omega_2}^N(\operatorname{rng}\sigma)$. Define $k_0$ by:

$k_0 : N_0 \overset{\sim}{\leftrightarrow} C$, where $N_0$ is transitive, $\sigma_0 = k_0^{-1} \circ \sigma$, $\theta_0, \mathbb{N}_0, s_0 = \sigma_0(\overline{\theta \mathbb{N}}, \overline{s})$.

Just as before we get:

(1) $\langle N_0, \sigma_0\rangle$ is the liftup of $\langle \overline{N}, \sigma \restriction H_{\omega_3}^{\overline{N}}\rangle$,
$k_0$ is the unique $k : N_0 \prec N$ s.t. $k_0\sigma_0 = \sigma$ and $k_0 \restriction \omega_3^{N_0} = \mathrm{id}$,
(where $\omega_3^{N_0} = \sup \sigma_0{}''\omega_3^{\overline{N}}$).

Now let $\alpha_0$ be the least $\alpha$ s.t. $L_\alpha(N_0)$ is admissible. We define a language $\mathcal{L}_0$ on $L_{\alpha_0}(N_0)$ as follows:

**Definition** $\mathcal{L}_0$ is the language on $L_{\alpha_0}(N_0)$ with:

**Predicate:** $\in$
**Constants:** $\mathring{\sigma}$, $\underline{x}$ $(x \in L_{\alpha_0}(N_0))$
**Axioms:**
- Basic axioms and ZFC$^-$
- $\mathring{\sigma} : \underline{\overline{N}} \prec \underline{N_0}$ $\underline{\omega_2^{\overline{N}}}$-cofinally
- $\mathring{\sigma}(\overline{\theta}, \overline{N}, \overline{s}) = \theta_0, \mathbb{N}_0, s_0$.

(2) $\mathcal{L}_0$ is consistent.

*Proof*. Let $\langle N_1, \sigma_1\rangle$ be the liftup of $\langle \overline{N}, \sigma \restriction H_{\omega_2}^{\overline{N}}\rangle$. Define $k_1 : N_1 \prec N_0$ by:

$$k_1\sigma_1 = \sigma_0, \quad k_1 \restriction \gamma_1 = \mathrm{id}, \text{ where } \gamma_1 = \sup \sigma'\omega_2^{\overline{N}} = \sigma_1(\omega_2^{\overline{N}}).$$

Let $\mathcal{L}_1$ be the corresponding language on $L_{\alpha_1}(N_1)$, where $\alpha_1 = \delta_{N_1}$. Just as before it suffices to show that $\mathcal{L}_1$ is consistent. This clear, however, since $\langle H_{\omega_2}, \sigma_1 \rangle$ is a model. QED(2)

Now let $S'$ be a Namba sequence. Work in $\mathbf{V}[S']$. Let $\mu$ be a regular cardinal in $\mathbf{V}[S']$ with $N \in H_\mu$. Set:

$$M = \langle H_\mu, N, \sigma, \mathbb{N}, s \rangle.$$

Let $\pi : \tilde{M} \prec M$, where $\tilde{M}$ is transitive and countable. Then $\tilde{M} \in H_{\omega_1} \subset \mathbf{V}$ in $V[S']$. Let

$$\pi(\tilde{\mathbb{N}}, \tilde{\sigma}, \tilde{N}, \tilde{\mathcal{L}}, \tilde{k}, \tilde{C}) = \mathbb{N}, \sigma, N, \mathcal{L}_0, k_0, C_0.$$

Let $\mathfrak{A} \in \mathbf{V}$ be a solid model of $\tilde{\mathcal{L}}$. Set $\tilde{\sigma} = \tilde{k} \circ \overset{\circ}{\sigma}{}^{\mathfrak{A}}$; $\sigma' = \pi \circ \tilde{\sigma}$. It follows easily that:

(3) (a) $\sigma' : \overline{N} \prec N$
(b) $\sigma'(\overline{\theta}, \overline{\mathbb{N}}, \overline{s}) = \theta, \mathbb{N}, s$
(c) $C^N_{\omega_1}(\operatorname{rng} \sigma') = C$

Now let $\overline{S} = S_{\overline{G}}$ and set: $S = \sigma' {}'' \overline{S}$. Then $S \in \mathbf{V}[S']$ is a cofinal $\omega$-sequence in $\omega_2^{\mathbf{V}}$; hence:

(4) $S$ is a Namba sequence and $\mathbf{V}[S] = \mathbf{V}[S']$. (Hence $\sigma' \in \mathbf{V}[S]$.)

But we know:

(5) $S = \sigma' {}'' \overline{S}$.

Let $G = G_S$. There is then a $p \in G$ which forces the existence of a $\sigma' \in \mathbf{V}[S]$ satisfying (3), (5). This proves the Main Claim. QED(Lemma 6.2)

Now let $\kappa > \omega_1$ be a regular cardinal. Let $A \subset \kappa$ be a stationary set of $\omega$-cofinal ordinals. Our final example is the forcing $\mathbb{P}_A$ which is designed to shoot a cofinal normal sequence of order type $\omega_1$ through $A$:

**Definition** $\mathbb{P}_A$ is the set of normal functions $p : \nu + 1 \to A$, where $\nu < \omega_1$. The extension relation is defined by:

$$p \leq q \quad \text{in} \quad \mathbb{P}_A \longleftrightarrow_{\text{Df}} q \subset p.$$

Clearly, if $G$ is $\mathbb{P}_A$-generic, then $\bigcup G : \omega_1 \to A$ is normal and cofinal in $\kappa$. $\mathbb{P}_A$ adds no new countable subsets of the ground model. If, however, $\{\lambda < \kappa \mid \operatorname{cf}(\lambda) = \omega \wedge \lambda \notin A\}$ is stationary, then $\mathbb{P}_A$ will not be a complete forcing.

**Lemma 6.3** *$\mathbb{P}_A$ is subcomplete.*

*Proof*. Clearly $\delta(\mathbb{P}_A) \geq \kappa$, since otherwise $\kappa$ would remain regular. Now let $\theta > 2^{2^\kappa}$. Let $N = L_\tau^B$ be a ZFC$^-$ model s.t. $\tau > \theta$ and $H_\theta \subset N$. Let $\sigma : \overline{N} \prec N$ where $\overline{N}$ is countable and full. Let $\sigma(\overline{\theta}, \overline{\mathbb{P}}, \overline{A}, \overline{\kappa}, \overline{s}) = \theta, \mathbb{P}_A, A, \kappa, s$. Let $\overline{G}$ be $\overline{\mathbb{P}}$-generic over $\overline{N}$. It suffices to show:

**Main Claim** There is $p \in \mathbb{P}$ s.t. whenever $G \ni p$ is $\mathbb{P}_A$-generic, there is $\sigma' \in \mathbf{V}[G]$ s.t.

(a) $\sigma' : \overline{N} \prec N$,
(b) $\sigma'(\overline{\theta}, \overline{\mathbb{P}}, \overline{\kappa}, \overline{A}, \overline{s}) = \theta, \mathbb{P}, \kappa, A, s$,
(c) $C_\kappa^N(\text{rng}\,\sigma') = C_\kappa^N(\text{rng}\,\sigma)$,
(d) $\sigma' {}''\overline{G} \subset G$.

(1) Let $\sigma' \in \mathbf{V}$ satisfy (a), (b), (c) and

(e) $\sup \sigma' {}''\overline{\kappa} \in A$.

Then the Main Claim holds.

*Proof*. Let $\overline{F} = \bigcup \overline{G}$. Then $\overline{F}$ is a cofinal normal map of $\omega_1^{\overline{N}}$ into $\overline{A}$, where $\sigma(\overline{A}) = A$. Define $p \in \mathbb{P}_A$ by:

$$p(\xi) = \begin{cases} \sigma' \overline{F}(\xi) & \text{for } \xi < \omega_1^{\overline{N}}, \\ \sup \sigma' {}''\overline{\kappa} & \text{for } \xi = \omega_1^{\overline{N}}. \end{cases}$$

Clearly $p \leq \sigma'(q)$ for $q \in \overline{G}$. Hence if $G \ni p$ is generic, then $\sigma' {}''\overline{G} \subset G$.

QED(1)

We must produce a $\sigma'$ satisfying (a), (b), (c) and (e). For $\xi < \kappa$ set: $C_\xi = C_\xi^N(\text{rng}\,\sigma)$. Set:

$$D = \{\tau < \kappa \mid \tau = \kappa \cap C_\tau\}.$$

Then $D$ is club in $\kappa$. Hence then is $\kappa_0 \in D \cap A$. Set:

**Definition** $k_0 : N_0 \overset{\sim}{\leftrightarrow} C_{\kappa_0}$, where $N_0$ is transitive; $\sigma_0 = k_0^{-1} \circ \sigma$; $\theta_0, \mathbb{P}_0, s_0, A_0 = \sigma_0(\overline{\theta}, \overline{\mathbb{P}}, \overline{s}, \overline{A})$. (Hence $\kappa_0 = \sigma_0(\overline{\kappa})$.)

We again let $\alpha_0 = \delta_{N_0}$ be least s.t. $L_{\alpha_0}(N_0)$ is admissible and define:

**Definition** $\mathcal{L}_0$ is the language on $L_{\alpha_0}(N_0)$ with:

**Predicate:** $\in$
**Constants:** $\mathring{\sigma}$, $\underline{x}$ $(x \in L_{\alpha_0}(N_0))$
**Axioms:**
- Basic axioms and ZFC$^-$
- $\mathring{\sigma} : \underline{\overline{N}} \prec \underline{N_0}$ $\underline{\overline{\kappa}}$-cofinally
- $\mathring{\sigma}(\underline{\overline{\theta}}, \underline{\overline{\mathbb{P}}}, \underline{\overline{s}}, \underline{\overline{\kappa}}, \underline{\overline{A}}) = \underline{\theta_0}, \underline{\mathbb{P}_0}, \underline{s_0}, \underline{\kappa_0}, \underline{A_0}$.

(2) $\mathcal{L}_0$ is consistent.

*Proof*. Let $\langle N_1, \sigma_1 \rangle$ be the liftup of $\langle \overline{N}, \sigma \restriction H_{\overline{\kappa}}^{\overline{N}} \rangle$. Note that $\sigma \restriction H_{\overline{\kappa}}^{\overline{N}} = \sigma_0 \restriction H_{\overline{\kappa}}^{\overline{N}}$. Hence there is $k_1 : N_1 \prec N_0$ defined by:

$$k_1 \sigma_1 = \sigma_0, \quad k_1 \restriction \kappa_1 = \mathrm{id},$$

where $\kappa_1 = \sigma_1(\overline{\kappa}) = \sup \sigma'' \overline{\kappa}$. Let $\mathcal{L}_1$ be the corresponding language on $L_{\alpha_1}(N_1)$, where $\alpha_1 = \delta_{N_1}$. By the usual argument it suffices to show that $\mathcal{L}_1$ is consistent: Since $N_0 = C_{\kappa_0}^{N_0}(\mathrm{rng}\sigma_0)$, we can conclude that $\sigma_0 : \overline{N} \prec N_0$ is $\overline{\kappa}^{+\overline{N}}$-cofinal. Hence:

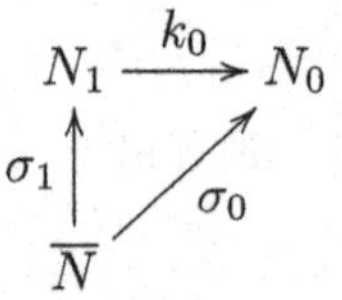

$$\begin{array}{ccc} N_1 & \xrightarrow{k_0} & N_0 \\ \sigma_1 \uparrow & \nearrow \sigma_0 & \\ \overline{N} & & \end{array}$$

where all maps are cofinal and all structures are almost full. $\mathcal{L}_1$ is trivially consistent, however, since $\langle H_\kappa, \sigma_1 \rangle$ models $\mathcal{L}_1$. QED(2)

Now let $M = \langle H_\kappa, N_0, \kappa_0, A_0, \sigma_0 \rangle$. Let $\pi : \tilde{M} \prec M$ s.t. $\tilde{M}$ is countable and transitive. Let $\pi(\tilde{\mathcal{L}}) = \mathcal{L}_0$. Then $\tilde{\mathcal{L}}$ is a consistent language on $L_{\tilde{\alpha}}(\tilde{N}) = \pi^{-1}(L_{\alpha_0}(N_0))$. Hence $\tilde{\mathcal{L}}$ has a solid model $\mathfrak{A}$. Set:

$$\sigma' = k_0 \circ \pi \circ \overset{\circ}{\sigma}^{\mathfrak{A}}.$$

Then $\sigma'$ satisfies (a), (b), (c), (e) of (1). QED(Lemma 6.3)

## 4. Iterating Subcomplete Forcing

The *two step iteration theorem* for subcomplete forcing says that if $\mathbb{A}$ is subcomplete and

$$\Vdash_{\mathbb{A}} \overset{\circ}{\mathbb{B}} \quad \text{is subcomplete},$$

then $\mathbb{A} * \overset{\circ}{\mathbb{B}}$ is subcomplete. Equivalently:

**Theorem 1** *Let $\mathbb{A} \subseteq \mathbb{B}$ where $\mathbb{A}$ is subcomplete and*

$$\Vdash_{\mathbb{A}} \check{\mathbb{B}}/\overset{\circ}{G} \quad \textit{is subcomplete.}$$

*Then $\mathbb{B}$ is subcomplete.*

(**Note** The definitions of $\mathbb{A} * \overset{\circ}{\mathbb{B}}$, $\check{\mathbb{B}}/\overset{\circ}{G}$ and other Boolean conventions employed here can be found in Section 0. $\overset{\circ}{G}$ is the canonical generic name – i.e. $\Vdash_{\mathbb{A}} \overset{\circ}{G}$ is $\check{\mathbb{A}}$-generic over $\check{\mathbf{V}}$, and $[[\check{a} \in \overset{\circ}{G}]] = a$ for $a \in \mathbb{A}$.)

*Proof of Theorem* 1. Let $\theta$ be big enough that $\theta$ verifies the subcompleteness of $\mathbb{A}$ and:

$$\Vdash_{\mathbb{A}} \check{\theta} \text{ verifies the subcompleteness of } \check{\mathbb{B}}/\mathring{G}.$$

Let $N = L_\tau^A$ be a ZFC$^-$ model s.t. $H_\theta \subset N$ and $\tau > \theta$. Let $\sigma : \overline{N} \prec N$ where $\overline{N}$ is countable and sound. Let:

$$\sigma(\overline{\theta}, \overline{\mathbb{A}}, \overline{\mathbb{B}}, \overline{s}) = \theta, \mathbb{A}, \mathbb{B}, s$$

where $s \in N$. Let $\overline{G}$ be $\overline{\mathbb{B}}$-generic over $\overline{N}$. We must find $b \in \mathbb{B} \setminus \{0\}$ s.t. whenever $G \ni b$ is $\mathbb{B}$-generic, there is $\sigma' \in \mathbf{V}[G]$ satisfying (a)–(d) in the definition of subcompleteness.

Let $\overline{G}_0 = \overline{G} \cap \overline{\mathbb{A}}$. Then $\overline{G}_0$ is $\overline{\mathbb{A}}$-generic over $\overline{N}$. Since $\theta$ verifies the subcompleteness of $\mathbb{A}$, there exist $a \in \overline{\mathbb{A}} \setminus \{0\}$, $\mathring{\sigma}_0 \in \mathbf{V}^{\mathbb{A}}$ s.t. whenever $G_0 \ni a$ is $\mathbb{A}$-generic and $\sigma_0 = \mathring{\sigma}_0^{G_0}$, then (a)–(d) hold with $\overline{\mathbb{A}}$, $\overline{G}_0$, $\mathbb{A}$, $G_0$, $\sigma_0$ in place of $\overline{\mathbb{B}}$, $\overline{G}$, $\mathbb{B}$, $G$, $\sigma'$. Let $\overline{\mathbb{B}}^* = \overline{\mathbb{B}}/\overline{G}_0$. Let $G_0 \ni a$ be $\mathbb{A}$-generic. Set: $\mathbb{B}^* = \mathbb{B}/G_0$. Clearly, $\sigma_0$ extends to $\sigma_0^*$ s.t

$$\sigma_0^* : \overline{N}[\overline{G}_0] \prec N[G_0] \quad \text{and} \quad \sigma_0^*(\overline{G}_0) = G_0.$$

In other words, $\sigma_0^* : \overline{N}^* \prec N^*$ where: $\overline{N} = L_{\overline{\tau}}^{\overline{A}}$, $\overline{N}^* = L_{\overline{\tau}}^{\overline{A}, \overline{G}_0}$, $N = L_\tau^A$, $N^* = L_\tau^{A, G_0}$. Note that $H_\theta^{\mathbf{V}[G_0]} = H_\theta^{\mathbf{V}}[G_0] \subset N^*$. Moreover $\overline{G}^*$ is $\overline{\mathbb{B}}^*$-generic over $\overline{N}^*$ where:

$$\overline{G}^* = \overline{G}/\overline{G}_0 = \{b/\overline{G}_0 \mid b \in \overline{G}\}.$$

Clearly

$$\sigma_0^*(\overline{\theta}, \overline{\mathbb{A}}, \overline{\mathbb{B}}, \overline{\mathbb{B}}^*, \overline{s}) = \theta, \mathbb{A}, \mathbb{B}, \mathbb{B}^*, s.$$

Since $\theta$ verifies the subcompleteness of $\mathbb{B}^*$ in $\mathbf{V}[G_0]$, we conclude that there is $b^* \in \mathbb{B}^*$ s.t. whenever $G^* \ni b^*$ is $\mathbb{B}^*$-generic over $\mathbf{V}[G_0]$, then there is $\sigma^* \in \mathbf{V}[G_0][G^*]$ with:

(a*) $\sigma^* : \overline{N}^* \prec N^*$,
(b*) $\sigma^*(\overline{\theta}, \overline{\mathbb{A}}, \overline{\mathbb{B}}, \overline{\mathbb{B}}^*, \overline{s}) = \theta, \mathbb{A}, \mathbb{B}, \mathbb{B}^*, s$
(c*) $C_{\delta^*}^{N^*}(\text{rng}(\sigma^*)) = C_{\delta^*}^{N^*}(\text{rng}(\sigma_0^*))$, where $\delta^* = \delta(\mathbb{B}^*)$.
(d*) $\sigma^* {}''\overline{G}^* \subset G^*$.

Note that $\overline{G} = \overline{G}_0 * \overline{G}^* =_{\text{Df}} \{b \in \overline{\mathbb{B}} \mid b/\overline{G}_0 \in \overline{G}^*\}$. Set $G = G_0 * G^*$. Set $\sigma' = \sigma^* \restriction \overline{N}$. Then $\sigma' \in \mathbf{V}[G] = \mathbf{V}[G_0][G^*]$. We show:

**Claim** $\sigma'$ satisfies:

(a) $\sigma' : \overline{N} \prec N$
(b) $\sigma'(\overline{\theta}, \overline{\mathbb{A}}, \overline{\mathbb{B}}, \overline{s}) = \theta, \mathbb{A}, \mathbb{B}, s$

(c) $C^N_\delta(\mathrm{rng}(\sigma')) = C^N_\delta(\mathrm{rng}(\sigma))$, where $\delta = \delta(\mathbb{B})$.
(d) $\sigma'{}''\overline{G} \subset G$.

We note first that the claim proves the theorem, since $G$ is $\mathbb{B}$-generic and there must, therefore, be a $b \in G$ which forces the existence of such a $\sigma'$. We now prove the claim. (a), (b), (d) are immediate. We prove (c). Note that $\delta \geq \delta^*$, we have:

$$C^{N^*}_\delta(\mathrm{rng}(\sigma^*)) = C^{N^*}_\delta(\mathrm{rng}(\sigma^*_0)).$$

Since $C^N_\delta(\mathrm{rng}(\sigma_0)) = C^N_\delta(\mathrm{rng}(\sigma))$, it suffices to show:
(1) $C^N_\delta(\mathrm{rng}(\sigma')) = N \cap C^{N^*}_\delta(\mathrm{rng}(\sigma^*))$
(2) $C^N_\delta(\mathrm{rng}(\sigma_0)) = N \cap C^{N^*}_\delta(\mathrm{rng}(\sigma^*_0))$.
We proof (1), the proof of (2) being virtually identical. ($\subset$) is trivial. We prove ($\supset$).
Let $x \in N \cap C^{N^*}_\delta(\mathrm{rng}(\sigma^*))$. Then $x$ is $N[G_0]$-definable in $\xi$, $\sigma^*(z)$, $G_0$, where $\xi < \delta$, $z \in \overline{N}$. But, letting $t \in \overline{N}^{\overline{\mathbb{A}}}$ s.t. $\langle z, \overline{G}_0\rangle = t^{\overline{G}_0}$, we have:

$$\langle \sigma^*(z), G_0\rangle = \sigma^*(\langle z, \overline{G}_0\rangle) = \sigma^*(t^{\overline{G}_0}) = \sigma'(t)^{G_0}.$$

Hence:

$$x = \text{ that } x \text{ s.t. } N[G_0] \models \varphi[x, \xi, \sigma'(t)^{G_0}].$$

But since $\sigma'(\overline{\mathbb{B}}) = \mathbb{B}$, we have: $\sigma'(\overline{\delta}) = \delta$, where $\overline{\delta} = \delta(\overline{\mathbb{B}})$. Since $\overline{\delta} \geq \delta(\overline{\mathbb{A}})$, there is $f \in \overline{N}$ mapping $\overline{\delta}$ onto a dense subset of $\overline{A}$. Hence $\sigma'(f)$ maps $\delta$ onto a dense subset of $\mathbb{A}$. Hence there is $\nu < \delta$ s.t. $\sigma'(f)(\nu) \in G_0$ and $\sigma'(f)(\nu)$ forces $\varphi(\check{x}, \check{\xi}, \sigma'(t))$. Thus: $x =$ that $x$ s.t. $\sigma'(f)(\nu) \Vdash^N_{\mathbb{A}} \varphi(\check{x}, \check{\xi}, \sigma'(t))$. Hence $x \in C^N_\delta(\mathrm{rng}(\sigma'))$. QED(Theorem 1)

The proof of Theorem 1 shows more than we have stated. We can omit the assumption that $\mathbb{A}$ is subcomplete and omit the map $\sigma$, assuming, however, that $\Vdash_{\mathbb{A}} \check{\mathbb{B}}/\mathring{G}_0$ is subcomplete, where $\mathring{G}_0$ is the canonical $\mathbb{A}$-generic name. Let $\theta$ be big enough that

$$\Vdash_{\mathbb{A}} \check{\theta} \quad \text{verifies the subcompleteness of } \check{\mathbb{B}}/\mathring{G}_0.$$

Let $N$ be as before and let $\overline{N}$ be countable and full. Suppose that $a \in \mathbb{A} \setminus \{0\}$ and $\mathring{\sigma} \in \mathbf{V}^{\mathbb{A}}$ are given s.t. whenever $G_0 \ni a$ is $\mathbb{A}$-generic and $\sigma_0 = \mathring{\sigma}_0^{G_0}$, then

- $\sigma_0 : \overline{N} \prec N$
- $\sigma_0(\overline{\theta}, \overline{\mathbb{A}}, \overline{\mathbb{B}}, \overline{s}) = \theta, \mathbb{A}, \mathbb{B}, s$
- $\sigma_0{}''\overline{G}_0 \subset G_0$.

Our proof then yields a $b^* \in \mathbb{B}^* = (\mathbb{B}/G_0) \setminus \{0\}$ s.t. if $G \supset G_0$ is $\mathbb{B}$-generic and $b^* \in G^* = G/G_0 = \{c/G_0 \mid c \in G\}$, then there is $\sigma' \in \mathbf{V}[G]$ s.t. (a), (b), (d) and:

(c′) $C^N_\delta(\text{rng}(\sigma')) = C^N_\delta(\text{rng}(\sigma_0))$

hold, where $\delta = \delta(\mathbb{B})$. We can improve on this still further. Suppose that $t \in \mathbf{V}^{\mathbb{A}}$ s.t. and $a \Vdash t \in \check{\overline{N}}$. This means that $t^{G_0} \in \overline{N}$ whenever $G_0 \ni a$ is $\mathbb{A}$-generic. We can then select our $b^*$ so as to force:

(e*) $\sigma^*(t^{G_0}) = \sigma_0(t^{G_0})$

in addition to (a*)–(d*). It then follows that:

(e′) $\sigma'(t^{G_0}) = \sigma_0(t^{G_0})$.

Since, whenever $G_0 \ni a$ is $\mathbb{A}$-generic, we can find a $b^* \in \mathbb{B}/G_0$ forcing (a), (b), (d), (c′), (e′), we conclude that there is $b^0 \in \mathbf{V}^{\mathbb{A}}$ s.t. $a$ forces $b^* = \mathring{b}^{G_0}$ to have these properties. We may assume w.l.o.g. that $\Vdash_{\mathbb{A}} \mathring{b} \in \check{\mathbb{B}}/\mathring{G}_0$ and $[[\mathring{b} \neq 0]]_{\mathbb{A}} = a$. By Section 0, Fact 4 there is then a unique $b \in \mathbb{B}$ s.t. $\Vdash_{\mathbb{A}} \check{b}/\mathring{G}_0 = \mathring{b}$. Letting $h = h_{\mathbb{A},\mathbb{B}}$ be defined as in Section 0 by $h(c) = \bigcap\{a \in \mathbb{A} \mid c \subset a\}$ for $a \in \mathbb{B}$, we conclude by Section 0, Fact 3 that:

$$h(b) = [[\check{b}/\mathring{G}_0 \neq 0]] = [[\mathring{b} \neq 0]] = a.$$

Clearly, if $G \ni b$ is $\mathbb{B}$-generic, then $G_0 \ni a$ is $\mathbb{A}$-generic, where $G_0 = G \cap \mathbb{A}$. Thus $b/G_0 = \mathring{b}^{G_0} = b^*$ has the above properties and (a), (b), (d), (c′),(e′) hold.

Putting all of this together, we get a very useful technical lemma:

**Lemma 1.1** *Let $\mathbb{A} \subseteq \mathbb{B}$ and let: $\Vdash_{\mathbb{A}} \check{\mathbb{B}}/\mathring{G}$ is subcomplete. Let $\theta$ be big enough that $\mathbb{B} \in H_\theta$ and: $\Vdash_{\mathbb{A}} \check{\theta}$ verifies the subcompleteness of $\check{\mathbb{B}}/\mathring{G}$. Let $N = L^A_\tau$ be a ZFC$^-$ model s.t. $H_\theta \subset N$ and $\theta < \tau$. Let $\overline{N}$ be countable and full. Let $\overline{\mathbb{A}} \subseteq \overline{\mathbb{B}}$ in $\overline{N}$, where $\overline{G}$ is $\overline{\mathbb{B}}$-generic over $\overline{N}$. Set: $\overline{G}_0 = \overline{G} \cap \overline{\mathbb{A}}$. Suppose that $a \in \mathbb{A} \setminus \{0\}$, $\mathring{\sigma}_0 \in \mathbf{V}^{\mathbb{A}}$ s.t. whenever $G_0 \ni a$ is $\mathbb{A}$-generic and $\sigma_0 = \mathring{\sigma}_0^{G_0}$, then:*

*(i)* $\sigma_0 : \overline{N} \prec N$
*(ii)* $\sigma_0(\overline{\theta}, \overline{\mathbb{A}}, \overline{\mathbb{B}}, \overline{s}) = \theta, \mathbb{A}, \mathbb{B}, s$
*(iii)* $\sigma_0{}''\overline{G}_0 \subset G_0$
*(iv)* $t^{G_0} \in \overline{N}$.

*Let $h = h_{\mathbb{A},\mathbb{B}}$. Then there are $b \in \mathbb{B} \setminus \{0\}$, $\overset{\circ}{\sigma} \in \mathbf{V}^{\mathbb{B}}$ s.t. $a = h(b)$ and whenever $G \ni b$ is $\mathbb{B}$-generic, $\sigma = \overset{\circ}{\sigma}^G$, and $G_0 = G \cap \mathbb{A}$, then*

*(a)* $\sigma : \overline{N} \prec N$
*(b)* $\sigma(\overline{\theta}, \overline{\mathbb{A}}, \overline{\mathbb{B}}, \overline{s}) = \theta, \mathbb{A}, \mathbb{B}, s$
*(c)* $C_\delta^N(\text{rng}(\sigma)) = C_\delta^N(\text{rng}(\sigma_0))$ $(\delta = \delta(\mathbb{B}))$
*(d)* $\sigma''\overline{G} \subset G$
*(e)* $\sigma(t^{G_0}) = \sigma_0(t^{G_0})$.

⋆ ⋆ ⋆ ⋆ ⋆

We now prove a theorem about iterations of length $\omega$.

**Theorem 2** *Let $\langle \mathbb{B}_i \mid i < \omega \rangle$ be s.t. $\mathbb{B}_0 = 2$, $\mathbb{B}_i \subseteq \mathbb{B}_{i+1}$ and $\Vdash_{\mathbb{B}_i} (\check{\mathbb{B}}_{i+1}/\overset{\circ}{G}$ is subcomplete) for $i < \omega$. Let $\mathbb{B}_\omega$ be the inverse limit of $\langle \mathbb{B}_i \mid i < \omega \rangle$. Then $\mathbb{B}_\omega$ is subcomplete.*

*Proof*. Let $\theta$ be big enough that $\Vdash_{\mathbb{B}_i} \check{\theta}$ verifies the subcompleteness of $\check{\mathbb{B}}_{i+1}/\overset{\circ}{G}$ for $i < \omega$. Let $N = L_\tau^A$ s.t. $H_\theta \subset N$, $\theta < \tau$, and $N$ is a ZFC$^-$ model. Let $\sigma : \overline{N} \prec N$ s.t. $\sigma(\overline{\theta}, \langle \overline{\mathbb{B}}_i \mid i \leq \omega \rangle, \overline{s}) = \theta, \langle \mathbb{B}_i \mid i \leq \omega \rangle, s$, and $\overline{N}$ is countable and full. Let $\overline{G} = \overline{G}_\omega$ be $\overline{\mathbb{B}}_\omega$-generic over $\overline{N}$ and set $\overline{G}_i = \overline{G} \cap \overline{\mathbb{B}}_i$. Then $\overline{G}_i$ is $\overline{\mathbb{B}}_i$-generic over $\overline{N}$. We claim that there is $b \in \mathbb{B}_\omega \setminus \{0\}$ s.t. whenever $G \ni b$ is $\mathbb{B}_\omega$-generic, there is $\sigma' \in \mathbf{V}[G]$ s.t.

(a) $\sigma' : \overline{N} \prec N$
(b) $\sigma'(\overline{\theta}, \langle \overline{\mathbb{B}}_i \mid i < \omega \rangle, \overline{s}) = \theta, \langle \mathbb{B}_i \mid i < \omega \rangle, s$
(c) $C_\delta^N(\text{rng}(\sigma')) = C_\delta^N(\text{rng}(\sigma))$, where $\delta = \delta(\mathbb{B}_\omega)$.
(d) $\sigma'{}''\overline{G} \subset G$.

We first construct a sequence $b_i$, $\overset{\circ}{\sigma}_i$ $(i < \omega)$ s.t. $b_i \in \mathbb{B}_i$, $h_i(b_{i+1}) = b_i$ (where $h_i = h_{\mathbb{B}_i,\mathbb{B}_{i+1}}$) and whenever $G_i \ni b_i$ is $\mathbb{B}_i$-generic, then, letting $\sigma_i = \overset{\circ}{\sigma}_i^{G_i}$, we have:

(a′) $\sigma_i : \overline{N} \prec N$
(b′) $\sigma_i(\overline{\theta}, \langle \overline{\mathbb{B}}_i \mid i \leq \omega \rangle, \overline{s}) = \theta, \langle \mathbb{B}_i \mid i \leq \omega \rangle, s$
(c′) $C_\delta^N(\text{rng}(\sigma_i)) = C_\delta^N(\text{rng}(\sigma))$
(d′) $\sigma_i''\overline{G} \subset G_i$.

Now let $\langle x_i \mid i < \omega \rangle$ enumerate $\overline{N}$. Set: $u_i =$ the $\overline{N}$-least $u$ s.t. $\sigma(x_i) \in \sigma_i(u)$ and $\overline{\overline{u}} \leq \overline{\delta} =_{\text{Df}} \delta(\overline{\mathbb{B}})$ in $\overline{N}$. (This exists, since $\text{rng}(\sigma) \subset C_\delta^N(\text{rng}(\sigma_i)) = \bigcup\{\sigma_i(u) \mid \overline{\overline{u}} \leq \overline{\delta} \text{ in } \overline{N}\}$ by Section 3, Lemma 5.5.) $\sigma_i$ will satisfy the additional requirements:

(e′) $\sigma_0 = \sigma$

(f′) $\sigma_i(x_h) = \sigma_h(x_h)$ for $h < i$, where $\sigma_h =_{\text{Df}} \overset{\circ}{\sigma}_h(G_i \cap \mathbb{B}_h)$.

(**Note** Then $\sigma_h = \overset{\circ}{\sigma}{}_h^{G_i}$, since we assume: $\mathbf{V}^{\mathbb{B}_h} \subseteq \mathbf{V}^{\mathbb{B}_i}$ (i.e. the identity is the natural injection of $\mathbf{V}^{\mathbb{B}_h}$ into $\mathbf{V}^{\mathbb{B}_i}$). Thus $t^{G_i \cap \mathbb{B}_h} = t^{G_i}$ for $t \in \mathbf{V}^{\mathbb{B}_h}$, $h < i$.)

(g′) $\sigma_i(u_h) = \sigma_h(u_h)$ for $h < i$.

Note that $u_i = \overset{\circ}{u}{}_i^{G_i}$ for a $\overset{\circ}{u}_i \in \mathbf{V}^{\mathbb{B}_i}$. We set: $b_0 = 1$, $\overset{\circ}{\sigma}_0 = \check{\sigma}$. Given $b_i$, $\overset{\circ}{\sigma}_i$, Lemma 1.1 then gives us $b_{i+1}$, $\overset{\circ}{\sigma}_{i+1}$. (Take $\sigma_{i+1}(t^{G_i}) = \sigma_i(t^{G_i})$ where $\Vdash_{\mathbb{B}_i} t = \langle \check{x}_0, \ldots, \check{x}_i, \overset{\circ}{u}_0, \ldots, \overset{\circ}{u}_i \rangle$.) Since $h_i(b_{i+1}) = b_i$, the sequence $\vec{b} = \langle b_i \mid i < \omega \rangle$ is a thread in $\langle \mathbb{B}_i \mid i < \omega \rangle$. Hence $b = \bigcap_i b_i \neq 0$ in $\mathbb{B}_\omega$, since $\mathbb{B}_\omega$ is the inverse limit. Now let $G \ni b$ be $\mathbb{B}_\omega$-generic. Set $G_i = G \cap \mathbb{B}_i$, $\sigma_i = \overset{\circ}{\sigma}{}_i^G = \overset{\circ}{\sigma}{}_i^{G_i}$. Then (a′)–(g′) hold for $i < \omega$. By (f′) we can define $\sigma' : \overline{N} \prec N$ by: $\sigma'(x) = \sigma_i(x)$ for $i$ s.t. $\sigma_i(x) = \sigma_j(x)$ for $i \leq j$. (a), (b) are then trivial. We prove:

(c) $C_\delta^N(\text{rng}(\sigma')) = C_\delta^N(\text{rng}(\sigma))$.

*Proof*. Set $C_i = C_\delta^N(\text{rng}(\sigma_i))$ for $i < \omega$. (Hence $C_0 = C_\delta^N(\text{rng}(\sigma))$.)
($\subset$) It suffices to show $\text{rng}(\sigma') \subset C_0$. But $\sigma'(x_i) = \sigma_i(x_i) \in C_i = C_0$.
($\supset$) We show $\text{rng}(\sigma) \subset C_\delta^N(\text{rng}(\sigma'))$.
$\sigma(x_i) \in \sigma_i(u_i) = \sigma'(u_i) \subset \bigcup\{\sigma'(u) \mid \overline{\overline{u}} \leq \overline{\delta} \text{ in } \overline{N}\} = C_\delta^N(\text{rng}(\sigma'))$. QED(c)

Finally we show:

(d) $\sigma' {}''\overline{G} \subset G$.

*Proof*. We first note that $\sigma' {}''\overline{G}_i \subset G$ for $i < \omega$, since if $a \in \overline{G}_i$, then $\sigma'(a) = \sigma_j(a) \in G_j \subset G$ for some $j \geq i$. Now let $\overline{a} \in G$. Since $\overline{\mathbb{B}}_\omega$ is the inverse limit of $\langle \overline{\mathbb{B}}_i \mid i < \omega \rangle$, we may assume w.l.o.g. that $\overline{a} = \bigcap_{i<\omega} \overline{a}_i$ where $\langle \overline{a} \mid i < \omega \rangle$ is a thread in $\langle \overline{\mathbb{B}}_i \mid i < \omega \rangle$. Let $\sigma'(\langle \overline{a}_i \mid i < \omega \rangle) = \langle a_i \mid i < \omega \rangle$. Then $\langle a_i \mid i < \omega \rangle$ is a thread in $\langle \mathbb{B}_i \mid i < \omega \rangle$ and $\sigma'(\overline{a}) = \sigma'(\bigcap_i \overline{a}_i) = \bigcap_i a_i \in G$ by the completeness of $G$ wrt. $\mathbf{V}$, since $a_i \in G$ for $i < \omega$. QED(Theorem 2)

**Note** Theorem 2 can be generalized to countable support iterations of length $< \omega_2$. At $\omega_2$ it can fail, however, since in a countable support iteration we are required to take a direct limit at $\omega_2$. If some earlier stage changed the cofinality of $\omega_2$ to $\omega$ (e.g. if $\mathbb{B}_1$ were Namba forcing), then the direct limit would not be subcomplete. Hence for longer iterations we must employ *revised countable support* iterations, which we discuss in the next section.

### *Revised Countable Support Iterations*

**Definition** By an *iteration* we mean a sequence $\langle \mathbb{B}_i \mid i < \alpha \rangle$ s.t.

- $\mathbb{B}_0 = 2$
- $\mathbb{B}_i \subseteq \mathbb{B}_j$ for $i \leq j < \alpha$
- If $\lambda < \alpha$ is a limit ordinal, then $\bigcup_{i<\lambda} \mathbb{B}_i$ generates $\mathbb{B}_\lambda$.

In dealing with an iteration we shall employ obvious notational simplifications, writing e.g. $\Vdash_i$ for $\Vdash_{\mathbb{B}_i}$, $[[\varphi]]_i$ for $[[\varphi]]_{\mathbb{B}_i}$ etc. We also write: $h_i(b) = h_{\mathbb{B}_i}(b) =_{\text{Df}} \bigcap\{a \in \mathbb{B}_i \mid b \subset a\}$ in $\mathbb{B}_i$, for $b \in \bigcup_{j<\alpha} \mathbb{B}_j$. Recall that:

- $h_i(b) \neq 0 \leftrightarrow b \neq 0$
- $h_i(\bigcup_{j \in I} b_j) = \bigcup_{i \in I} h_i(b_j)$
- $a \cap h_i(b) = h_i(a \cap b)$ for $a \in \mathbb{B}_i$
- $h_i(b) = [[\check{b}/\mathring{G} \neq 0]]_i$.

Our definition of "iteration" permits great leeway in defining $\mathbb{B}_\lambda$ at limit $\lambda$. In practice people usually employ one of a number of standard limiting procedures, such as *finite support* (FS), *countable support* (CS) or *revised countable support* (RCS) iterations. RCS iterations are particulary suited to subcomplete forcing. The definition of RCS iteration is given in Section 0. For present purposes all we need to know is:

**Fact** Let $\mathbb{B} = \langle \mathbb{B}_i \mid i < \alpha \rangle$ be an RSC iteration. Then:

(a) If $\lambda < \alpha$ and $\langle \xi_i \mid i < \omega \rangle$ is monotone and cofinal in $\lambda$, then:

 (i) If $\langle b_i \mid i < \omega \rangle$ is a thread through $\langle \mathbb{B}_{\xi_i} \mid i < \omega \rangle$, then $\bigcap_{i<\omega} b_i \neq \emptyset$ in $\mathbb{B}_i$.

 (ii) The set of such $\bigcap_i b_i$ is dense in $\mathbb{B}_\lambda$.

(b) If $\lambda < \alpha$ and $\Vdash_i \text{cf}(\check{\lambda}) > \omega$ for $i < \lambda$, then $\bigcup_{i<\lambda} \mathbb{B}_i$ is dense in $\mathbb{B}_\lambda$.

(c) If $i < \lambda$ and $G$ is $\mathbb{B}_i$-generic, then the iteration $\langle \mathbb{B}_{i+j}/G \mid j < \alpha - i \rangle$ satisfies (a), (b) in $\mathbf{V}[G]$.

(**Note** By a "thread" through $\langle \mathbb{B}_i \mid i < \omega \rangle$ we mean a sequence $\langle b_i \mid i < \omega \rangle$ wrt. $b_0 \neq 0$, $b_i \in \mathbb{B}_i$, and $h_i(b_j) = b_i$ for $i \leq j < \omega$.)

**Theorem 3** *Let $\mathbb{B} = \langle \mathbb{B}_i \mid i < \alpha \rangle$ be an RCS-iteration s.t. for all $i+1 < \alpha$:*

*(a)* $\mathbb{B}_i \neq \mathbb{B}_{i+1}$

*(b)* $\Vdash_i (\check{\mathbb{B}}_{i+1}/\mathring{G}$ *is subcomplete)*
*(c)* $\Vdash_{i+1} (\delta(\check{\mathbb{B}}_i)$ *has cardinality* $\leq \omega_1)$.

*Then every* $\mathbb{B}_i$ *is subcomplete.*

*Proof*. Set: $\delta_i = \delta(\mathbb{B}_i)$. Then

(1) $\delta_i \leq \delta_j$ for $i \leq j < \alpha$,

since if $X$ is dense in $\mathbb{B}_j$, then $\{h_i(a) \mid a \in X\}$ is dense in $\mathbb{B}_i$.

(2) $\overline{\overline{\nu}} \leq \delta_\nu$ for $\nu < \alpha$.

*Proof*. Suppose not. Let $\nu$ be the least counterexample. Then $\nu > 0$ is a cardinal. If $\nu < \omega$, then $\delta_\nu < \omega$ and hence $\mathbb{B}_\nu$ is atomic with $\delta_\nu$ the number of atoms. Let $\nu = n+1$. Then $\delta_n < \delta_\nu < n+1$ by (a). Hence $\delta_n < n$. Contradiction! Hence $\nu \geq \omega$ is a cardinal. If $\nu$ is a limit cardinal, then $\delta_\nu \geq \sup_{i<\nu} \delta_\nu \geq \nu$. Contradiction! Thus $\nu$ is a successor cardinal. Let $X \subset \mathbb{B}_\nu$ be dense in $\mathbb{B}_\nu$ with $\overline{\overline{X}} = \delta_\nu < \nu$. Then $X \subset \mathbb{B}_\eta$ for an $\eta < \nu$ by the regularity of $\nu$. Hence $\mathbb{B}_\eta = \mathbb{B}_\nu$, contradicting (a).

QED(2)

By induction on $i$ we prove:

**Claim** Let $G$ be $\mathbb{B}_h$-generic, $h \leq i$. Then $\mathbb{B}_i/G$ is subcomplete in $\mathbf{V}[G]$. (Hence $\mathbb{B}_i \simeq \mathbb{B}_i/\{1\}$ is subcomplete in $\mathbf{V}$, taking $h = 0$.) The case $h = i$ is trivial, since then $\mathbb{B}_i/G \simeq 2$. Hence $i = 0$ is trivial. Now let $i = j+1$. Then $\mathbb{B}_j/G \subset \mathbb{B}_i/G$. Let $\tilde{G}$ be $\mathbb{B}_j/G$-generic over $\mathbf{V}[G]$. Then $G' = G * \tilde{G} = \{b \in \mathbb{B}_j \mid b/G \in \tilde{G}\}$ is $\mathbb{B}_j$-generic over $\mathbf{V}$. But then $(\mathbb{B}_i/G)/\tilde{G} \simeq \mathbb{B}_i/G'$ is subcomplete in $\mathbf{V}[G'] = \mathbf{V}[G][\tilde{G}]$ by (b). Hence we have shown: $\Vdash_{\mathbb{B}_j/G} ((\mathbb{B}_i\check{/}G)/\mathring{G}$ is subcomplete). But $\mathbb{B}_j/G$ is subcomplete in $\mathbf{V}[G]$ by the induction hypothesis, so it follows by the two step theorem that $\mathbb{B}_i/G$ is subcomplete in $\mathbf{V}[G]$. There remains the case that $i = \lambda$ is a limit ordinal. By our induction hypothesis $\mathbb{B}_j/G_h$ is subcomplete in $\mathbf{V}[G_h]$ for $h \leq j < \lambda$. But then $\langle \mathbb{B}_{h+i}/G_h \mid i < \lambda - h\rangle$ satisfies the same induction hypothesis, since if $i \leq k < \lambda - h$ and $\hat{G}$ is $\mathbb{B}_{h+i}/G_h$-generic over $\mathbf{V}[G_h]$, then $G = G_h * \tilde{G}$ is $\mathbb{B}_{h+i}$-generic over $\mathbf{V}$ and $(\mathbb{B}_{h+k}/G_h)/\tilde{G} \simeq \mathbb{B}_{h+k}/G$ is subcomplete in $\mathbf{V}$.

**Case 1** $\mathrm{cf}(\lambda) \leq \delta_i$ for an $i < \lambda$.
Then $\mathrm{cf}(\lambda) \leq \omega_1$ in $\mathbf{V}[G_j]$ for $i < j < \lambda$ whenever $G_j$ is $\mathbb{B}_j$-generic. It suffices to prove the claim for such $j$, since if $h < j$ and $G_h$ is $\mathbb{B}_h$-generic, we can then use the two step theorem to show – exactly as in the successor case – that $\mathbb{B}_\lambda/G_h$ is subcomplete in $\mathbf{V}[G_h]$. Hence it will suffice to prove:

**Claim** Assume $\mathrm{cf}(\lambda) \le \omega_1$ in $\mathbf{V}$. Then $\mathbb{B}_\lambda$ is subcomplete, since the same proof can then be carried out in $\mathbf{V}[G_j]$ to show that $\mathbb{B}_\lambda/G_j$ is subcomplete. Fix $f : \omega_1 \to \lambda$ s.t. $\sup f''\omega_1 = \lambda$. Let $\theta > \lambda$ be a cardinal s.t. $\overline{\overline{\mathbb{B}}} < \theta$ and $\theta$ is big enough that:

$$\Vdash_i (\check{\theta} \text{ witnesses the subcompleteness of } \check{\mathbb{B}}_j/\mathring{G})$$

for $i \le j < \lambda$. Let $N = L_\tau^A$ be a $\mathrm{ZFC}^-$ model s.t. $H_\theta \subset N$, $\theta < \tau$. Let $\sigma : \overline{N} \prec N$ s.t. $\overline{N}$ is countable and full. Suppose also that: $\sigma(\overline{\theta}, \overline{\mathbb{B}}, \overline{\lambda}, \overline{f}, \overline{s}) = \theta, \mathbb{B}, \lambda, f, s$.

**Claim** There is $b \in \mathbb{B}_\lambda \setminus \{0\}$ s.t. whenever $G \ni b$ is $\mathbb{B}_\lambda$-generic, there is $\sigma' \in \mathbf{V}[G]$ s.t.

(a) $\sigma' : \overline{N} \prec N$
(b) $\sigma'(\overline{\theta}, \overline{\mathbb{B}}, \overline{\lambda}, \overline{f}, \overline{s}) = \theta, \mathbb{B}, \lambda, f, s$
(c) $C_\delta^N(\mathrm{rng}(\sigma')) = C_\delta^N(\mathrm{rng}(\sigma))$, where $\sigma = \sup\{\delta_i \mid i < \lambda\}$.
(d) $\sigma''\overline{G} \subset G$.

Set: $\tilde{\lambda} = \sup \sigma''\overline{\lambda}$. It is easily verified that there is a sequence $\langle \nu_i \mid i < \omega \rangle$ in $\omega_1^{\overline{N}}$ s.t., setting $\overline{\xi}_i = \overline{f}(\nu_i)$, we have: $\overline{\xi}_0 = 0$, and $\langle \overline{\xi} \mid i < \omega \rangle$ is monotone and cofinal in $\overline{\lambda}$. (We can assume w.l.o.g. that $\overline{f}(0) = 0$.) Set $\xi_i = f(\nu_i)$. Then $\xi_i = \sigma(\overline{\xi}_i)$ and $\langle \xi_i \mid i < \omega \rangle$ is monotone and cofinal in $\tilde{\lambda}$. Moreover:

(3) $\sigma'(\overline{\xi}_i) = \xi_i$ whenever $\sigma' : \overline{N} \prec N$ s.t. $\sigma'(\overline{f}) = f$.

We now closely imitate the proof of Theorem 2, constructing a sequence $b_i$, $\mathring{\sigma}_i$ $(i < \omega)$ s.t. $b_i \in \mathbb{B}_{\xi_i}$, $h_{\xi_i}(b_{i+1}) = b_i$, and whenever $G_i \ni b_i$ is $\mathbb{B}_{\xi_i}$-generic, then, letting $\sigma_i = \mathring{\sigma}_i^{G_i}$, we have:

(a′) $\sigma_i : \overline{N} \prec N$
(b′) $\sigma_i(\overline{\theta}, \overline{\mathbb{B}}, \overline{f}, \overline{s}) = \theta, \mathbb{B}, f, s$
(c′) $C_{\delta_i}^N(\mathrm{rng}(\sigma_i)) = C_\delta^N(\mathrm{rng}(\sigma))$
(d′) $\sigma_i''\overline{G}_i \subset G_i$
(e′) $\sigma_0 = \sigma$
(f′) $\sigma_i(x_h) = \sigma_h(x_h)$ $(h \le i)$ where $\sigma_h = \mathring{\sigma}_h^{G_i}$
($\langle x_\ell \mid \ell < \omega \rangle$ being an arbitrarily chosen enumeration of $\overline{N}$.)
(g′) $\sigma_i(u_h) = \sigma_h(u_h)$ $(h \le i)$, where $u_i =$ the $\overline{N}$-least $u$ s.t. $\sigma(x_i) \in \sigma_i(u)$ and

$$\overline{\overline{u}} < \overline{\delta} =_{\mathrm{Df}} \sigma^{-1}(\delta) \text{ in } \overline{N}.$$

The construction is exactly as before using that $\sigma_i(\overline{\mathbb{B}}_{\overline{\xi}_j}) = \mathbb{B}_{\xi_j}$ for all $j$ and that $\Vdash_{\mathbb{B}_{\xi_j}} (\check{\mathbb{B}}_{\xi_{j+1}}/\mathring{G}$ is subcomplete). As before set: $\sigma'(x) = \sigma_i(x)$, where $i$

is big enough that $\sigma_i(x) = \sigma_j(x)$ for $i \leq j$. The verification of (a)–(c) is exactly as before. To verify (d), we first note that, as before,

(4) $\sigma'{}''G_i \subset G$ for $i < \omega$.

We then consider two cases: If $\mathrm{cf}(\lambda) = \omega$ in $N$, then $\mathrm{cf}(\overline{\lambda}) = \omega$ in $\overline{N}$ and $\tilde{\lambda} = \lambda$. $\overline{\mathbb{B}}_{\overline{\lambda}}$ is then the inverse limit of $\langle \overline{\mathbb{B}}_{\overline{\xi}_i} \mid i < \omega \rangle$ and $\mathbb{B}_\lambda$ is the inverse limit of $\langle \mathbb{B}_{\xi_i} \mid i < \omega \rangle$. We then proceed exactly as before. If $\mathrm{cf}(\lambda) = \omega_1$, $\overline{\mathbb{B}}_\lambda$ is the direct limit – i.e. $\bigcup\limits_{i<\omega} \overline{\mathbb{B}}_i$ is dense in $\overline{\mathbb{B}}_\lambda$. The conclusion then follows by (4). QED(Case 1)

**Case 2** Case 1 fails.
Then $\lambda$ is regular and $\delta_i < \lambda$ for all $i < \lambda$. Hence $\lambda = \sup\limits_{i<\lambda} \delta_i$. Let $N$, $\overline{N}$, $\theta$, $\sigma$, $\overline{G}$ be as before with $\sigma(\overline{\theta}, \overline{\mathbb{B}}, \overline{s}, \overline{\lambda}) = \theta, \mathbb{B}, s, \lambda$. (However, there is now nothing corresponding to the function $f$.) As before set: $\tilde{\lambda} = \sup \sigma''\lambda$. It suffices to show:

**Claim** There is $c \in \mathbb{B}_\lambda$ s.t. whenever $G \ni c$ is $\mathbb{B}_\lambda$-generic, there is $\sigma' \in \mathbf{V}[G]$ with:

(a) $\sigma' : \overline{N} \prec N$
(b) $\sigma'(\overline{\theta}, \overline{\mathbb{B}}, \overline{\lambda}, \overline{s}) = \theta, \mathbb{B}, \lambda, s$
(c) $C^N_\lambda(\mathrm{rng}(\sigma')) = C^N_\lambda(\mathrm{rng}(\sigma))$
(d) $\sigma'{}''\overline{G} \subset G$.

Choose a sequence $\langle \overline{\xi}_i \mid i < \omega \rangle$ which is monotone and cofinal in $\overline{\lambda}$ with $\overline{\xi}_0 = 0$. Set: $\xi_i = \sigma(\overline{\xi}_i)$. As before, our strategy is to construct $c_i$, $\overset{\circ}{\sigma}_i$ $(i < \omega)$ s.t. $c_0 = 1$, $\overset{\circ}{\sigma}_0 = \check{\sigma}$, $\langle c_i \mid i < \omega \rangle$ is a thread in $\langle \mathbb{B}_{\xi_i} \mid i < \omega \rangle$, and $c_i$ forces $\overset{\circ}{\sigma}_i : \check{\overline{N}} \prec \check{N}$. The intention is, again, that if $c = \bigcap\limits_i c_i \in G$ and $G$ is $\mathbb{B}_\lambda$-generic, then we can define the embedding $\sigma' \in \mathbf{V}[G]$ from the sequence $\sigma_i = \overset{\circ}{\sigma}{}^G_i$ $(i < \omega)$. However, since we no longer have the function $f$ available in defining $\langle \xi_i \mid i < \omega \rangle$, we shall *not* be able to enforce: $\sigma_i(\overline{\xi}_j) = \xi_j$ for $j < \omega$. Nonetheless we can enforce: $\sup \sigma_i{}''\overline{\lambda} = \tilde{\lambda}$, and shall have to make do with that. We inductively construct $c_i \in \mathbb{B}_{\xi_i}$, $\overset{\circ}{\sigma}_i \in \mathbf{V}^{\mathbb{B}_{\xi_i}}$ with the properties:

(I) (a) $c_0 = 1$
  (b) $h_{\xi_h}(c_i) = c_h$ for $i = h + 1$.

(II) Let $G \ni c_i$ be $\mathbb{B}_{\xi_i}$-generic. Set: $G_\eta = G \cap \mathbb{B}_\eta$ $(\eta \leq \xi_i)$, $\overline{G}_\eta = \overline{G} \cap \overline{\mathbb{B}}_\eta$ $(\eta \leq \overline{\xi}_i)$, $\sigma_h = \overset{\circ}{\sigma}{}^G_h = \overset{\circ}{\sigma}{}^{G_{\xi_h}}_h$ for $h \leq i$. Then:
  (a) $\sigma_i : \overline{N} \prec N$
  (b) $\sigma_i(\overline{\theta}, \overline{\mathbb{B}}, \overline{\lambda}, \overline{s}) = \theta, \mathbb{B}, \lambda, s$

(c) $C_\lambda^N(\mathrm{rng}(\sigma_i)) = C_\lambda^N(\mathrm{rng}(\sigma))$
(d) Let $\sigma_i(\overline{\xi}_m) \leq \xi_i < \sigma_i(\overline{\xi}_{m+1})$. Then $\sigma_i{}''\overline{G}_{\overline{\xi}_m} \subset G$.
(e) $\sigma_i(x_h) = \sigma_h(x_h)$ for $h < i$, $\langle x_\ell \mid \ell < \omega\rangle$ being a fixed enumeration of $\overline{N}$.
(f) $\sigma_i(u_h) = \sigma_h(u_h)$ for $h < i$, where $u_h =$ the $\overline{N}$-least $u$ s.t. $\sigma(x_h) \in \sigma_h(u)$.
(g) $\sigma_i = \sigma_h$ if $\sigma_h(\overline{\xi}_m) \leq \xi_h < \xi_i < \sigma_h(\overline{\xi}_{m+1})$

(I), (II) are easily seen to imply the claim. Set $c = \bigcap\limits_i c_i$. Then $c \neq 0$, since $c$ is a thread in $\langle \mathbb{B}_{\xi_i} \mid i < \omega\rangle$. Let $G \ni c$ be $\mathbb{B}_\lambda$-generic. Define $\sigma_i = \overset{\circ}{\sigma}{}_i^G$ $(i < \omega)$ and define $\sigma'(x) = \sigma_j(x)$ where $\sigma_j(x) = \sigma_k(x)$ for all $k \geq j$. (a)–(c) follow exactly as before. We prove (d). Since $\overline{\mathbb{B}}_{\overline{\lambda}}$ is the direct limit of $\langle \overline{\mathbb{B}}_{\overline{\xi}_i} \mid i < \omega\rangle$, it suffices to show:

(d') $\sigma'{}''\overline{G} \subset G$ for $i < \omega$, where $\overline{G}_\eta = \overline{G} \cap \overline{B}_\eta$.

*Proof*. Let $a \in \overline{G}_{\overline{\xi}_i}$. We first note that for $j \geq i$ sufficiently large we have: $\sigma_j(\overline{\xi}_m) \leq \xi_j < \sigma_j(\overline{\xi}_{m+1})$ for an $m \geq i$, since otherwise $\xi_j < \sigma_j(\overline{\xi}_i)$ for arbitrarily large $j$. But $\sigma_j(\overline{\xi}_i) = \sigma'(\overline{\xi}_i)$ for sufficient large $j$. Hence $\sigma'(\overline{\xi}_i) \geq \sup\limits_j \xi_j = \lambda$. Contradiction! If we also pick $j$ large enough that $\sigma_j(a) = \sigma'(a)$, then $\sigma'(a) = \sigma_j(a) \in G$, since $a \in G_{\xi_m}$. QED(d)

It remains only to construct $c_i$, $\overset{\circ}{\sigma}_i$ and verify (I), (II). This will be somewhat trickier than the construction in Theorem 2. We shall also have to add further induction hypotheses to (I), (II). Before defining $c_i$ we define a $b_i \in \mathbb{B}_{\xi_i}$ s.t.

(III) (a) $b_0 = 1$, $\overset{\circ}{\sigma}_0 = \check{\sigma}$
(b) $h_{\xi_j}(b_i) = c_j$ if $i = j+1$
(c) (II)(a)–(g) hold whenever $b_i \in G$.

$\overset{\circ}{\sigma}_i$ will be defined simultaneously with $b_i$, before defining $c_i$. Our next induction hypothesis states an important property of $b_i$:

**Definition** Let $\nu \leq \xi_i < \mu < \tilde{\lambda}$ s.t. $\xi_h < \nu$ for $h < i$,

$$a^{j\nu\mu} =_{\mathrm{Df}} b_i \cap [[\overset{\circ}{\sigma}_i(\check{\overline{\xi}}_j) = \check{\nu} \wedge \overset{\circ}{\sigma}_i(\check{\overline{\xi}}_{j+1}) = \check{\mu}]]_{\xi_i}.$$

It follows easily that:

(5) $a^{j\nu\mu} \cap a^{j'\nu'\mu'} = 0$ if $\langle j, \nu, \mu\rangle \neq \langle j', \nu', \mu'\rangle$.

*Proof*. Suppose $a^{j\nu\mu} \cap a^{j'\nu'\mu'} \in G$ where $G$ is $\mathbb{B}_{\xi_i}$-generic. Then $j = j'$, since if e.g. $j < j'$, then $\mu \leq \sigma_i(\overline{\xi}_{j+1}) \leq \sigma_i(\overline{\xi}_{j'}) = \nu' \leq \xi_i$. Contradiction! But then $\nu = \sigma_i(\xi_j) = \nu'$, $\mu = \sigma_i(\overline{\xi}_{j+1}) = \mu'$. Contradiction! QED(1)

Our final induction hypothesis reads:

(IV) $a^{j\nu\mu} \cap [[\overset{\circ}{\sigma}_i(\check{x}) = \check{y}]]_{\xi_i} \in \mathbb{B}_\nu$ if $\sup_{h<i} \xi_h < \nu \leq \xi_i < \mu$.

Hence $a^{j\nu\mu} = a^{j\nu\mu} \cap [[\overset{\circ}{\sigma}_i(\check{0}) = \check{0}]] \in \mathbb{B}_\nu$.

**Definition** $A = A_i =$ the set of all $a^{j\nu\mu} \neq 0$ s.t. $\sup_{h<i} \xi_h < \nu \leq \xi_i < \mu$.

By (IV) we see that for each $a = a^{j\nu\mu} \in A$ there is $\overset{\circ}{\sigma}_a \in \mathbf{V}^{\mathbb{B}_\nu}$ s.t.

(6) $\overset{\circ}{\sigma}{}_a^{G_\nu} = \sigma_i^G$ for $\mathbb{B}_{\xi_i}$-generic $G \ni a$.

But:

(7)
If $G \ni a$ is $\mathbb{B}_\nu$-generic, then $G$ extends to a $\mathbb{B}_{\xi_i}$-generic $G'$ s.t. $G = G' \cap \mathbb{B}_\nu$. Hence: $\overset{\circ}{\sigma}{}_a^G = \overset{\circ}{\sigma}{}_a^{G'} = \overset{\circ}{\sigma}{}_i^{G'}$.

Thus we have:

(8)
Let $G \ni a$ be $\mathbb{B}_\nu$-generic, where $a = a^{j\nu\mu} \in A_i$. Then (II) holds with $\sigma_a = \overset{\circ}{\sigma}{}_a^G$ in place of $\sigma_i$, $\sigma_h = \overset{\circ}{\sigma}{}_h^G = \overset{\circ}{\sigma}{}_h^{G_{\xi_h}}$ for $h < i$, where $G_\eta =_{\text{Df}} G \cap \mathbb{B}_\eta$ $(\eta \leq \nu)$.

**Note** Since $a \in G$, (d) then reduces to: $\sigma_a{}''\overline{G}_{\overline{\xi}_j} \subset G$.

**Note** We then have: $\sigma_a(x_h) = \sigma_h(x_h)$, $\sigma_a(u_h) = \sigma_h(u_h)$ for $h < i$.

Whenever $\nu < \mu < \lambda$ and $G$ is $\mathbb{B}_\nu$-generic, we know that $\mathbb{B}_\mu/G$ is subcomplete in $\mathbf{V}[G]$. Then, using (8), Lemma 1.1, and repeating the construction of $b_{i+1}$, $\overset{\circ}{\sigma}_{i+1}$ from $b_i$, $\overset{\circ}{\sigma}_i$ in the proof of Theorem 2, we get:

(9)
Let $a \in A_i$, $a = a^{j\nu\mu}$. There are $\tilde{a} \in \mathbb{B}_\mu$, $\overset{\circ}{\sigma}{}'_a \in \mathbf{V}^{\mathbb{B}_\mu}$ s.t. $h_\nu(\tilde{a}) = a$ and whenever $G \ni \tilde{a}$ is $\mathbb{B}_\mu$-generic, $\sigma_a = \overset{\circ}{\sigma}{}^G$, and $\sigma'_a = \overset{\circ}{\sigma}{}'^G_a$, then we have:

(a) $\sigma'_a : \overline{N} \prec N$
(b) $\tau'_a(\overline{\theta}, \overline{\mathbb{B}}, \overline{\lambda}, \overline{s}) = \theta, \mathbb{B}, \lambda, s$
(c) $C_\lambda^N(\text{rng}(\sigma'_a)) = C_\lambda^N(\text{rng}(\sigma_a))$
(d) $\sigma'_a{}''\overline{G}_{\overline{\xi}_{j+1}} \subset G$

(e) Let $r$ be least s.t. $\mu \leq \xi_r$. Then $\sigma'_a(x_h) = \sigma_a(x_h)$ for $h < r$.
(f) Let $r$ be as above. Then $\sigma'_a(u^a_h) = \sigma_a(u^a_h)$ for $h < r$, where $u^a_h =$ the $\overline{N}$-least $u \in \overline{N}$ s.t. $\overline{\overline{u}} \leq \overline{\lambda}$ in $\overline{N}$ and $\sigma(x_h) \in \sigma_a(u_h)$.
(g) $\sigma'_a(\overline{\xi}_\ell) = \sigma_a(\overline{\xi}_\ell)$ for $\ell \leq j+1$.

For each $a \in A_i$ fix such a pair $\tilde{a}$, $\overset{\circ}{\sigma}{}'_a$, which can be regarded as an instruction to be used later in forming $b_r$, where $r$ is least s.t. $\mu \leq \xi_r$. If $G$ is $\mathbb{B}_{\xi_r}$-generic and $a \cap b_r \in G$, we shall want: $\tilde{a} \in G$ and $\sigma_r = \overset{\circ}{\sigma}{}'^G_a$ (where $\sigma_r = \overset{\circ}{\sigma}{}^G_r$). In particular, we want: $a \cap b_r = \tilde{a}$. But we shall also require: $h_{\xi_i}(b_r) = c_i$. Hence we need: $a \cap c_i = h_\xi(a \cap b_r) = h_\xi(\tilde{a})$. This is why $b_i$ must be "shrunk" to $c_i$. Accordingly we define $c_i$ as follows:

**Definition** Let $b_i$ be given. Set $\overline{b} = b_i \setminus \bigcup A_i$. Then:

$$c_i =_{\mathrm{Df}} \overline{b} \cup \bigcup_{a \in A_i} h_{\xi_i}(\tilde{a}).$$

We are working by induction on $i$. We assume (I)–(IV) to hold below $i$ and (III), (IV) to hold at $i$. We must now verify (I), (II) at $i$. (II) is immediate by (III)(c), since $c_i \subset b_i$. (I)(b) holds, since $h_{\xi_j} h_{\xi_i}(\tilde{a}) = h_{\xi_j}(\tilde{a}) = h_{\xi_j} h_\nu(\tilde{a}) = h_{\xi_j}(a)$ for $i = j+1$ and $a = a^{\ell\mu\nu} \in A_i$. Hence:

$$h_{\xi_j}(c_i) = h_{\xi_j}(\overline{b}) \cup \bigcup_{a \in A} h_{\xi_j}(\tilde{a}) = h_{\xi_j}(\overline{b} \cup \bigcup_{a \in A} h_{\xi_i}(\tilde{a})) = h_{\xi_j}(b_i) = c_j.$$

For (I)(a) note that $A_0 = \{a\}$ where $a = a^{0,0,\xi_1} = 1$, since $\sigma_0 = \sigma$ by (III)(a). Hence $c_0 = h_0(\tilde{a}) = 1$. This completes the verification of (I)–(IV) at $i$, given (III), (IV) at $i$ and (I)–(IV) below $i$. Now let (I)–(IV) hold below $i$. We must define $b_i$, $\overset{\circ}{\sigma}_i$ and verify (III), (IV) at $i$. For $i = 0$ set: $b_0 = 1$, $\overset{\circ}{\sigma}_0 = \check{\sigma}$. The verifications are trivial.
Now let $i = j+1$. Note that $A_\ell$, $\langle \tilde{a} \mid a \in A_\ell \rangle$, $\langle \sigma'_a \mid a \in A_\ell \rangle$ have been defined for $\ell \leq j$. Set:

**Definition** $\hat{A}_j =$ the set of $a = a^{h\nu\mu} \in \bigcup_{\ell \leq j} A_\ell$ s.t. $\xi_j < \mu$.

(10) Let $a, a' \in \hat{A}_j$, $a = a^{h\nu\mu}$, $a' = a^{h'\nu'\mu'}$. Then $a \cap a' = 0$ if $\langle h, \nu, \mu \rangle \neq \langle h', \nu', \mu' \rangle$.

*Proof*. Suppose not. Let $a \in A_\ell$, $a' \in A_{\ell'}$. Then $\ell \neq \ell'$ by (4). Let e.g. $\ell < \ell'$ Let $a \cap a' \in G$, where $G$ is $\mathbb{B}_j$-generic. Set $\sigma_\ell = \overset{\circ}{\sigma}{}^G_\ell$ for $\ell \leq j$. Then:

$$\sigma_\ell(\overline{\xi}_h) = \nu \leq \xi_\ell < \nu' \leq \xi_{\ell'} \leq \xi_j < \mu = \sigma_\ell(\overline{\xi}_{h+1}).$$

Hence $\sigma_{\ell'} = \sigma_\ell$ by (II)(g). Hence $h < h'$, since $\sigma_\ell(\overline{\xi}_{h'}) = \nu' > \sigma_\ell(\overline{\xi}_h)$. Hence $\sigma_\ell(\overline{\xi}_{h+1}) \leq \nu' < \mu$. Contradiction! QED(10)

We now define:

**Definition** $b_i = \bigcup\{h_{\xi_i}(\tilde{a}) \mid a \in \hat{A}_j\}$ for $i = j+1$.

To define $\sigma_i$ we set: $\tilde{A}$ = the set of $a^{i\nu\mu} \in \hat{A}_j$ s.t. $\mu \leq \xi_i$. $\overset{\circ}{\sigma}_i \in \mathbf{V}^{\mathbb{B}_i}$ is then a name s.t. $[[\overset{\circ}{\sigma}_i = \overset{\circ}{\sigma}'_a]] = \tilde{a}$ if $a \in \tilde{A}$, $[[\overset{\circ}{\sigma}_i = \overset{\circ}{\sigma}_j]] \cap b_i = b_i \setminus \bigcup \tilde{A}$.

Then:

(11) (III)(c) holds at $i$.

*Proof*. Let $G \ni b_i$ be $\mathbb{B}_{\xi_i}$-generic.
**Case 1** $\tilde{a} \in G$ for an $a \in \tilde{A}$.
Let $a = a^{h\nu\mu} \in A_\ell$, $\mu \leq \xi_i$ (hence $\xi_j < \mu \leq \xi_i$). Thus $\sigma_i = \sigma'_a$. (II)(a)–(d) hold by (9)(a)–(d). Note that the $r$ in (9)(e), (f) is $r = i$. But, if $a \in A_\ell$, $\ell \leq j$, then $\sigma_\ell = \sigma_a$. Hence $\sigma_\ell(\overline{\xi}_h) = \nu \leq \xi_\ell \leq \xi_{\ell'} < \sigma_\ell(\overline{\xi}_{h+1}) = \mu$ for $\ell \leq \ell' \leq j$. Hence: $\sigma_a = \sigma_{\ell'}$ for $\ell \leq \ell' \leq j$. But then (II)(e), (f) hold by (9)(e), (f). Finally (II)(g) holds vacuously, since $\xi_j < \mu = \sigma_i(\overline{\xi}_{h+1}) \leq \xi_i$, hence $\xi_j < \sigma_i(\overline{\xi}_m)$ where $\sigma_i(\overline{\xi}_m) \leq \xi_i < \sigma_i(\overline{\xi}_{m+1})$.
**Case 2** Case 1 fails.
Then $\sigma_i = \sigma_j$. (II)(a)–(g) then follow trivially. QED(11)

(III)(a) holds vacuously at $i = j+1$. We prove:

(12) (III)(b) holds at $i$.

*Proof*. Clearly $h_{\xi_j}(b_i) = \bigcup_{a \in \hat{A}_j} h_{\xi_j}(\tilde{a})$. Hence we need:

**Claim** $c_j = \bigcup_{a \in \hat{A}_j} h_{\xi_j}(\tilde{a})$.

For $j = 0$ this is trivial, so let $j = \ell + 1$. Recall that $c_j = \overline{b} \cup \bigcup_{a \in A_j} h_{\xi_j}(\tilde{a})$, where $\overline{b} = b_j \setminus \bigcup_{a \in A_j} a$, so it suffices to show:

**Claim** $\overline{b} = \bigcup_{a \in A'} h_{\xi_j}(\tilde{a})$ where $A' = \hat{A}_j \setminus A_j$.

($\supset$) Let $a' \in A'$. Then $a' \in \hat{A}_\ell$. Hence $h_{\xi_j}(\tilde{a}') \subset \bigcup_{a \in \hat{A}_\ell} h_j(\tilde{a}) = b_j$. But for all $a \in A_j$ we have $a \cap a' = 0$ by (11). Hence $h_{\xi_j}(\tilde{a}) \cap h_{\xi_j}(\tilde{a}') = a \cap h_{\xi_j}(\tilde{a}') = h_{\xi_j}(a \cap \tilde{a}') = 0$, since $a \cap \tilde{a}' \subset a \cap a' = 0$. Hence $h_{\xi_j}(\tilde{a}') \subset \overline{b}$.
($\subset$) Suppose not. Then there is $a \in \hat{A}_j \setminus A'$ s.t. $\overline{b} \cap h_{\xi_j}(\tilde{a}) \neq 0$. But then $a \in A_j$ and $h_{\xi_j}(\tilde{a}) = a$. Hence $a \cap \overline{b} = 0$ by the definition of $\overline{b}$. QED(12)

It remains only to show:

(13) (IV) holds at $i$.

*Proof*. Let $a = a^{h,\nu,\mu} \in A_i$. Then $\xi_j < \nu \leq \xi_i$. $a \cap [[\mathring{\sigma}_i(\check{x}) = \check{y}]] = b_i \cap d$, where $d = [[\mathring{\sigma}_i(\check{\overline{\xi}}_h) = \check{\nu} \wedge \mathring{\sigma}_i(\overline{\xi}_{h+1}) = \check{\mu} \wedge \mathring{\sigma}_i(\check{x}) = \check{y}]]_{\mathbb{B}_{\xi_i}} = [[\varphi(\mathring{\sigma}_i)]]$, where the formula $\varphi(v)$ is $\Sigma_0$ in the parameters $\check{\overline{\xi}}_h$, $\check{\overline{\xi}}_{h+1}$, $\check{\nu}$, $\check{\mu}$, $\check{x}$, $\check{y}$, all of which lie in $\mathbf{V}^2$. Recall that we are assuming $\mathbf{V}^{\mathbb{B}_\eta} \subseteq \mathbf{V}^{\mathbb{B}_\tau}$ for $\eta \leq \tau$ (i.e. $\mathbb{B}_\eta$ is completely contained in $\mathbb{B}_\tau$ and the identity is the natural embedding of $\mathbf{V}^{\mathbb{B}_\eta}$ in $\mathbf{V}^{\mathbb{B}_\tau}$). As mentioned in Section 0, this has the consequence that if $\psi$ is a $\Sigma_0$ formula and $t_1, \ldots, t_m \in \mathbf{V}^{\mathbb{B}_\eta}$, then:

$$a \Vdash_{\mathbb{B}_\tau} \psi(\vec{t}) \longleftrightarrow a \Vdash_{\mathbb{B}_\eta} \psi(\vec{t}) \quad \text{for } a \in \mathbb{B}_\eta,$$

or in other words: $[[\psi(\vec{t})]]_{\mathbb{B}_\tau} = [[\psi(\vec{t})]]_{\mathbb{B}_\eta} \in \mathbb{B}_\eta$.
We shall make strong use of this. We know: $b_i = \bigcup_{e \in \hat{A}_j} h_{\xi_i}(\tilde{e})$. Hence it suffices to assign to each $e \in \hat{A}_j$ an $e^* \in \mathbb{B}_\nu$ s.t.

$$h_{\xi_i}(\tilde{e}) \cap d = e^*,$$

since then we have:

$$b_i \cap d = \bigcup_{e \in \hat{A}_j} e^* \in \mathbb{B}_\nu.$$

For $h_{\xi_i}(\tilde{e}) \cap d = 0$ we, of course, set $e^* = 0$. Now let $h_{\xi_i}(\tilde{e}) \cap d \neq 0$. Let $e = a^{\overline{h},\overline{\nu},\overline{\mu}} \in \hat{A}_j$. Let $G \ni h_{\xi_i}(\tilde{e}) \cap d$ be $\mathbb{B}_{\xi_i}$-generic. Set: $\sigma_i = \mathring{\sigma}^G$, $\mathring{\sigma}_j = \mathring{\sigma}_j^G$.

**Case 1** $\overline{\mu} \leq \xi_i$.
Then $\tilde{e} = h_{\xi_i}(\tilde{e}) \in \mathbb{B}_{\overline{\mu}} \wedge G$. Hence $\sigma_i = \sigma'_e =_{\mathrm{Df}} \mathring{\sigma}'^G_e$. Hence $[[\varphi(\mathring{\sigma}'_e)]] \in G$. Conversely, if $\tilde{e} \cap [[\varphi(\mathring{\sigma}'_e)]] \in G$, then $\sigma_i = \sigma'_e$ and hence $\tilde{e} \cap d \in G$. Since this holds for all $G$, we conclude:

$$\tilde{e} \cap d = \tilde{e} \cap [[\varphi(\mathring{\sigma}'_e)]] \in \mathbb{B}_{\overline{\mu}}.$$

However, $\overline{\mu} \leq \nu$, since otherwise we would have $\sigma_i(\overline{\xi}_\ell) = \sigma_j(\overline{\xi}_\ell)$ for $\ell \leq \overline{h}+1$ and $\sigma_i(\overline{\xi}_h) = \nu < \overline{\mu} = \sigma_i(\overline{\xi}_{\overline{h}+1})$. Hence $h \leq \overline{h}$ and $\sigma_i(\overline{\xi}_h) \leq \sigma_j(\overline{\xi}_{\overline{h}}) = \overline{\nu} \leq \xi_j < \nu$. Contradiction! QED(Case 1)

**Case 2** $\overline{\mu} > \xi_i$.
We show that this cannot occur. Clearly, if $G \ni h_{\xi_i}(\tilde{e}) \cap d$ is $\mathbb{B}_{\xi_i}$-generic, then $\sigma_i = \sigma_j = \sigma'_e$ by the definition of $\sigma'_e$. But then $\tilde{e} \cap d = 0$, since if $G \ni \tilde{e} \cap d$ were $\mathbb{B}_{\overline{\mu}}$-generic, then $\sigma_i(\overline{\xi}_{\overline{h}}) = \overline{\nu} \leq \xi_j < \nu \leq \xi_i < \overline{\mu} = \sigma_i(\overline{\xi}_{\overline{h}+1})$. Hence $\nu = \sigma_i(\overline{\xi}_h)$ is impossible. Contradiction!
Since $d \in \mathbb{B}_{\xi_i}$, we conclude: $h_{\xi_i}(\tilde{e}) \cap d = h_{\xi_i}(\tilde{e} \cap d) = 0$. Contradiction.

QED(13)

This completes the proof of Theorem 3.

The above theorem can be adapted to iterations which allow more freedom in the formation of limit algebras.

**Definition** An iteration $\mathbb{B} = \langle \mathbb{B}_i \mid i < \alpha \rangle$ is *nicely subcomplete* iff the following hold:

(a) For all $i + 1 < \alpha$:

  (i) $\Vdash_i \check{\mathbb{B}}_{i+1}/\mathring{G}$ is subcomplete,

  (ii) $\Vdash_{i+1} \text{card}(\delta(\check{\mathbb{B}}_i)) \leq \omega_1$.

(b) If $\lambda < \alpha$ and $\langle \xi_n \mid n < \omega \rangle$ is monotone and cofinal in $\lambda$, then

  (i) $\bigcap_n b_n \neq 0$ in $\mathbb{B}_\lambda$ whenever $b = \langle b_n \mid n < \omega \rangle$ is a thread in $\langle \mathbb{B}_{\xi_n} \mid n < \omega \rangle$,

  (ii) $\mathbb{B}_\lambda$ is subcomplete if $\mathbb{B}_i$ is subcomplete for $i < \lambda$.

(c) If $\lambda < \alpha$ and $\Vdash_i \text{cf}(\check{\lambda}) > \omega$ for all $i < \lambda$, then $\bigcup_{i<\lambda} \mathbb{B}_i$ is dense in $\mathbb{B}_\lambda$.

(d) If $i < \alpha$ and $G$ is $\mathbb{B}_i$-generic, then (a)–(c) hold for $\langle \mathbb{B}_{i+j}/G \mid j < \alpha - i \rangle$ in $\mathbf{V}[G]$.

(This allows greater freedom in forming limit algebras at points which acquire cofinality $\omega$, but requires us to take direct limits at other points.)

**Theorem 4** *Let $\mathbb{B} = \langle \mathbb{B}_i \mid i < \alpha \rangle$ be nicely subcomplete. Then every $\mathbb{B}_i$ is subcomplete.*

*Proof.* (sketch) By induction on $i$ we again prove:

**Claim** Let $h \leq i$. Let $G$ be $\mathbb{B}_h$-generic. Then $\mathbb{B}_i/G$ is subcomplete in $\mathbf{V}[G]$.

The cases $h = i$, $i = j + 1$ are again trivial, so assume that $i = \lambda$ is a limit ordinal. We again have the two cases:

**Case 1** $\text{cf}(\lambda) \leq \overline{\overline{\delta(\mathbb{B}_h)}}$ for an $h < \lambda$.

**Case 2** Case 1 fails.

In Case 1 it again suffices to prove the claim for sufficiently large $h < \lambda$, so we assume $\text{cf}(\lambda) \leq \omega_1$ in $\mathbf{V}[G]$ whenever $G$ is $\mathbb{B}_h$-generic. But then we can assume $\text{cf}(\lambda) \leq \omega_1$ in $\mathbf{V}$, since the same proof can be carried out in $\mathbf{V}[G]$ for $\langle \mathbb{B}_{h+j}/G \mid j < \alpha - h \rangle$. This splits into two subcases:

**Case 1.1** $\text{cf}(\lambda) = \omega$.

Then $\mathbb{B}_\lambda$ is subcomplete by (b)(ii).

**Case 1.2** $\text{cf}(\lambda) = \omega_1$.

We then literally repeat the argument in the proof of Theorem 3, using that $\mathbb{B}_\lambda$ is the direct limit of $\langle \mathbb{B}_i \mid i < \lambda \rangle$.

(**Note** If we instead assumed $\mathrm{cf}(\lambda) = \omega$, the proof in Theorem 3 would no longer work, since the set of $\bigcap_n b$ s.t. $b = \langle b_n \mid n < \omega \rangle$ is a thread in $\langle \mathbb{B}_{\xi_n} \mid n < \omega \rangle$ may not be dense in $\mathbb{B}_\lambda$.) QED(Case 1)

In Case 2 we literally repeat the proof in Theorem 3, using that if $\tilde{\lambda} = \sup \sigma '' \overline{\lambda}$, then by (b)(i), if $c = \bigcap_n c_n$ is a thread in $\langle \mathbb{B}_{\xi_n} \mid n < \omega \rangle$ ($\xi_n = \sigma(\overline{\xi}_n)$, where $\langle \overline{\xi}_n \mid n < \omega \rangle$ is monotone and cofinal in $\overline{\lambda}$), then $c \in \mathbb{B}_{\tilde{\lambda}}$. Just as before we utilize the fact that we can ensure that $\sup \sigma_n '' \overline{\lambda} = \lambda$, even though we cannot fix the values of $\sigma_n(\overline{\xi}_i)$ $(i < \omega)$. QED(Theorem 4)

## *Forcing Axioms*

We say that a complete BA $\mathbb{B}$ *satisfies Martin's Axiom* iff whenever $\langle \Delta_i \mid i < \omega_1 \rangle$ is a sequence of dense sets in $\mathbb{B}$, there is a filter $G$ on $\mathbb{B}$ s.t. $G \cap \Delta_i \neq \emptyset$ for $i < \omega$. The original Martin's Axiom said that this holds for all $\mathbb{B}$ satisfying the countable chain condition. This axiom is consistent relative to ZFC. It was later discovered that very strong versions of Martin's Axiom can be proven consistent relative to a supercompact cardinal. The best known of these are the *proper forcing axiom* (PFA), which posits Martin's Axiom for proper forcings and *Martin's Maximum* (MM) which is equivalent to Martin's Axiom for semiproper forcings. Both of these strengthen the original Martin's Axiom, hence imply the negation of CH. Here we shall consider the *subcomplete forcing axiom* (SCFA), which says that Martin's Axiom holds for subcomplete forcings. This, it turns out, is compatible with CH, hence cannot be a strengthening of the original Martin's Axiom (though it is, of course, a strengthening of Martin's Axiom for complete forcings). Nonetheless it turns out that SCFA has some of the more striking consequences of MM. A fuller account of this can be found in [FA].

We recall from Section 3.1 that the notion of subcompleteness is "locally based" in the sense that, if $\theta$, $\theta'$ are cardinals with $\overline{\overline{H_\theta}} < \theta'$, then we need only consider $N = L_\tau^A$ of size less than $\theta'$, in order to determine whether $\theta$ verifies the subcompleteness of a given $\mathbb{B}$. In other words, $\mathfrak{P}(H_\theta)$ contains all the information needed to determine this. As a consequence we get Section 3, Corollary 2.3, which says that, if $W$ is an inner model and $\mathfrak{P}(H_\theta) \subset W$, then the question, whether $\theta$ verifies the subcompleteness of $\mathbb{B}$, is absolute in $W$.

Using this we prove:

**Theorem 5** *Let $\kappa$ be supercompact. There is a subcomplete $\mathbb{B} \subset \mathbf{V}_\kappa$ s.t. whenever $G$ is $\mathbb{B}$-generic, then:*

*(a)* $\kappa = \omega_2$,
*(b)* CH *holds,*
*(c)* SCFA *holds.*

*Proof*. Let $f$ be a Laver function (i.e. for each $x$ and each cardinal $\beta$ there is a supercompact embedding $\pi : \mathbf{V} \to W$ s.t. $x = \pi(f)(\kappa)$ and $W^\beta \subset W$). We define and RCS iteration $\langle \mathbb{B}_i \mid i \leq \kappa \rangle$ by:

- $\mathbb{B}_0 = 2$.
- If $\Vdash_i$ $f(i)$ is a subcomplete forcing, then $\Vdash_i$ $\check{\mathbb{B}}_{i+1}/\mathring{G} \simeq f(i) * \text{coll}(w_1, \overline{\overline{f(i)}})$.
- If $\nVdash_i f(i)$ is a subcomplete forcing, then $\Vdash_i \check{\mathbb{B}}_{i+1}/\mathring{G} \simeq \text{coll}(w_1, w_2)$.

Let $G$ be $\mathbb{B} = \mathbb{B}_\kappa$-generic. Then CH holds in $\mathbf{V}[G]$, since there will be a stage $\mathbb{B}_{i+1}$ which makes CH true by collapsing. But then it remains true at later stages, since no reals are added. We now show that SCFA holds in $\mathbf{V}[G]$. Let $\mathbb{A} \in \mathbf{V}[G]$ be subcomplete. Let $\mathbb{A} = \mathring{\mathbb{A}}^G$. Let $U \in \mathbf{V}$, $U \subset \mathbf{V}^{\mathbb{B}_\kappa}$ s.t.

(1) $[[x \in \mathring{\mathbb{A}}]] \subset \bigcup_{z \in U} [[x = z]]$ for $x \in \mathbf{V}^{\mathbb{B}_\kappa}$.

We may also assume w.l.o.g. that $\mathring{\mathbb{A}}$ is forced to be subcomplete and in fact:

(2) $\Vdash_\kappa \check{\theta}$ verifies the subcompleteness of $\mathring{\mathbb{A}}$.

Let $\beta = \overline{\overline{\mathbf{V}}}_\beta$ where $\mathring{\mathbb{A}}, U, \theta \in \mathbf{V}_\beta$. Let $\pi : \mathbf{V} \to W$ be a supercompact embedding s.t. $\mathring{\mathbb{A}} = \pi(f)(\kappa)$ and $W^\beta \subset W$. (Hence $\mathbf{V}_{\beta+1} \subset W$.) Then:

(3) $\theta$ verifies the subcompleteness of $\mathbb{A}$ in $W[G]$,

since this depends only on $\mathfrak{P}(H_\theta) \subset W$. Now let: $\langle \mathbb{B}'_i \mid i \leq \kappa' \rangle = \pi(\langle \mathbb{B}_i \mid i \leq \kappa \rangle)$. Then $\mathbb{B}_\kappa = \mathbb{B}'_\kappa$ and $G$ is $\mathbb{B}'_\kappa$-generic over $W$. Hence we can form $G' \supset G$ which is $\mathbb{B}'_\kappa$-generic over $W$. Since $\mathbb{B}'_{\kappa+1} \simeq \mathbb{A} * \text{coll}(\omega_1, \overline{\overline{\mathbb{A}}})$, there is $A \in W[G']$ which is $\mathbb{A}$-generic over $W[G]$. Now let $\pi^*$ be the unique $\pi^* \supset \pi$ s.t.

(4) $\pi^* : \mathbf{V}[G] \prec W[G'] \wedge \pi^*(G) = G'$.

Then

(5) $\pi^*(\mathbb{A}) = \mathbb{A}'$, where $\mathbb{A}' = \pi(\mathring{\mathbb{A}})^{\mathbb{B}'_{\kappa'}}$.

But

(6)
$\pi^* \upharpoonright \mathbb{A} \in W[G']$, since $\pi \upharpoonright U \in W$ and $\pi^* \upharpoonright \mathbb{A}$ is definable as that $\tilde{\pi}$ s.t. $\tilde{\pi}(t^G) = \pi(t)^{G'}$ whenever $t \in U$ and $t^G \in \mathbb{A}$.

Let $A'$ be the filter on $\mathbb{A}'$ generated by $\pi^{*\prime\prime}\mathbb{A}$. Let $\langle \Delta_i \mid i < \omega_1 \rangle$ be a sequence of dense sets in $\mathbb{A}$ in $\mathbf{V}[G]$. Let $\langle \Delta'_i \mid i < \omega_1 \rangle = \pi^*(\langle \Delta_i \mid i < \omega_1 \rangle)$. Obviously $A' \cap \Delta'_i \neq \emptyset$ for $i < \omega_1$. Since $\pi : \mathbf{V}[G] \prec W[G]$, we conclude that there is a filter $\tilde{A}$ on $\mathbb{A}$ in $\mathbf{V}[G]$ s.t. $\tilde{A} \cap \Delta_i \neq \emptyset$ for all $i < \omega_2$. QED(Theorem 5)

In [FA] we show that subcomplete forcings are $\diamondsuit$-preserving – i.e. if $\diamondsuit$ holds in $\mathbf{V}$, it continues to hold in $\mathbf{V}[G]$. It follows easily from this that $\diamondsuit$ holds in the model $\mathbf{V}[G]$ just constructed. If we do a prior application of Silver forcing to make GCH true, then the ultimate model will also satisfy GCH. Hence we have, in fact, shown the consistency of

$$\text{SCFA} + \diamondsuit + \text{GCH}$$

relative to a supercompact cardinal.

SCFA has two of the more striking consequences of MM: Friedman's principle and the singular cardinal hypothesis at singular strong limit cardinals. Friedman's principle at a regular cardinal $\tau > \omega_1$ says that if $A \subset \tau$ is any stationary set of $\omega$-cofinal ordinals, then there is a normal function $f : \omega_1 \to A$ (i.e. $f$ is monotone and continuous at limits). It is easily seen that Friedman's principle at $\beta^+$ implies the negation of $\Box_\beta$.

**Lemma 6** *Assume SCFA. Let $\kappa > \omega_1$ be regular. Then Friedman's principle holds at $\kappa$.*

*Proof*. Let $\mathbb{P}_A$ be as in the final example of Section 3.3, where $A \subset \kappa$ is a stationary set of $\omega$-cofinal ordinals. Let

$$\Delta_i = \text{ the set of } p \in \mathbb{P}_A \text{ s.t. } i+1 \subset \text{dom}(p) \text{ for } i < \omega_1.$$

Then $\Delta_i$ is dense in $\mathbb{P}_A$. By SCFA there is a set $G$ of mutually compatible conditions s.t. $G \cap \Delta_i = \emptyset$ for $i < \omega_1$. But then the function $f = \bigcup G$ has the desired property. QED(Lemma 6)

By essentially the same proof we get.

**Lemma 7.1** *Assume SCFA. Let $\tau > \omega_1$ be regular. Let $A_i \subset \tau$ be a stationary set of $\omega$-cofinal points for $i < \omega_1$. Let $\langle D_i \mid i < \omega_1 \rangle$ be a partition of $\omega_1$ into disjoint stationary sets. Then there is a normal function $f : \omega_1 \to \tau$ s.t. $f(j) \in A_i$ for $j \in D_i$.*

*Proof*. We need only to show that the appropriate forcing $\mathbb{P}$ is subcomplete The proof is exactly like Section 3.3, Lemma 6.3. QED(Lemma 7.1)

The singular cardinal hypothesis for strong limit cardinals then follows by a well known argument of Solovay:

**Corollary 7.2** *Assume SCFA. Let $\tau$ be as above. Then $\tau^{\omega_1} = \tau$.*

*Proof*. Let $\langle A_\xi \mid \xi < \tau \rangle$ partition $\{\lambda < \tau \mid \mathrm{cf}(\lambda) = \omega\}$ into disjoint stationary sets. For each $a \in [\tau]^{\omega_1}$ let $\langle \xi_i \mid i < \omega_1 \rangle$ enumerate $a$. Let $f : \omega_1 \to \bigcup\limits_{i<\omega_1} A_{\xi_i}$ be normal s.t. $f(j) \in A_{\xi_i}$ if $j \in D_i$, where $\langle D_i \mid i < \omega_1 \rangle$ partitions $\omega_1$ into stationary sets. Let $\lambda = \sup f''\omega_1$. Then

$$a = B_\lambda =_{\mathrm{Df}} \{\xi \mid A_\xi \cap \lambda \text{ is stationary in } \lambda\}.$$

Hence $[\tau]^{\omega_1} \subset \{B_\lambda \mid \lambda < \tau\}$. QED(Corollary 7.2)

**Corollary 7.3** *Assume SCFA. If $\mathrm{cf}(\beta) \leq \omega_1 < \beta$ and $2^{\underset{\smile}{\beta}} \leq \beta^+$, then $2^\beta = \beta^+$.*

*Proof*. $2^\beta = (2^{\underset{\smile}{\beta}})^{\mathrm{cf}(\beta)} \leq (\beta^+)^{\omega_1} = \beta^+$.

Using Silver's Theorem we conclude:

**Corollary 7.4** *Assume SCFA. If $\beta$ is a singular strong limit cardinal, then $2^\beta = \beta^+$.*

## 5. $\mathcal{L}$-Forcing

In the following, assume CH. Let $\beta > \omega_1$ be a cardinal and assume: $2^{\underset{\smile}{\beta}} = \beta$ (i.e. $2^\alpha \leq \beta$ for $\alpha < \beta$). Let $M = L^A_\beta =_{\mathrm{Df}} \langle L_\beta[A], \in, A \cap L_\beta[A] \rangle$ s.t. $L_\beta[A] = H_\beta$ and $A \subset H_\beta$. Suppose we have forcing conditions which do not collapse $\omega_1$, but do add a map collapsing $\beta$ onto $\omega_1$. The existence of such a map is equivalent to the existence of a commutative "tower" $\langle M_i \mid i < \omega_1 \rangle$, $\langle \pi_{ij} \mid i \leq j < \omega_1 \rangle$ s.t. each $M_i$ is countable and transitive, $\pi_{ij} : M_i \to M_j$ for $i \leq j < \omega_1$, and the tower converges to $M$ (i.e. there are $\langle \pi_i \mid i < \omega_1 \rangle$ s.t. $(M, \langle \pi_i \mid i < \omega_1 \rangle)$ is the direct limit of $\langle M_i \mid i < \omega_1 \rangle$, $\langle \pi_{ij} \mid i \leq j < \omega_1 \rangle$).

In $\mathcal{L}$-forcing we attempt to collapse $\beta$ onto $\omega_1$ by conditions which directly describe such a tower (or at least a commutative directed system converging to $M$). The "$\mathcal{L}$" in "$\mathcal{L}$-forcing" refers to an infinitary language on a structure of the form:

$$N = \langle H_{\beta^+}, \in, M, \ldots \rangle$$

in the ground model $\mathbf{V}$. $\mathcal{L}$ then determines a set of conditions $\mathbb{P}_{\mathcal{L}}$. $\mathcal{L}$-forcing has been used to add new reals with interesting properties. In these notes, however, we shall concentrate wholly on a form of $\mathcal{L}$-forcing which does *not* add new reals. This means, of course, that $H_{\omega_1}$ is absolute. Hence all countable initial segments of our "tower" will lie in $\mathbf{V}$.

The theory of $\mathcal{L}$-forcing is developed in [LF]. In that paper, however, we dealt only with forcings which literally added a tower converging to $M$ in the aforementioned sense. In later applications we found it better to replace the tower by other sorts of convergence systems. We therefore adopt a more general approach here. The proofs in [LF] can be readily adapted to this approach.

Recall that we are working in first order set theory, so we cannot literally quantify over arbitrary classes. Instead we work with "virtual classes", which are expressions of the form $\{x \mid \varphi(x)\}$ where $\varphi = \varphi(x)$ is a formula of ZF. Normally we suppose $x$ to be the only variable occurring free in $\varphi$. We define:

**Definition** An *approximation system* is a pair $\langle \Gamma, \Pi \rangle$ of virtual classes s.t. (I)–(VII) below are provable in ZFC$^-$.

(I) $\Gamma$ is a class of pairs $\langle M, C \rangle$ s.t.

(a) $M = L_\tau^{A_1,\ldots,A_n}$ for some $A_1, \ldots, A_n, \tau$.

(b) $C \subset M$.

**(Definition** For $u \in \Gamma$ set: $u = \langle M_u, C_u \rangle$.)

(II) $\Pi$ is a class of triples $\langle \pi, u, v \rangle$ s.t. $u, v \in \Gamma$, $\pi : M_u \prec M_v$, $C_u = \pi^{-1}{}''C_v$.

**(Definition** $\pi : u \lhd v \leftrightarrow_{\mathrm{Df}} \langle \pi, u, v \rangle \in \Pi$, $\;u \lhd v \leftrightarrow_{\mathrm{Df}} \bigvee \pi \; \pi : u \lhd v$.)

(III) There is at most one $\pi$ s.t. $\pi : u \lhd v$.

**(Definition** $\pi_{uv} \simeq_{\mathrm{Df}}$ that $\pi$ s.t. $\pi : u \lhd v$.)

(IV) (a) $u \lhd u \wedge \pi_{uu} = \mathrm{id}$ for $u \in \Gamma$.

(b) $u \lhd v \lhd w \to (u \lhd w \wedge \pi_{uw} = \pi_{vw} \circ \pi_{uv})$.

(c) If $u, v \lhd w$ and $\mathrm{rng}(\pi_{uw}) \subset \mathrm{rng}(\pi_{vw})$, then $u \lhd v$ and $\pi_{uv} = \pi_{vw}^{-1} \circ \pi_{uw}$.

We say that a set $X \subset \Gamma$ is $\lhd$-*directed* iff for all $u, v \in X$ there is $w$ s.t. $u, v \lhd w$. In this case we can form a direct limit $v$, $\langle \pi_u \mid u \in X \rangle$ of $\langle u \mid u \in X \rangle$, $\langle \pi_{uu'} \mid u \lhd u' \wedge u, u' \in X \rangle$. Then $v = \langle \mathfrak{A}, C \rangle$, where $\mathfrak{A}$ is a (possibly ill founded) ZFC$^-$ model. If $\mathfrak{A}$ is well founded, we can take it as transitive. Clearly the transitivized direct limit of $X$, if it exists, is uniquely determined by $X$.

(V) Let $X \subset \Gamma$ be $\lhd$-directed. Let $v, \langle \pi_u \mid u \in X \rangle$ be the transitivized direct limit. Then $v \in \Gamma$ and $\pi_u = \pi_{uv}$ for $u \in X$. Moreover, if $u \lhd w$ for all $u \in X$, then $v \lhd w$. (Hence $\pi_{vw}$ is uniquely determined by: $\pi_{vw}\pi_u = \pi_{uw}$ for $u \in X$.)

If $t = \{x \mid \varphi(x)\}$ is a virtual class and $W$ is any set or class, we can form $t^W$ (the interpretation of $t$ in $W$) by relativizing all quantifiers in $\varphi$ to $W$.

(VI) If $M$ is an admissible set, then $\Gamma \cap M = \Gamma^M$ and $\Pi \cap M = \Pi^M$.

If $\mathfrak{A} = \langle |\mathfrak{A}|, \in_{\mathfrak{A}} \rangle$ is any binary structure we can form the relativization $t^{\mathfrak{A}}$ by relativizing quantifiers to $|\mathfrak{A}|$ and simultaneously replacing $\in$ by $\in_{\mathfrak{A}}$ in $\varphi$.

(VIII) If $\mathfrak{A}$ is a solid model of ZFC$^-$ and $A = \text{wfc}(\mathfrak{A})$, then $\Gamma \cap A = \Gamma^{\mathfrak{A}} \cap A$, and $\Pi \cap A = \Pi^{\mathfrak{A}} \cap A$.

Hence:

(1) If $\mathbf{V}[G]$ is a generic extension of $\mathbf{V}$, then $\Gamma^{\mathbf{V}[G]} \cap \mathbf{V} = \Gamma^{\mathbf{V}}$, $\Pi^{\mathbf{V}[G]} \cap \mathbf{V} = \Pi^{\mathbf{V}}$.

*Proof*. Let $x \in \mathbf{V}$. Then $x \in M \in \mathbf{V}$, where $M$ is admissible. Hence $x \in \Gamma^{\mathbf{V}[G]} \leftrightarrow x \in \Gamma^M \leftrightarrow x \in \Gamma^{\mathbf{V}}$, applying (VI) first in $\mathbf{V}[G]$, then in $\mathbf{V}$.

QED(1)

**Remark** In practice (I)–(VII) will follow readily from the definitions given for $\Gamma$, $\Pi$, so we shall not bother to verify them in detail. In all cases $\Gamma$ and $\Pi$ will also be provably primitive recursive in ZFC$^-$, so the absoluteness properties (VI), (VII) will follow by Section 1.3. However, it will also be easy to verify these properties directly without going through the theory of pr functions.

⋆ ⋆ ⋆ ⋆ ⋆

A simple example of an approximation system is:

$\Gamma$ is the set of all $\langle M, C \rangle$ s.t.

- $M = L_\tau^A$ for some $A$, $\tau$.
- $M$ models ZFC$^-$ and $\omega_1$ exists and CH.
- $C$ maps $\omega_1^M$ onto $M$.

$\Pi$ is then the set of all $\langle \pi, u, v \rangle$ s.t. $u, v \in \Gamma$, $\pi : M_u \prec M_v$ and $\pi \circ C_u \subset C_v$. (Note that in this example we have $\Gamma, \Pi \subset H_{\omega_2}$.) The absoluteness properties are straightforward, since if $M, N, \pi \in A$ and $A$ is admissible, then $\pi : M \prec N$ is uniformly expressible by a $\Sigma_1$ formula in any solid $\mathfrak{A}$ extending $A$.

Now let an approximation system $\langle \Gamma, \Pi \rangle$ be given. Let $M = L^A_\beta$ be as described at the outset with $\beta > \omega_1$ and $H_\beta = L_\beta[A]$. Our aim in $\mathcal{L}$-forcing is to generically add $C \subset M$ in the extension $\mathbf{V}[C]$ s.t. $\langle M, C \rangle \in \Gamma^{\mathbf{V}[C]}$ and $\langle M, C \rangle$ is the limit of a directed $X \subset \Gamma \cap H_{\omega_1}$. At the same time we want to add no reals, so that $\Gamma \cap H_{\omega_1}$ remains absolute. Since we are assuming CH it follows easily that $\text{card}(M) = \omega_1$ in $\mathbf{V}[G]$. (In the above example we would accomplish this explicitly, since $C$ would map $\omega_1$ onto $M$.)

$\mathcal{L}$ is a language on $N = \langle H_{\beta^+}, \in, M, <, \ldots \rangle$, where $<$ is a well ordering of $H_{\beta^+}$.

(**Note** $N$ remains a ZFC$^-$ model, hence admissible, no matter which predicates and constants we adjoin to it.)

The only nonlogical predicate of $\mathcal{L}$ is $\in$. In addition to the constants $\underline{x}$ $(x \in N)$ there will be one further constant $\mathring{C}$. We always suppose $\mathcal{L}$ to contain the following *core axioms*:

- ZFC$^-$ (here the usual finite axioms are meant, so we could write them as a single $M$-finite conjunction).
- $\bigwedge v (v \in \underline{x} \leftrightarrow \bigvee\!\!\!\bigvee_{z \in x} v = \underline{z})$ for $x \in N$.
- $H_{\omega_1} = \underline{H_{\omega_1}}$ (or equivalently $\mathfrak{P}(\omega) = \underline{\mathfrak{P}(\omega)}$).
- $\langle \underline{M}, \mathring{C} \rangle \in \Gamma$.
- For all countable $X \subset \underline{M}$ there is $u \in \Gamma \cap H_{\omega_1}$ s.t. $X \subset \text{rng}(\pi_{u, \langle \underline{M}, \mathring{C} \rangle})$.

  ($\mathcal{L}$ might, of course, contain further axioms as well.)

Assume that $\mathcal{L}$ is consistent. Then it is forced to be consistent by the forcing collapsing the cardinality of $N$ to $\omega$ with finite conditions. Hence, in such generic extensions, $\mathcal{L}$ has solid models. In what follows, without confusion, when we claim any solid model of $\mathcal{L}$, it's understood that we work in such generic extensions.

**Definition** Let $\mathfrak{A}$ be a solid model of $\mathcal{L}$. $\Gamma^{\mathfrak{A}}$, $\Pi^{\mathfrak{A}}$, $\lhd^{\mathfrak{A}}$, $\pi^{\mathfrak{A}}_{uv}$ $(u \lhd^{\mathfrak{A}} v)$ are defined in the obvious way. Set:

$$\tilde{\Gamma} = \tilde{\Gamma}^{\mathfrak{A}} =_{\text{Df}} \{ e \in \Gamma \cap H_{\omega_1} \mid e \lhd \langle M, \mathring{C}^{\mathfrak{A}} \rangle \text{ in } \mathfrak{A} \}.$$

For $e \in \tilde{\Gamma}$ set: $\pi^{\mathfrak{A}}_e =_{\text{Df}} \pi^{\mathfrak{A}}_{e, \langle M, \mathring{C}^{\mathfrak{A}} \rangle}$.

**Lemma 1.1** *Let $\mathfrak{A}$ be as above. Then $\tilde{\Gamma}$ is a $\lhd$-directed system with limit $\langle M, \mathring{C}^{\mathfrak{A}} \rangle$, $\langle \pi^{\mathfrak{A}}_e \mid e \in \tilde{\Gamma} \rangle$.*

*Proof*. $M = \bigcup_{e\in\tilde{\Gamma}} \mathrm{rng}(\pi_e^{\mathfrak{A}})$ is trivial. We show that $\tilde{\Gamma}$ is directed. Let $e_0, e_1 \in \tilde{\Gamma}$. Let $u \in \tilde{\Gamma}$ s.t. $\mathrm{rng}(\pi_{e_0}^{\mathfrak{A}}) \cup \mathrm{rng}(\pi_{e_1}^{\mathfrak{A}}) \subset \mathrm{rng}(\pi_u^{\mathfrak{A}})$. Then $e_0, e_1 \lhd u$ and $\pi_{e_h u} = (\pi_u^{\mathfrak{A}})^{-1} \circ \pi_{e_h}^{\mathfrak{A}}$.

QED(Lemma 1.1)

**Lemma 1.2** *Let $\mathfrak{A}$ be as above. Let $A \in \mathfrak{A}$ s.t. $A \subset M$. There is $e \in \tilde{\Gamma}$ s.t. $\mathrm{rng}(\pi_e^{\mathfrak{A}}) \prec \langle M, A\rangle$.*

*Proof*. In $\mathfrak{A}$ construct $\langle e_i \mid i < \omega\rangle$, $\langle X_i \mid i < \omega\rangle$ s.t. $\bigcup_{h<i} \mathrm{rng}(\pi_{e_h}^{\mathfrak{A}}) \subset X_i \subset \mathrm{rng}(\pi_{e_i}^{\mathfrak{A}})$ and $X_i \prec \langle M, A\rangle$. It follows easily that $e_h \lhd e_i$ for $h \le i < \omega$ and $\{e_h \mid h < \omega\}$ has a direct limit $e$, $\langle \pi_{e_i e} \mid i < \omega\rangle$. But then $e \lhd \langle M, \mathring{C}^{\mathfrak{A}}\rangle$ and $\mathrm{rng}(\pi_e^{\mathfrak{A}}) = \bigcup_{i<\omega} \mathrm{rng}(\pi_{e_i}^{\mathfrak{A}}) = \bigcup_{i<\omega} X_i \prec \langle M, A\rangle$. QED(Lemma 1.2)

**Corollary 1.3** *Let $\mathfrak{A}$ be as above. Let $U \subset \mathfrak{P}(M)$ s.t. $U \in \mathfrak{A}$ is countable in $\mathfrak{A}$. There is $e \in \tilde{\Gamma}$ s.t. $\mathrm{rng}(\pi_e^{\mathfrak{A}}) \prec \langle M, A\rangle$ for all $A \in U$.*

*Proof*. Let $\langle A_i \mid i < \omega\rangle \in \mathfrak{A}$ enumerate $U$ and apply Lemma 1.2 to $A = \{\langle x, i\rangle \mid x \in A_i\}$. QED(Corollary 1.3)

**Corollary 1.4** *Let $\mathfrak{A}$ be as above. Let $U, V$ be countable in $\mathfrak{A}$ s.t. $U \subset M$, $V \subset \mathfrak{P}(M)$. There is $e \in \tilde{\Gamma}$ s.t. $U \subset \mathrm{rng}(\pi_e^{\mathfrak{A}}) \prec \langle M, A\rangle$ for $A \in V$.*

*Proof*. Apply Corollary 1.3 to $U \cup V$. QED(Corollary 1.4)

If $\mathcal{L}$ is consistent, we can define a set $\mathbb{P} = \mathbb{P}_{\mathcal{L}}$ of conditions as follows:

**Definition** Let $\tilde{\mathbb{P}}$ be the set of $p = \langle p_0, p_1\rangle$ s.t. $p_0 \in \Gamma \cap H_{\omega_1}$ and

$$p_1 \subset \mathfrak{P}(M) \times \mathfrak{P}(M_{p_0}) \quad \text{is countable.}$$

For $p \in \tilde{\mathbb{P}}$ let $\varphi_p$ be the conjunction of the $\mathcal{L}$ statements:

- $\underline{p_0} \lhd \langle \underline{M}, \mathring{C}\rangle$
- If $\pi = \pi_{\underline{p_0}, \langle \underline{M}, \mathring{C}\rangle}$, then $\pi : \langle \underline{M}_{p_0}, \overline{a}\rangle \prec \langle \underline{M}, a\rangle$ for all $\langle a, \overline{a}\rangle \in \underline{p_1}$.

Set $\mathcal{L}(p) = \mathcal{L} + \varphi_p$. We set: $\mathbb{P} = \{p \in \tilde{\mathbb{P}} \mid \mathrm{con}(\mathcal{L}(p))\}$, where $\mathrm{con}(\mathcal{L}(p))$ is the statement that "$\mathcal{L}(p)$ is consistent". The extension relation on $\mathbb{P}$ is then defined by:

**Definition** Let $p, q \in \mathbb{P}$

$$p \le q \longleftrightarrow_{\mathrm{Df}} (q_0 \lhd p_0 \wedge \mathrm{rng}(q_1) \subset \mathrm{rng}(p_1) \wedge \pi_{q_0 p_0} : \langle M_{q_0}, \overline{a}\rangle \prec \langle M_{p_0}, a'\rangle$$
$$\text{whenever } \langle a, \overline{a}\rangle \in q_1,\ \langle a, a'\rangle \in p_1).$$

**Lemma 2.1** *$\leq$ is a partial ordering.*

*Proof*. Transitivity is immediate. Now let $p \leq q$, $q \leq p$. We claim that $p = q$, $p_0 = q_0$ is immediate. But if $\langle a, \overline{a} \rangle \in q_1$, $\langle a, a' \rangle \in p_1$, then $\overline{a} = a'$, since $\pi_{q_0 p_0} = \text{id}$. Hence $q_1 = p_1$. QED(Lemma 2.1)

**Definition** Let $p \in \mathbb{P}$. $M_p = M_{p_0}$, $C_p = C_{p_0}$, $F^p = p_1$, $R^p = \text{rng}(p_1)$, $D^p = \text{dom}(p_1)$.

**Lemma 2.2** *Let $p \in \mathbb{P}$. Then*

*(a) $F^p$ is a function.*
*(b) If $R^p$ is closed under set difference, then $F^p : D^p \leftrightarrow R^p$.*
*(c) $F^p \upharpoonright M_p$ injects $M_p$ into $M$.*

*Proof*. Let $\mathfrak{A}$ be a solid model of $\mathcal{L}(p)$. Let $\pi = \pi_p^{\mathfrak{A}} =_{\text{Df}} (\pi_{p_0, \langle M, \mathring{C}^{\mathfrak{A}} \rangle})^{\mathfrak{A}}$.
(a) Let $\langle a, \overline{a} \rangle, \langle a, \overline{a}' \rangle \in F^p$. Then $\overline{a} = \overline{a}' = \pi^{-1}{}''a$.
(b) Let $\langle a, \overline{a} \rangle, \langle b, \overline{b} \rangle \in F^p$. It suffices to show:
**Claim:** $\overline{a} \subset \overline{b} \to a \subset b$.
Set $c = a \setminus b$, $\overline{c} = \overline{a} \setminus \overline{b} = \emptyset$. Then $(F^p)^{-1}(c) = \pi^{-1}{}''c = \pi^{-1}{}''a \setminus \pi^{-1}{}''b = \overline{b} \setminus \overline{a} = \emptyset$. Hence $c = \emptyset$, since $\pi : \langle M_{p_0}, \overline{c} \rangle \prec \langle M, c \rangle$. QED(b)
(c) Let $x \in M_{p_0}$, $\langle x, \overline{x} \rangle \in F^p$. Then $\pi(\overline{x}) = x \in M$ since $\pi : \langle M_p, \overline{x} \rangle \prec \langle M, x \rangle$.

QED(Lemma 2.2)

We define:

**Definition** $\pi^p = F^p \upharpoonright M_p$.

**Note** By the proof of (c) we have: $\mathcal{L}(p) \vdash \underline{\pi^p} \subset \pi_{\underline{p}, \langle \underline{M}, \mathring{C} \rangle}$.

We now prove the main lemma on extendability of conditions.

**Lemma 3.1** $\mathbb{P} \neq \emptyset$. *Moreover, if $p, q \in \mathbb{P}$ and $\mathcal{L}(p) \cup \mathcal{L}(q)$ is consistent, there is $r$ s.t. $r \leq p, q$. Moreover, if $X \subset \mathfrak{P}(M)$ is any countable set, we may choose $r$ s.t. $X \subset R^r$.*

*Proof*. To see $\mathbb{P} \neq \emptyset$ let $\mathfrak{A}$ be any solid model of $\mathcal{L}$. Let $e \lhd \langle M, \mathring{C}^{\mathfrak{A}} \rangle$ in $\mathfrak{A}$ where $e \in \Gamma \cap H_{\omega_1}$. Then $\mathfrak{A} \models \mathcal{L}(p)$ where $p = \langle e, \emptyset \rangle$. Hence $p \in \mathbb{P}$.
Now let $\mathfrak{A} \models \mathcal{L}(p) \cup \mathcal{L}(q)$. Let $X \subset \mathfrak{P}(M)$ be countable in $\mathbf{V}$. Let $Y = X \cup R^p \cup R^q$. There is $e \in H_{\omega_1} \cap \Gamma$ s.t. $e \lhd \langle M, \mathring{C}^{\mathfrak{A}} \rangle$ in $\mathfrak{A}$ and $\pi_e^{\mathfrak{A}} \prec \langle M, A \rangle$ for all $A \in Y$. For $A \in Y$ set $\overline{A} = (\pi_e^{\mathfrak{A}})^{-1}{}''A$. Letting $\langle A_i \mid i < \omega \rangle$ be an enumeration of $Y$ in $\mathbf{V}$, we see that $\langle \overline{A}_i \mid i < \omega \rangle \in H_{\omega_1}$. Hence $F \in \mathbf{V}$

where $F = \{\langle A, \overline{A}\rangle \mid A \in Y\} = \{\langle A_i, \overline{A}_i\rangle \mid i < \omega\}$. Set $r = \langle e, F\rangle$. Then $\mathfrak{A} \models \mathcal{L}(r)$ and $r \leq p, q$. QED(Lemma 3.1)

**Corollary 3.2** *$p$, $q$ are compatible in $\mathbb{P}$ iff $\mathcal{L}(p) \cup \mathcal{L}(q)$ is consistent.*

*Proof*. ($\leftarrow$) Lemma 3.1.
($\rightarrow$) If $r \leq p, q$, then $\mathcal{L}(r) \vdash \mathcal{L}(p) \cup \mathcal{L}(q)$. QED(Corollary 3.2)

**Corollary 3.3** *Let $p \in \mathbb{P}$, $X \subset \mathfrak{P}(M)$ where $X$ is countable. There is $r \leq p$ with $X \subset R^r$.*

**Corollary 3.4** *Let $p \in \mathbb{P}$, $u \subset M$, $u$ is countable. There is $r \leq p$ with $u \subset \mathrm{rng}(\pi^r)$.*

**Lemma 3.5** *Let $p \in \mathbb{P}$, $u \subset M_p$, $u$ finite. There is $r \leq p$ s.t. $r_0 = p_0$ and $u \subset \mathrm{dom}(\pi^r)$.*

*Proof*. Let $\mathfrak{A}$ be a solid model of $\mathcal{L}(p)$. Set: $r_0 = p_0$, $F^r = F^p \cup (\pi_p^{\mathfrak{A}} \upharpoonright u)$. Then $\mathfrak{A} \models \mathcal{L}(r)$. QED(Lemma 3.5)

Using these extension lemmas we get:

**Lemma 3.6** *Let $G$ be $\mathbb{P}$-generic. For $p \in G$ set: $\pi_p^G = \bigcup\{\pi^q \mid q \in G \wedge p_0 = q_0\}$. Then:*

*(a) $\{p_0 \mid p \in G\}$ is a $\lhd$-directed system with limit $\langle M, C^G\rangle$, $\langle \pi_p^G \mid p \in G\rangle$, where $C^G = \bigcup\limits_p \pi_p^G {}''C_p$.*

*(b) $\pi_p^G : \langle M_p, \overline{a}\rangle \prec \langle M, a\rangle$ for $\langle a, \overline{a}\rangle \in F^p$.*

**Note** $\pi_p^G : p_0 \lhd \langle M, C^G\rangle$ in $\mathbf{V}[G]$ by (a).

The proof is straightforward. Now let $\kappa > (2^\beta)$ be regular in $\mathbf{V}$. Then $\kappa$ remains regular in $\mathbf{V}[G]$, since $\mathbb{P} \in H_\kappa$. $\langle H_\kappa^{\mathbf{V}[G]}, C^G\rangle$ then models all of the core axioms except possibly the axiom: $H_{\omega_1} = \underline{H_{\omega_1}}$.

We now state a condition called *revisability* which will guarantee that no reals are added – hence that all core axioms hold in $\langle H_\kappa^{\mathbf{V}[G]}, C^G\rangle$.

We first define:

**Definition** Let $N^* = \langle H_\delta, M, <, \ldots\rangle$ be a model of countable or finite type, where $\delta > 2^\beta$ is a cardinal and $<$ well orders $H_\delta$. Let $p \in \mathbb{P}$. $p$ *conforms* to $N^*$ iff whenever $a_1, \ldots, a_n \in R^p$ ($n \geq 0$) and $b \subset M$ is $N^*$-definable in $a_1, \ldots, a_n$, then $b \in R^p$.

**Note** If $p$ conforms to $N^*$ then $R^p \neq \emptyset$ and $F^p : D^p \leftrightarrow R^p$ by Lemma 2.2.

**Note** $\{p \mid p$ conforms to $N^*\}$ is dense in $\mathbb{P}$ by the extension lemmas.

Before defining revisability we prove a theorem:

**Lemma 4** *Let p conforms to $N^*$. There is a unique $\overline{N}^* = \overline{N}^*(p, N^*)$ s.t.*

*(i) $\overline{N}^*$ is transitive and of the same type as $N^*$.*
*(ii) If $a_1, \ldots, a_n \in R^p$ $(n \geq 0)$ and $b \subset M$ is $N^*$-definable in $a_1, \ldots, a_n$, then $\overline{a}_1^p, \ldots, \overline{a}_n^p \in \overline{N}^*$ (where $\overline{a}_i^p = (F^p)^{-1}(a_i)$) and $\overline{b}^p (= (F^p)^{-1}(b))$ is $\overline{N}^*$-definable in $\overline{a}_1^p, \ldots, \overline{a}_n^p$ by the same definition.*
*(iii) Each $x \in \overline{N}^*$ is $\overline{N}^*$-definable from parameters in $M_p \cup D^p$.*

*Moreover, if $\mathfrak{A}$ is a solid model of $\mathcal{L}(p)$, then $\pi_p^{\mathfrak{A}} \cup F^p$ extends uniquely to a $\pi \supset \pi_p^{\mathfrak{A}} \cup F^p$ s.t. $\pi : \overline{N}^* \prec N^*$.*

*Proof*. We use the following:

**Fact** For any $X \subset M$ the following are equivalent:

(a) $X \prec \langle M, a \rangle$ for all $a \in R^p$.
(b) Let $Y =$ the smallest $Y \prec N^*$ s.t. $X \cup R^p \subset Y$. Then $Y \cap M = X$.

((b) $\rightarrow$ (a) is trivial. (a) $\rightarrow$ (b) follows from the fact that each $z \in Y$ is $N^*$-definable from parameters in $X \cup R^p$.)

Let $\tilde{Y} =$ the smallest $\tilde{Y} \prec N^*$ s.t. $M \cup R^p \subset \tilde{Y}$. Then $\tilde{Y}$ has cardinality $\beta$ in $\mathbf{V}$. Hence, if – in some extension $\mathbf{V}[G]$ – $\mathfrak{A}$ is a solid model of $\mathcal{L}(p)$, then $\tilde{N}^* \in N \subset \mathfrak{A}$, where $\tilde{\pi} : \tilde{N}^* \overset{\sim}{\leftrightarrow} \tilde{Y}$ is the transitivation of $\tilde{Y}$. Working in $\mathfrak{A}$, we now form $Z =$ the smallest $Z \prec \tilde{N}^*$ s.t. $X \cup R^p \subset Z$, where $X = \mathrm{rng}(\pi_p^{\mathfrak{A}})$. Transitivize $Z$ to get $\overline{\pi} : \overline{N}^* \overset{\sim}{\leftrightarrow} Z$. Then $\overline{N}^* \in H_{\omega_1}^{\mathfrak{A}} = H_{\omega_1}^{\mathbf{V}}$.

**Claim 1** $\overline{N}^*$ satisfies (i)–(iii).

*Proof*. Let $\pi = \tilde{\pi}\overline{\pi} : \overline{N}^* \overset{\sim}{\leftrightarrow} Y =$ the smallest $Y \prec N^*$ s.t. $X \cup R^p \subset Y$. Then $\pi \restriction M_p = \pi_p^{\mathfrak{A}}$, since $X = Y \cap M$ by the above Fact. For $a \in R^p$ we have $\pi^{-1}(a) = \pi^{-1}{}''(X \cap a) = (\pi_p^{\mathfrak{A}})^{-1}{}''(X \cap a) = \overline{a}^p$. Thus $\pi \supset \pi_p^{\mathfrak{A}} \cup F^p$. Using this, (i)–(iii) follow easily.

**Claim 2** At most one $\overline{N}^*$ satisfies (i)–(iii).

*Proof*. Let $\overline{N}_0^*, \overline{N}_1^*$ be two different ones. Then

(1) Let $x_1, \ldots, x_n \in M_p$, $b_1, \ldots, b_m \in D^p$. Then $\overline{N}_0^* \models \varphi(\vec{x}, \vec{b}) \leftrightarrow \overline{N}_1^* \models \varphi(\vec{x}, \vec{b})$.

*Proof*. Let $b_i = \overline{a}_i^p$, $a_i \in R^p$. Set: $c = \{\langle \vec{x} \rangle \in M \mid N^* \models \varphi(\vec{x}, \vec{a})\}$. Then by (ii):

$$\overline{c}^p = \{\langle \vec{x} \rangle \in M_p \mid \overline{N}_h^* \models \varphi(\vec{x}, \vec{b})\}. \qquad \text{QED(1)}$$

But it then follows straightforwardly that $\mathrm{id} \restriction (M_p \cup D^p)$ extends to a $\sigma : \overline{N}_0^* \overset{\sim}{\leftrightarrow} \overline{N}_1^*$. Hence $\sigma = \mathrm{id}$, since the models are transitive. QED(Claim 2)

In the proof of Claim 1, we have shown that, if $\mathfrak{A}$ is a solid model of $\mathcal{L}(p)$, then $\pi_p^{\mathfrak{A}} \cup F^p$ extends to a $\pi : \overline{N}^* \prec N^*$. It remains only to note that $\pi$ is unique, since every $z \in \text{rng}(\pi)$ is $N^*$-definable from elements of $X \cup R^p = \text{rng}(\pi_p^{\mathfrak{A}} \cup F^p)$.

QED(Lemma 4)

**Note** Clearly $M_p = \overline{M}$, where $\overline{N}^* = \langle \overline{H}, \overline{M}, <, \ldots \rangle$.

We now define:

**Definition** $\mathbb{P} = \mathbb{P}_{\mathcal{L}}$ is *revisable* iff for sufficiently large cardinals $\Omega > 2^\beta$: Let $N^* = \langle H_\Omega, M, <, \mathbb{P}, \ldots \rangle$ where $<$ well orders $H_\Omega$. Let $p$ conform to $N^*$ and set $\overline{N}^* = \overline{N}^*(p, N^*)$. Let $\overline{G}$ be $\overline{\mathbb{P}}$-generic over $\overline{N}^*$, where $\overline{N}^* = \langle \overline{H}, \overline{M}, <, \overline{\mathbb{P}}, \ldots \rangle$. Then there is $q \in \mathbb{P}$ s.t. $M_q = M_p$, $C_q = C^{\overline{G}}$, and $F^q = F^p$.
(In other words $q = \langle \langle M_p, C^{\overline{G}} \rangle, F^p \rangle \in \mathbb{P}$.)

**Lemma 5.1** *Let $\mathbb{P}$ be revisable. Then $\mathbb{P}$ adds no new reals.*

*Proof*. Let $\Vdash \overset{\circ}{f} : \omega \to 2$.

**Claim** $\Delta = \{p \mid \bigvee f \; p \Vdash \overset{\circ}{f} = \check{f}\}$ is dense in $\mathbb{P}$.

Let $r \in \mathbb{P}$. Pick $\Omega$ big enough to verify revisability and set $N^* = \langle H_\Omega, M, <, \mathbb{P}, \overset{\circ}{f}, r, \ldots \rangle$. Let $p$ conform to $N^*$. Set $\overline{N}^* = \overline{N}^*(p, N^*)$. Let $\overline{N}^* = \langle \overline{H}, \overline{M}, <, \overline{\mathbb{P}}, \overline{f}, \overline{r}, \ldots \rangle$. Let $\overline{G} \ni \overline{r}$ be $\overline{\mathbb{P}}$-generic over $\overline{N}^*$. Let $f = \overline{f}^{\overline{G}}$. Let $q = \langle \langle \overline{M}, C^{\overline{G}} \rangle, F^p \rangle \in \mathbb{P}$.

**Claim** $q \le r$ and $q \Vdash \overset{\circ}{f} = \check{f}$.

*Proof*. Let $\mathfrak{A}$ be a solid model of $\mathcal{L}(q)$. Let $\sigma \supset \pi_q^{\mathfrak{A}} \cup F^q$ s.t. $\sigma : \overline{N}^* \prec N^*$.

(1) $q \le r$.

*Proof*. Let $\overline{C} = C^{\overline{G}}$, $r_0 = \overline{r}_0 \lhd \langle \overline{M}, \overline{C} \rangle = q_0$ and $\pi_{r_0,q_0} = \pi_{\overline{r}}^{\overline{G}}$. But $R^r \subset R^q$, since $r$ is $\overline{N}^*$-definable. Let $\langle a, \overline{a} \rangle \in F^r$, $\langle a, a' \rangle \in F^q$.

**Claim** $\pi_{r_0,q_0} : \langle M_r, \overline{a} \rangle \prec \langle M_q, a' \rangle$.

This is clear, since $a' = (F^q)^{-1}(a) = \sigma^{-1}(a)$ and hence $\langle \overline{a}, a' \rangle \in \sigma^{-1}{}''F^r = F^{\overline{r}}$).

QED(1)

(2) Let $\overline{s} \in \overline{G}, s = \sigma(\overline{s})$. Then $\mathfrak{A} \models \mathcal{L}(s)$.

*Proof*. $s_0 = \overline{r}_0 \lhd q_0 = \langle \overline{M}, \overline{C} \rangle \lhd \langle M, \dot{C}^{\mathfrak{A}} \rangle$, and $\pi_{s_0}^{\mathfrak{A}} = \sigma \circ \pi_{s_0,q_0}$.

Let $\langle a, \overline{a} \rangle \in F^s$. Then $a = \sigma(a')$, where $\langle a', \overline{a} \rangle \in F^{\overline{s}}$. Hence,

$$\pi^{\mathfrak{A}}_{s_0} : \langle M_s, \overline{a} \rangle \prec \langle M, a \rangle.$$

QED(2)

(3) $q \Vdash \overset{\circ}{f} = \check{f}$.

Suppose not. Then there is $i$ s.t. $f(i) = h$ and $q \not\Vdash \overset{\circ}{f}(\check{i}) = \check{h}$. Let $q' \leq q$ s.t. $q' \Vdash \overset{\circ}{f}(\check{i}) \neq \check{h}$. Let $\mathfrak{A}$ be a solid model of $\mathcal{L}(q')$, hence of $\mathcal{L}(q)$. Let $\overline{s} \in \overline{G}$ s.t. $\overline{s} \Vdash_{\overline{\mathbb{P}}} \overline{f}(\check{i}) = \check{h}$. Let $\sigma$ be as above. Let $s = \sigma(\overline{s})$. Then $s \Vdash \overset{\circ}{f}(\check{i}) = \check{h}$. Hence $q'$, $s$ are incompatible. But $\mathfrak{A} \models \mathcal{L}(q') \cup \mathcal{L}(s)$. Contradiction! by Lemma 3.1.

QED(Lemma 5.1)

Now let $\mathcal{L}^c$ be $\mathcal{L}$ with its axioms reduced to the core axioms. (Thus $\mathcal{L}^c$ is uniquely determined by $\Gamma$, $\Pi$.) By Lemma 5.1 we have:

**Lemma 5.2** *Let $\mathbb{P}$ be revisable. Let $G$ be $\mathbb{P}$-generic. Let $p \in G$. Set: $\mathfrak{A} = \langle H_\kappa^{\mathbf{V}[G]}, C^G \rangle$, where $\kappa > 2^\beta$ is regular. Then $\mathfrak{A}$ models $\mathcal{L}^c(p)$.*

An examination of the proof of Lemma 4 shows, however, the proof of the final clause in that Lemma used only that $\mathfrak{A}$ models $\mathcal{L}^c(p)$. Hence:

**Corollary 5.3** *Let $\mathbb{P}$ be revisable. Let $G$ be $\mathbb{P}$-generic. Let $p \in G$ where $p$ conforms to $N^* = \langle H_\Omega, M, <, \ldots \rangle$. Let $\overline{N}^* = \overline{N}^*(p, N^*)$. There is a unique $\sigma \supset \pi_p^G \cup F^p$ s.t. $\sigma : \overline{N}^* \prec N^*$.*

*Proof*. $\pi_p^G = \pi_p^{\mathfrak{A}}$ where $\mathfrak{A}$ is as in Lemma 5.2. QED(Corollary 5.3)

Combining this with the proof of Lemma 5.1 we get:

**Lemma 5.4** *Let $\mathbb{P}$ be revisable. Let $N^* = \langle H_\Omega, M, <, \mathbb{P}, \ldots \rangle$ where $\Omega$ verifies revisability. Let $p$ conform to $N^*$. Set:*

$$\overline{N}^* = \overline{N}^*(p, N^*) = \langle \overline{H}, \overline{M}, <, \overline{\mathbb{P}}, \ldots \rangle.$$

*Let $\overline{G}$ be $\overline{\mathbb{P}}$-generic over $\overline{N}^*$ and set: $q = \langle \langle M_p, C^{\overline{G}} \rangle, F^p \rangle$. Let $G \ni q$ be $\mathbb{P}$-generic. Let $\sigma \supset \pi_p^G \cup F^p$ s.t. $\sigma : \overline{N}^* \prec N^*$. Then $\sigma''\overline{G} \subset G$. (Hence $\sigma$ extends uniquely to $\sigma^* : \overline{N}^*[\overline{G}] \prec N^*[G]$ with $\sigma^*(\overline{G}) = G$.)*

*Proof*. The proof of (2) in Lemma 5.1 made use of a solid model $\mathfrak{A}$ of $\mathcal{L}(q)$. An examination of this proof shows, however, that it is enough that $\mathfrak{A}$ models $\mathcal{L}^c(q)$. Hence we can take $\mathfrak{A} = A$, where $A = \langle H_\kappa^{\mathbf{V}[G]}, C^G \rangle$ is as above. Hence $\pi_q^A = \pi_q^G$ and if $\sigma \supset \pi_q^G \cup F^q$ is s.t. if $\sigma : \overline{N}^* \prec N^*$ then $A \models \mathcal{L}(s)$ whenever $\overline{s} \in \overline{G}$ and $s = \sigma(\overline{s})$. If $s \notin G$, there would be $p \in G$

incompatible with $s$. But $A \models \mathcal{L}(p) \cup \mathcal{L}(s)$. Contradiction!

QED(Lemma 5.4)

We say that $\mathcal{L}$ is *modest* if all of its axioms can be forced by $\mathbb{P}_{\mathcal{L}}$-more precisely:

**Definition** Let $\mathcal{L}$ satisfy the core axioms. $\mathcal{L}$ is *modest* iff whenever $G$ is $\mathbb{P}_{\mathcal{L}}$-generic there is a regular $\kappa > 2^{\beta}$ s.t. $A = \langle H_{\kappa}^{\mathbf{V}[G]}, C^G \rangle$ satisfies $\mathcal{L}$.

Lemma 5.2 says that $\mathcal{L}^c$ is modest. Assuming modesty, we have a simple criterion for deciding whether a given condition lies in a generic set $G$:

**Lemma 5.5** *Let $\mathbb{P} = \mathbb{P}_{\mathcal{L}}$ where $\mathcal{L}$ is modest. Let $G$ be $\mathbb{P}$-generic. Let $p \in \mathbb{P}$. Then $p \in G$ iff the following hold:*

- $p_0 \lhd \langle M, C^G \rangle$,
- $\pi_{p_0}^G : \langle M_p, \overline{a} \rangle \prec \langle M, a \rangle$ *whenever* $\langle a, \overline{a} \rangle \in F^p$.

*Proof.* ($\rightarrow$) is trivial. We prove ($\leftarrow$). Let $\kappa$ be regular s.t. $\kappa > 2^{\beta}$ and $A = \langle H_{\kappa}, C^G \rangle$ satisfies $\mathcal{L}$. Then $A \models \mathcal{L}(p)$. Iff $p \notin G$ there would be a $q \in \mathbb{P}$ s.t. $p$, $q$ are incompatible. But $A \models \mathcal{L}(p) \cup \mathcal{L}(q)$. QED(Lemma 5.5)

**Note** In [LF], §4 we have shown that the assumption of modesty can be omitted from Lemma 5.5 assuming that $\mathbb{P}$ adds no reals. This is because $\mathbb{P} = \mathbb{P}_{\mathcal{L}^*}$, where $\mathcal{L}^*$ is the set of $\mathcal{L}$ statements forced to hold in $A = \langle H_{\kappa}^{\mathbf{V}[G]}, C^G \rangle$, where $\kappa > 2^{\beta}$ is regular. We shall not use that here, however, since our languages will always be modest. (We are unlikely to adopt an axiom without the expectation that it will be forced.)

Finally, we note that there is an apparently weaker notion of revisability relative to a parameter:

**Definition** $\mathbb{P}$ is *weakly revisable* iff there exist a cardinal $\Omega > 2^{\beta}$ and an $s \in H_{\Omega}$ s.t. whenever $N^* = \langle H_{\Omega}, M, <, \mathbb{P}, s, \ldots \rangle$ and $p$ conforms to $N^*$, then, letting $\overline{N}^* = \overline{N}^*(p, N^*) = \langle \overline{H}, \overline{M}, <, \overline{\mathbb{P}}, \overline{s}, \ldots \rangle$, we have: Let $\overline{G}$ be $\overline{\mathbb{P}}$-generic over $\overline{N}^*$. Then $q = \langle \langle \overline{M}, C^{\overline{G}} \rangle, F^p \rangle \in \mathbb{P}$.

It turns out that this is equivalent to full revisability. This fact is useful (and may be used tacitly) in verifying revisability.

**Lemma 5.6** *Let $\mathbb{P}$ be weakly revisable. Then it is fully revisable.*

*Proof.* Let $\Omega$ be the smallest cardinal verifying weak revisability. Let $\Omega' > \overline{\overline{H}}_{\Omega}$ be a cardinal. Let $N'^* = \langle H_{\Omega'}, M, <', \mathbb{P}, \ldots \rangle$. Let $p$ conform to $N'^*$ and let $\overline{N}'^* = \overline{N}^*(p, N'^*) = \langle \overline{H}', \overline{M}', <, \overline{\mathbb{P}}', \ldots \rangle$. Let $\overline{G}$ be $\overline{\mathbb{P}}'$-generic over $\overline{N}'^*$.

**Claim** $q = \langle\langle \overline{M}', C^{\overline{G}}\rangle, F^p\rangle \in \mathbb{P}$.

Note that $\Omega$, $s$ are $N'^*$-definable, where $s =$ the $<'$-least $s$ s.t. $\langle\Omega, s\rangle$ verifies weak revisability. Let $\mathfrak{A}$ be a solid model of $\mathcal{L}(p)$ and let $\sigma' \supset \pi_p^{\mathfrak{A}} \cup F^p$ s.t. $\sigma' : \overline{N}'^* \prec N'^*$. Let

(1) $\sigma'(\overline{\Omega}, \overline{s}) = \Omega, s$.

Set: $N^* = \langle H_\Omega, M, <, \mathbb{P}, s, \ldots\rangle$ where $<=<' \cap H_\Omega^2$. Then $p$ conforms to $N^*$. Set: $\overline{N}^* = \overline{N}^*(p, N^*) = \langle \overline{H}, \overline{M}, <, \overline{\mathbb{P}}, \overline{s}, \ldots\rangle$. Let $\sigma \supset \pi_p^{\mathfrak{A}} \cup F^p$ s.t. $\sigma : \overline{N}^* \prec N^*$. Then each $x \in \text{rng}(\sigma)$ is $N^*$-definable in parameters from $\text{rng}(\pi_p^{\mathfrak{A}} \cup F^p)$. Hence it is $N'^*$-definable in these parameters. Hence:

(2) $\text{rng}(\sigma) \subset \text{rng}(\sigma')$.

But:

(3) $\overline{M} = M_p = \overline{M}'$; $\sigma \upharpoonright \overline{M} = \pi_p^{\mathfrak{A}} = \sigma' \upharpoonright \overline{M}$.

Moreover, each $a \in \mathfrak{P}(\overline{M}) \cap \overline{N}^*$ is $\langle \overline{M}, b\rangle$-definable from parameters from $\overline{M}$, where $b \in D^p$. Similarly for $\mathfrak{P}(\overline{M}) \cap \overline{N}^*$. Hence:

(4) $\mathfrak{P}(\overline{M}) \cap \overline{N}^* = \mathfrak{P}(\overline{M}) \cap \overline{N}'^*$.

Since $\sigma \upharpoonright D^p = F^p = \sigma' \upharpoonright D^p$ and $\sigma \upharpoonright \overline{M} = \sigma' \upharpoonright \overline{M}'$, we conclude

(5) $\sigma \upharpoonright \mathfrak{P}(\overline{M}) = \sigma' \upharpoonright \mathfrak{P}(\overline{M})$.

$\overline{\mathbb{P}} = \langle |\overline{\mathbb{P}}|, \leq_{\overline{\mathbb{P}}}\rangle$ is canonically codable as a subset of $\mathfrak{P}(\overline{M})$. Similarly for $\overline{\mathbb{P}}'$. But $\sigma(\overline{\mathbb{P}}) = \sigma'(\overline{\mathbb{P}}') = \mathbb{P}$. It follows easily that.

(6) $\overline{\mathbb{P}} = \overline{\mathbb{P}}'$ and $\sigma \upharpoonright \mathbb{P} = \sigma' \upharpoonright \overline{\mathbb{P}}'$.

But if $\Delta \in \mathfrak{P}(\overline{\mathbb{P}}) \cap \overline{N}^*$, then $\Delta = (\sigma^{-1}) \cdot \sigma(\Delta) \in \overline{N}'^*$. Hence:

(7) $\mathfrak{P}(\overline{\mathbb{P}}) \cap \overline{N}^* \subset \overline{N}'^*$.

Hence $\overline{G}$ is generic over $\overline{N}^*$ and we conclude:

(8) $q = \langle\langle \overline{M}, C^{\overline{G}}\rangle, F^p\rangle \in \mathbb{P}$. QED(Lemma 5.6)

In conclusion we say a few words about the difference between the present approach and that taken in [LF]. There too we approximated $M = L_\tau^A$ s.t. $L_\tau^A = H_\beta$ for some $\beta > \omega_1$. Our intention, however, was simply to make $M$ the limit of a tower of countable models. In place of an approximation system $\Gamma, \Pi$ we worked with a collection $T$ of *tower segments* $\langle\langle M_i \mid i \leq \alpha\rangle, \langle \pi_{ij} \mid i \leq j \leq \alpha\rangle\rangle$ satisfying:

- $M_i = L_{\beta_i}^{A_i}$, $i \leq \omega_1^{M_i}$, $M_h \in H_{\omega_1}^{M_i}$ for $h < i$.

- $\pi_{ij} : M_i \prec M_j$ $(i \leq j)$ with $\pi_{ii} = \text{id}$.
- $\pi_{ij}\pi_{hi} = \pi_{hj}$.
- If $\lambda \leq \alpha$ is a limit ordinal, then $M_\lambda = \bigcup_{i<\lambda} \text{rng}(\pi_{i\lambda})$.

We sometimes imposed further requirements on $T$, but $T$ was always primitive recursive. For $t \in T$ we set:

$$t = \langle\langle M_i^t \mid i \leq \alpha_t\rangle, \langle \pi_{ij}^t \mid i \leq j \leq \alpha_t\rangle\rangle.$$

Call $t$ a segment of $s$ iff $\alpha_t \leq \alpha_s$ and

$$M_i^t = M_i^s, \quad \pi_{ij}^t = \pi_{ij}^s \quad \text{for } i \leq j \leq \alpha_t.$$

Our language contained a single constant $\overset{\circ}{t}$ in addition to $\underline{x}$ $(x \in N)$ and the core axioms:

$$\text{ZFC}^-, \quad H_{\omega_1} = \underline{H}_{\omega_1}, \quad \bigwedge v\Big(v \in \underline{x} \leftrightarrow \bigvee_{z \in x} w = \underline{z}\Big), \quad \overset{\circ}{t} \in T,$$

$$\alpha_{\overset{\circ}{t}} = \underline{\omega}_1, \quad M_{\underline{\omega}_1}^t = \underline{M}, \quad \bigwedge i < \underline{\omega}_1 \, M_i^t \in H_{\omega_1}.$$

We now show how to convert this approach into our present one. For each $t \in T$ set:

$$e_t = \langle M_{\alpha_t}, \{\langle y, x, i\rangle \mid i < \alpha_t \wedge \pi_{i\alpha_t}(x) = y\}\rangle.$$

Set $\Gamma = \{e_t \mid t \in T\}$. Note that $t$ is uniquely recoverable from $e_t$. We set:

$$e_t \lhd e_s \;\text{ iff }\; e_t \text{ is a segment of } e_1,$$

$$\pi : e_t \lhd e_s \;\text{ iff }\; (e_t \lhd e_s \text{ and } \pi = \pi_{\alpha_t,\alpha_s}^s).$$

Then $\Gamma, \Pi$ is an approximation system and the above core axioms translate into our usual core axioms.

## 6. Examples

We now display some specific examples of $\mathcal{L}$-forcing. All of them are revisable and will turn out to be subcomplete as well.

### 6.1. *Example 1*

Assume CH and $2^{\omega_1} = \omega_2$. Without adding reals we wish to make $\omega_2$ become $\omega$-cofinal. We first define our approximation system:

**Definition** $\Gamma =$ the set of $\langle M, C\rangle$ s.t.

- $M = L_\tau^A$ models $\text{ZFC}^-$ and "$\omega_1$ is the largest cardinal".
- $C$ is a cofinal subset of $On_M$ of order type $\omega$.

**Definition** For $u \in \Gamma$ set $u = \langle M_u, C_u \rangle$.

**Definition** $\Pi =$ the set of $\langle \pi, u, v \rangle$ s.t. $u, v \in \Gamma$, $\pi : M_u \prec M_v$, $\pi'' C_u = C_v$.

We again write $\pi : u \lhd v$ for $\langle \pi, u, v \rangle \in \Pi$ and $u \lhd v$ for $\bigvee \pi \ \pi : u \lhd v$.

**Definition** $\alpha_u = \omega_1^{M_u}$ for $u \in \Gamma$.

We note that:

(1) Let $v \in \Gamma$. Let $\alpha \leq \alpha_v$. There is at most one $u \in \Gamma$ s.t. $u \lhd v$ and $\alpha = \alpha_u$. Thus $\lhd$ is a tree.

Now let $M = L^A_{\omega_2}$, where $L_{\omega_2}[A] = H_{\omega_2}$. Set: $N = \langle H_{\omega_3}, M, <, \ldots \rangle$ where $<$ well orders $H_{\omega_3}$.

Let $\mathcal{L}$ be the language on $N$ constaining exactly the core axioms (wrt. $\Gamma$, $\Pi$).

**Lemma 1** *$\mathcal{L}$ is consistent.*

*Proof*. Let $\theta > 2^\beta$ be a regular cardinal. Let $H = H_\theta$ and $\sigma : \overline{H} \prec H$, where $\overline{H}$ is countable and transitive. Let $\sigma(\overline{M}, \overline{N}) = M, N$, $\sigma(\tilde{\mathcal{L}}) = \mathcal{L}$. Set $\tilde{M} = \bigcup\limits_{u \in \overline{M}} \sigma(u)$. Then $\sigma \upharpoonright \overline{M} : \overline{M} \prec \tilde{M}$ cofinally. Let $\langle \tilde{H}, \tilde{\sigma} \rangle$ be the liftup of $\langle \overline{M}, \sigma \upharpoonright \overline{M} \rangle$. Then $\tilde{\sigma} : \overline{H} \prec \tilde{H}$ $\omega_2^{\overline{M}}$-cofinally. Let $\tilde{k} : \tilde{H} \prec H$ s.t. $k\tilde{\sigma} = \sigma$, $k \upharpoonright \omega_2^{\tilde{H}} = \mathrm{id}$. Let $k(\tilde{\mathcal{L}}) = \mathcal{L}$. Then $\tilde{\mathcal{L}}$ is a language on $\tilde{N}$ and it suffices to show:

**Claim** $\tilde{\mathcal{L}}$ is consistent.

*Proof*. Let $\overline{C} \subset \overline{M}$ be cofinal in $On_{\overline{M}}$ with order type $\omega$. Set $\tilde{C} = \sigma'' \overline{C}$. Then $\langle H^{\tilde{H}}_{\omega_2}, \tilde{C} \rangle$ models $\tilde{\mathcal{L}}$. QED(Lemma 1)

Now let $\mathbb{P} = \mathbb{P}_{\mathcal{L}}$. We show that $\mathbb{P}$ satisfies a particularly strong form of revisability.

**Lemma 2** *Let $p \in \mathbb{P}$. Let $C$ be cofinal in $On_{M_p}$ with order type $\omega$. Then $q = \langle \langle M_p, C \rangle, F^p \rangle \in \mathbb{P}$.*

*Proof*. Let $\mathfrak{A}$ be a solid model of $\mathcal{L}(p)$. We shall "resection" $\mathfrak{A}$ to get a solid model $\mathfrak{A}'$ of $\mathcal{L}(q)$. Let $\mathfrak{A} = \langle |\mathfrak{A}|, C^{\mathfrak{A}} \rangle$. Set $\mathfrak{A}' = \langle |\mathfrak{A}|, C' \rangle$ where $C' = \pi_p^{\mathfrak{A}}{}'' C$. Since $C'$ is defined in $\mathfrak{A}$, we have $\mathfrak{A}' \models (\mathrm{ZFC}^- \wedge \underline{H}_{\omega_1} = H_{\underline{\omega}_1})$. Since $H^{\mathfrak{A}}_{\omega_1} = H_{\omega_1}$ it follows easily that whenever $X \subset M$ is countable in $\mathfrak{A}$, then there is $u \lhd \langle M, C' \rangle$ s.t. $u \in H_{\omega_1}$ and $X \subset \mathrm{rng}(\pi_u^{\mathfrak{A}})$. Hence all core axioms hold.

QED(Lemma 2)

An immediate corollary is:

**Corollary 2.1** *$\mathbb{P}$ is revisable.*

Thus, if $G$ is $\mathbb{P}$-generic and $\kappa > 2^{\beta}$ is regular, $\langle H_{\kappa}^{G}, C^{c}\rangle$ satisfies all core axioms. But these are exactly the axioms of $\mathcal{L}$. Hence $\mathcal{L}$ is modest.

Making use of Lemma 2 we now prove:

**Lemma 3** *$\mathbb{P}$ is subcomplete.*

*Proof.* Let $\theta > 2^{2^{\omega_2}}$. Let $W = L_{\tau}^{A}$ be a ZFC$^{-}$ model s.t. $H_{\theta} \subset W$ and $\theta < \tau$. Let $\pi : \overline{W} \prec W$, where $\overline{W}$ is countable and full. Let $\pi(\overline{\theta}, \overline{\mathbb{P}}, \overline{s}) = \theta, \mathbb{P}, s$. Since $\omega_2 \leq \delta(\mathbb{P})$ it suffices to show:
**Claim** Let $\overline{G}$ be $\overline{\mathbb{P}}$-generic over $\overline{W}$. There is $q \in \mathbb{P}$ s.t. whenever $G \ni q$ is $\mathbb{P}$-generic, then there is $\sigma \in \mathbf{V}[G]$ s.t.

(a) $\sigma : \overline{W} \prec W$
(b) $\sigma(\overline{\mathbb{P}}, \overline{\theta}, \overline{s}) = \mathbb{P}, \theta, s$
(c) $C_{\omega_2}^{W}(\text{rng}(\sigma)) = C_{\omega_2}^{W}(\text{rng}(\pi))$
(d) $\sigma''\overline{G} \subset G$.

Now let $C = C_{\omega_2}^{W}(\text{rng}(\pi))$, $k : \tilde{W} \overset{\sim}{\leftrightarrow} C$, where $\tilde{W}$ is transitive. Set $\tilde{\pi} = k^{-1}\cdot\pi$. Then $\tilde{\pi} : \overline{W} \prec \tilde{W}$ is $\omega_3^{\overline{W}}$-cofinal. If $\sigma$ satisfies (a)–(d) and we set: $\tilde{\sigma} = k^{-1}\sigma$, then $\tilde{\sigma} : \overline{W} \prec \tilde{W}$ is also $\omega_3^{\overline{W}}$-cofinal. But since $\tilde{\sigma}$ takes $\omega_2^{\overline{W}}$ cofinally to $\omega_2 = \omega_2^{\tilde{W}}$, it follows that $\tilde{\sigma}$ is $\omega_2$-cofinal.

The following lemma hints at the possibility of such a $\tilde{\sigma}$: Let $\tilde{\pi}(\overline{\theta}, \overline{\mathbb{P}}, \overline{s}) = \tilde{\theta}, \tilde{\mathbb{P}}, \tilde{s}$.

**Sublemma 3.1** *Let $\delta = \delta_{\tilde{W}} =$ the least $\delta$ s.t. $L_{\delta}(\tilde{W})$ is admissible. Then the following language $\tilde{\mathcal{L}}$ on $L_{\delta}(\tilde{W})$ is consistent:*
***Predicate:*** $\in$
***Constants:*** $\mathring{\sigma}$, $\underline{x}$ $(x \in L_{\delta}(\tilde{W}))$
***Axioms:*** *ZFC$^{-}$*, $\bigwedge v(v \in \underline{x} \leftrightarrow \bigvee\!\!\!\bigvee_{z\in x} v = \underline{z})$, $\mathring{\sigma} : \underline{\overline{W}} \prec \underline{\tilde{W}}$ *$\omega_2^{\tilde{W}}$-cofinally,*

$$\sigma(\underline{\overline{\mathbb{P}}}, \underline{\overline{\theta}}, \underline{\overline{s}}) = \tilde{\mathbb{P}}, \tilde{\theta}, \tilde{s}.$$

*Proof.* Let $\langle \hat{W}, \hat{\pi}\rangle$ be the liftup of $\langle \overline{W}, \pi \upharpoonright \overline{H}\rangle$, where $\overline{H} = (H_{\omega_2})^{\overline{W}}$. (Hence $\pi \upharpoonright \overline{H} = \tilde{\pi} \upharpoonright \overline{H}$.) Let $\hat{k} : \hat{W} \prec \tilde{N}$ s.t. $\hat{k}\hat{\pi} = \tilde{\pi}$, $\hat{k} \upharpoonright \omega_2^{\hat{W}} = \text{id}$. Then $\hat{k}$ is cofinal in $\tilde{W}$. Let $\hat{\delta} = \delta_{\hat{W}}$ be least s.t. $L_{\hat{\delta}}(\hat{W})$ is admissible. Let $\hat{\mathcal{L}}$ be defined on $L_{\hat{\delta}}(\hat{W})$ as $\tilde{\mathcal{L}}$ was defined on $L_{\delta}(\tilde{W})$ with $\hat{W}$, $\hat{\mathbb{P}}$, $\hat{\theta}$, $\hat{s}$ in place of $\tilde{W}$, $\tilde{\mathbb{P}}$, $\tilde{\theta}$, $\tilde{s}$, where $\hat{\pi}(\overline{\mathbb{P}}, \overline{\theta}, \overline{s}) = \hat{\mathbb{P}}, \hat{\theta}, \hat{s}$. It suffices to show:

**Claim** $\hat{\mathcal{L}}$ is consistent.
This is trivial, however, since $\langle \hat{W}, \hat{\pi}\rangle$ models $\hat{\mathcal{L}}$. QED(Sublemma 3.1)

Now let $\Omega > 2^\beta$ be a cardinal. Set: $N^* = \langle H_\Omega, M, W, \mathbb{P}, \theta, s, \pi, \ldots \rangle$. Let $p$ conform to $N^*$ and set:

$$\overline{N}^* = \overline{N}^*(p, N^*) = \langle H', M', W', \mathbb{P}', \theta', s', \pi', \ldots \rangle.$$

Let $\tilde{W}'$, $\tilde{\pi}'$, $\tilde{\mathcal{L}}'$ be defined in $\overline{N}^*$ as $\tilde{W}, \tilde{\pi}, \tilde{\mathcal{L}}$ were defined in $N^*$. Since $\overline{N}^*$ is countable, there is a solid model $\mathfrak{A}$ of $\tilde{\mathcal{L}}$. Set $\tilde{\sigma} = \mathring{\sigma}^{\mathfrak{A}}$. Then

$$\tilde{\sigma} : \overline{W} \prec \tilde{W}' \quad \omega_2^{\overline{W}}\text{-cofinally}.$$

Hence $\tilde{W}' = C^{\tilde{W}'}_{\omega_2^{W'}}(\mathrm{rng}(\tilde{\sigma}))$. Set: $\overline{C} = C^{\overline{G}}$, $C' = \pi' {}''\overline{C}$. Then $C'$ is cofinal in $\omega_2^{W'}$ and has order type $\omega$. Set: $q = \langle \langle M', C' \rangle, F^p \rangle$. Then $q \in \mathbb{P}$ by the strong revisability lemma. Let $G \ni q$ be $\mathbb{P}$-generic. Let $\pi^* \supset \pi_q^G \cup F^q$ s.t. $\pi^* : \overline{N}^* \prec N^*$. Let $\pi^*(k') = k$. Set: $\sigma' = k'\tilde{\sigma}$, $\sigma = \pi^*\sigma'$. Then $\sigma \in \mathbf{V}[G]$.

**Claim** $\sigma$ satisfies (a)–(d).

*Proof*. (a), (b) are trivial. We prove (c). Set $\omega_2' = \omega_2^{W'}$.

(1) $C^{W'}_{\omega_2'}(\mathrm{rng}(\sigma')) = C^{W'}_{\omega_2'}(\mathrm{rng}(\pi'))$,

since $k' {}''\tilde{W}' = C^{W'}_{\omega_2'}(\mathrm{rng}(\pi'))$ by definition and $k' {}''\tilde{W}' = C^{W'}_{\omega_2'}(\mathrm{rng}(\sigma'))$, since $\tilde{W}' = C^{\tilde{W}'}_{\omega_2'}(\mathrm{rng}(\tilde{\sigma}))$, $k' {}''\mathrm{rng}(\tilde{\sigma}) = \mathrm{rng}(\sigma')$, and $k' \upharpoonright \omega_2' = \mathrm{id}$.

(2) $C^W_{\omega_2}(\mathrm{rng}(\sigma)) \subset C^W_{\omega_2}(\mathrm{rng}(\pi))$,

since $\mathrm{rng}(\sigma) = \pi^* {}''\mathrm{rng}(\sigma') \subset \pi^* {}''C^{W'}_{\omega_2'}(\mathrm{rng}(\pi')) \subset \pi^*(C^{W'}_{\omega_2'}(\mathrm{rng}(\pi')) = C^W_{\omega_2}(\mathrm{rng}(\pi))$.

(3) $C^W_{\omega_2}(\mathrm{rng}(\pi)) \subset C^W_{\omega_2}(\mathrm{rng}(\sigma))$,

since $\mathrm{rng}(\pi) = \pi^* {}''\mathrm{rng}(\pi') \subset \pi^* {}''C^{W'}_{\omega_2}(\mathrm{rng}(\sigma')) \subset C^W_{\omega_2}(\mathrm{rng}(\sigma))$, since $\pi^* {}''\mathrm{rng}(\sigma') = \mathrm{rng}(\sigma)$ and $\pi^* {}''\omega_2' \subset \omega_2$. QED(c)

We now prove (d). Since $\mathcal{L}$ is modest we have:

**Sublemma 3.2** *Let $C = C^G$. Then $G = G^C =$ the set of $p \in \mathbb{P}$ s.t. $p_0 \lhd \langle M, C \rangle$ and $\pi : \langle M_p, \overline{a} \rangle \prec \langle M, a \rangle$ whenever $\langle a, \overline{a} \rangle \in F^p$, when $\pi = \pi_{p_0, \langle M, a \rangle}$.*

Now let $\overline{r} \in \overline{G}$, $r = \sigma_0(\overline{r})$. Then $r_0 = \overline{r}_0 \lhd \langle \overline{M}, \overline{C} \rangle$ and $\pi_{\overline{r}_0, \langle \overline{M}, \overline{C} \rangle} = \pi^{\overline{G}}_{\overline{r}}$, where $\overline{C} = C^{\overline{G}}$. Obviously,

$$\sigma' \upharpoonright \overline{M} : \langle \overline{M}, \overline{C} \rangle \lhd \langle M', C' \rangle$$

and

$$\pi_q^G : \langle M', C' \rangle \lhd \langle M, C \rangle,$$

where $\pi_q^G = \pi^* \restriction M'$. Hence

$$\sigma \restriction \overline{M} : \langle \overline{M}, \overline{C} \rangle \lhd \langle M, C \rangle.$$

Let $r = \sigma(\overline{r})$. Then $r_0 = \overline{r}_0$ and $F^r = \{\langle \pi^*(a), \overline{a} \rangle \mid \langle a, \overline{a} \rangle \in F^{\overline{r}}\}$. Clearly $r_0 \lhd \langle M, C \rangle$ and: $\pi_{r_0, \langle M, C \rangle} = \sigma \circ \pi_{\overline{r}}^G$. Now let $\langle a, \overline{a} \rangle = \langle \pi^*(a'), \overline{a} \rangle \in F^r$. Then $\pi_r^{\overline{G}} : \langle M_r, \overline{a} \rangle \prec \langle M', a' \rangle$ and $\sigma(\langle M', a' \rangle) = \langle M, a \rangle$. QED(Lemma 3)

**Note** We could in this case have omitted the predicate $C$ and simply taken $\Gamma$ as the set of $\langle M, \emptyset \rangle$ s.t. $M = L_\tau^A$ models ZFC$^-$ and "$\omega_1$ is the largest cardinal". $\Pi$ would then be defined as the set of $\langle \pi, u, v \rangle$ s.t. $u, v \in \Gamma$, $\pi'' M_u \prec M_v$ cofinally. If we call $\mathbb{P}'$ the resulting set of conditions, then it is the "same" as $\mathbb{P}$ in the sense that $\mathrm{BA}(\mathbb{P}) \simeq \mathrm{BA}(\mathbb{P}')$.

**Note** $\mathbb{P}$ is, in fact, equivalent to Namba forcing in the sense that $\mathrm{BA}(\mathbb{P}) \simeq \mathrm{BA}(\mathbb{N})$. This is surprising, since $\mathbb{P}$ not only looks different and has a different motivation, but the combinatorics involved in the proofs are quite different.

### 6.2. *Example 2*

Now let $\beta > \omega_2$ be a cardinal and assume: $2^\omega = \omega_1$, $2^{\omega_1} = \omega_2$, $2^{\beta} = \beta$. We shall develop a forcing very much like the previous forcing which, however, gives cofinality $\omega$ not only to $\omega_2$ but to every regular $\tau \in [\omega_2, \beta]$. There will be some variation in the definition of the forcing, depending on whether $\mathrm{cf}(\beta) = \omega_1$. Thus, in this example, we assume $\mathrm{cf}(\beta) = \omega_1$. In Example 3 we shall then detail the changes which must be made if $\mathrm{cf}(\beta) \neq \omega_1$. Let $M = L_\beta^A$ where $H_{\omega_2} = L_{\omega_2}[A]$ and $H_\beta = L_\beta[A]$. $M$ is then *smooth* in the sense defined in Section 3.2.

⋆ ⋆ ⋆ ⋆ ⋆

**Definition** Relabel the classes $\Gamma$, $\lhd$ defined in Example 1 as $\Gamma_0$, $\lhd_0$. Set: $\Gamma$ = the collection of $\langle M, C \rangle$ s.t.

- $M = L_\beta^A$ is smooth.
- $\gamma = \omega_2^M$ exists and $L_\gamma[A] = H_{\omega_2}$ in $M$.
- $C \subset \gamma$, $\sup C = \gamma$, $\mathrm{otp}\, C = \omega$.

For $u = \langle M_u, C_u \rangle \in \Gamma$ set:

$$\alpha_u = \alpha_{M_u} = \omega_1^{M_u}, \quad \gamma_u = \gamma_{M_u} = \omega_2^{M_u}, \quad M_u = L_{\beta_u}^{A_u},$$
$$M_u^0 = L_{\gamma_u}^{A_u}, \quad u^0 = \langle M_u^0, C_u \rangle.$$

Hence: $u^0 \in \Gamma_0$.

**Definition** Let $u, v \in \Gamma$. $\pi : u \lhd v$ iff the following hold:

- $\pi^0 : u^0 \lhd_0 v^0$ where $\pi^0 = \pi \upharpoonright M_u^0$.
- $\pi : M_u \prec M_v$.
- Let $\pi : M_u \to_{\Sigma_0} M_{uv}$ cofinally. Then $\langle M_{uv}, \pi \rangle$ is the liftup of $\langle M_u, \pi^0 \rangle$. (In other words $\pi : M_u \to_{\Sigma_0} M_{uv}$ $\gamma_u$-cofinally.)

$\langle \Gamma, \Pi \rangle$ is easily seen to be an approximation system.

⋆⋆⋆⋆⋆

We return to $M = L_\beta^A$ as stated at the outset. Let $N = \langle H_{\beta+}, M, <, \ldots \rangle$ where $<$ well orders $H_{\beta+}$. Let $\mathcal{L}$ be the language on $N$ containing only the core axioms (wrt. $\Gamma$, $\Pi$).

**Lemma 4** *$\mathcal{L}$ is consistent.*

*Proof*. Let $\theta > 2^\beta$ be a regular cardinal. Let $\pi : \overline{H} \prec H_\theta$ s.t. $\overline{H}$ is countable and transitive and $\pi(\overline{M}, \overline{N}, \overline{\mathcal{L}}) = M, N, \mathcal{L}$. Let $\hat{H} = H_{\omega_2}^{\overline{M}}$ and set

$$\langle \tilde{H}, \tilde{\pi} \rangle = \text{ the liftup of } \langle \overline{H}, \pi \upharpoonright \hat{H} \rangle.$$

Let $k : \tilde{H} \prec H_\theta$ s.t. $k\tilde{\sigma} = \sigma$. $k \upharpoonright \omega_2^{\tilde{H}} = \mathrm{id}$. Set $\tilde{M}, \tilde{N}, \tilde{\mathcal{L}} = \tilde{\pi}(\overline{M}, \overline{N}, \overline{\mathcal{L}})$. Then $k(\tilde{\mathcal{L}}) = \mathcal{L}$ and it suffices to show:

**Claim** $\tilde{\mathcal{L}}$ is consistent.

Let $\overline{C} \subset \omega_0^{\overline{M}}$ cofinally s.t. $\mathrm{otp}(\overline{C}) = \omega$. Set $\tilde{C} = \pi''\overline{C} = \tilde{\pi}''\overline{C}$. We prove:

**Claim** $\langle H_{\omega_2}, \tilde{C} \rangle$ models $\tilde{\mathcal{L}}$.

*Proof*. All axioms are trivial except for the last one. We show that if $X \subset M$ is countable, then there is $u \in \Gamma \cap H_{\omega_1}$ s.t. $u \lhd \langle M, \tilde{C} \rangle$ and $X \subset \mathrm{rng}(\pi_{u,\langle M, \tilde{C} \rangle})$. We construct such a $u$: Let $Z \prec \tilde{H}$ be countable s.t. $X \cup \mathrm{rng}(\tilde{\pi}) \subset Z$. Let $\pi' : H' \overset{\sim}{\leftrightarrow} Z$. Set: $M' = \pi'^{-1}(\tilde{M})$, $C' = \pi'^{-1}{}''\tilde{C}$, $\pi'' = \pi' \upharpoonright M'$. Then $X \subset \mathrm{rng}(\pi')$ and it suffices to show:

**Claim** $\pi'' : \langle M', C' \rangle \lhd \langle \tilde{M}, \tilde{C} \rangle$.

$\pi' \upharpoonright M^0 : \langle M^0, C' \rangle \lhd^0 \langle \tilde{M}^0, \tilde{C} \rangle$ is obvious. We therefore need only to show:

**Claim** Let $\pi'' : M' \to_{\Sigma_0} M^*$ cofinally. Then the map $\pi''$ is $\omega_2 M'$-cofinal into $M^*$.

*Proof*. First note that $\pi' : H' \prec \tilde{H}$ $\omega_2^{H'}$-cofinally, since if $x \in \tilde{H}$, then $x \in \tilde{\pi}(u)$, where $\overline{\overline{u}} < \omega_2$ in $\overline{H}$. Set $u' = \pi'^{-1}\tilde{\pi}(u)$. Then $x \in \pi'(u')$, $\overline{\overline{u}}' < \omega_2$ in $H'$. Now let $x \in M^*$. By cofinality there is $v \in M'$ s.t. $x \in \pi''(v)$. Let $u \in \tilde{H}$ s.t. $x \in \pi'(u)$ and $\overline{\overline{u}} < \omega_2$ in $H'$. Set: $w = u \cap v$. Then $x \in \pi''(w)$ where $\overline{\overline{w}} < \omega$ in $M'$, since $M' = H_{\beta'}$ in $H'$, where $\beta' = (\pi')^{-1}(\tilde{\beta})$.QED(Lemma 4)

We then define $\mathbb{P} = \mathbb{P}_{\mathcal{L}}$ as before. Exactly as before we get:

**Lemma 5** *Let $p \in \mathbb{P}$. Let $C \subset \gamma_p$ be cofinal in $\gamma_p$ with order type $\omega$, where $\gamma_p =_{\mathrm{Df}} \gamma_{p_0} = \omega_2^{M_p}$. Then $q = \langle\langle M_p, C\rangle, F^p\rangle \in \mathbb{P}$.*

Hence:

**Corollary 5.1** *$\mathbb{P}$ is revisable.*

Hence $\langle H_\kappa[G], C^G\rangle$ models the core axioms whenever $G$ is $\mathbb{P}$-generic and $\kappa > 2^\beta$. But $\mathcal{L}$ has only the core axioms and is, therefore, modest. Using this we obtain:

**Lemma 6** *$\mathbb{P}$ is subcomplete.*

The proof is virtually identical to that of Lemma 3. However, in the verification of (d) at the end of the proof we need additional justification for:

$$\sigma' \upharpoonright \overline{M} : \langle \overline{M}, \overline{C}\rangle \lhd \langle M', C'\rangle.$$

Letting $\sigma' \upharpoonright \overline{M}$ map $\overline{M}$ cofinally to $M^*$, we must show:

$$\sigma' \upharpoonright \overline{M} : \overline{M} \longrightarrow_{\Sigma_0} M^* \quad \omega_2^{\overline{M}}\text{-cofinally.}$$

This follows from

$$\sigma' : \overline{N} \prec \tilde{N}' \quad \omega_2^{\overline{N}}\text{-cofinally}$$

by the argument used in Lemma 4 to get:

$$\pi'' : M' \longrightarrow_{\Sigma_0} M^* \quad \omega_2^{M'}\text{-cofinally}$$

from: $\pi' : H' \prec \tilde{H}' \quad \omega_2^{H'}$-cofinally. QED(Lemma 6)

$\mathbb{P}$ obviously collapses $\beta$ to $\omega_1$. We now show that its successor is not collapsed:

**Lemma 7** *Let $G$ be $\mathbb{P}$-generic. Then $\beta^+$ is regular in $\mathbf{V}[G]$.*

This is immediate from:

**Sublemma 7.1** *$\mathbb{B} = \mathrm{BA}(\mathbb{P})$ has a dense subset of size $\beta$.*

*Proof.* We defined a collection $S$ of statements in the forcing language s.t. $\overline{\overline{S}} \leq \beta$ (in $\mathbf{V}$), and for each $p \in \mathbb{P}$ there is a $\psi \in S$ s.t. $0 \neq [[\psi]] \subset [p]$. ($[p]$ being the smallest $a \in \mathbb{B}$ s.t. $p \in a$.) Let $\mathring{C}$ be the canonical term s.t. $\mathring{C}^G = C^G$ for $\mathbb{P}$-generic $G$. For each triple $\langle u, \overline{a}, a\rangle$ s.t.

$$u = \langle M_u, C_u\rangle \in \Gamma \cap H_{\omega_1}, \quad \overline{a} : \omega \to \mathfrak{P}(M_u), \quad a : \omega \to M,$$

let $\psi_{u\bar{a}a}$ be the statement:

$$\check{u} \lhd \langle \check{M}, \overset{\circ}{C} \rangle \wedge \bigwedge i < \omega \bigwedge z (z \in \check{\bar{a}}(i) \longleftrightarrow \overset{\circ}{\pi}(z) \in \check{a}(i))$$

where $\overset{\circ}{\pi} = \pi_{\check{u}\langle \check{M}, \overset{\circ}{C} \rangle}$. All such triples are elements of $M$, so the set $S$ of such statements has at most cardinality $\beta$. We now show that for each $p \in \mathbb{P}$ there is $\psi \in S$ with $0 \neq [[\psi]] \subset [p]$. It suffices to prove this for a dense subset of $\mathbb{P}$, so assume w.l.o.g. that $p$ conforms to $N^* = \langle H_{(2^\beta)^+}, M, < \rangle$. Let $G \ni p$ be $\mathbb{P}$-generic. Let $\tilde{\beta} = \sup \pi_p^G{}''\beta_p$. Then $\tilde{\beta} < \beta$. Set $\tilde{M} = L^A_{\tilde{\beta}}$, where $M = L^A_\beta$. For each $a \in R^p$ set $\tilde{a} = a \cap \tilde{M}$. Let $\langle \langle a_i, \overline{a}_i \rangle \mid i < \omega \rangle$ enumerate $F^p$ in $\mathbf{V}$. Set $\overline{a} = \langle \overline{a}_i \mid i < \omega \rangle$, $\tilde{a} = \langle \tilde{a}_i \mid i < \omega \rangle$. Let $\psi = \psi_{p_0, \overline{a}, \tilde{a}}$. Then $[[\psi]] \neq 0$, since $\psi$ is true in $\mathbf{V}[G]$. We claim that $[[\psi]] \subset [p]$, or equivalently:

**Claim** Let $G$ be $\mathbb{P}$-generic. Then $G \cap [[\psi]] \neq \emptyset \rightarrow p \in G$.

Then $p_0 \lhd \langle M, C \rangle$, since $\psi$ is true in $\mathbf{V}[G]$. Let $\langle a, \overline{a} \rangle \in F^p$. We must show:

**Claim** $\pi : \langle M_p, \overline{a} \rangle \prec \langle M, a \rangle$, where $\pi = \pi_{p_0, \langle M, C \rangle}$.

Set: $b = \{ \langle z_1, \ldots, z_n \rangle \mid \langle M, a \rangle \models \mathcal{X}(z_1, \ldots, z_n) \}$. Then $b \in R^p$, since $p$ conforms to $N^*$. Moreover $\langle b, \overline{b} \rangle \in F^p$ where $\overline{b}$ has the same definition over $\langle M_p, \overline{a} \rangle$. Hence:

$$\langle M_p, \overline{a} \rangle \models \mathcal{X}(z_1, \ldots, z_n) \longleftrightarrow \langle z_1, \ldots, z_n \rangle \in \overline{b} \longleftrightarrow \pi(\langle z_1, \ldots, z_n \rangle) \in \tilde{b} = \tilde{M} \cap b$$
$$\longrightarrow \pi(\langle z_1, \ldots, z_n \rangle) \in b \longrightarrow \langle M, a \rangle \models \mathcal{X}(\pi(z_1), \ldots, \pi(z_n)).$$

Since this holds for all $\mathcal{X}$ we have: $\pi : \langle M_p, \overline{a} \rangle \prec \langle M, a \rangle$. QED(Lemma 7)

### 6.3. *Example 3*

We now assume $2^\omega = \omega_1$, $2^{\omega_1} = \omega_2$, $2^{\beta} = \beta$, and $\mathrm{cf}(\beta) \neq \omega_1$. We again want to give cofinality $\omega$ to all regular cardinals $\tau \in [\omega_2, \beta]$. It is clear that $\beta$ will also acquire cofinality $\omega$, since it either already has cofinality $\omega$, or its cofinality lies in $[\omega_2, \beta)$. The simplest way of handling this is to revise the definition of $\lhd$ to:

**Definition** Let $u, v \in \Gamma$. $\pi : u \lhd v$ iff the following hold:

- $\pi^0 : u^0 \lhd_0 v^0$ where $\pi^0 = \pi \upharpoonright M^0_u$.
- $\pi : M_u \prec M_v$ $\gamma_u$-cofinally.

Let $M = L^A_\beta$ where $L_\beta[A] = H_\beta$. As before set $N = \langle H_{\beta^+}, <, M, \ldots \rangle$. Let $\mathcal{L}$ be the language on $N$ with only the core axioms. Exactly as before we prove:

**Lemma 8** *$\mathcal{L}$ is consistent.*

(Note that if $\overline{N}$ is countable and transitive, $\pi : \overline{N} \prec N$, $\pi(\overline{M}) = M$, and $\overline{M}^0 = H^{\overline{M}}_{\omega_2}$, then if $\langle \tilde{N}, \tilde{\pi} \rangle$ is the liftup of $\langle \overline{N}, \pi \upharpoonright \overline{M}^0 \rangle$, then $\pi \upharpoonright \overline{M} : \overline{M} \prec \tilde{M}$ $\omega_2^{\overline{M}}$-cofinally, where $\tilde{M} = \tilde{\pi}(\overline{M})$.)

We then set $\mathbb{P} = \mathbb{P}_{\mathcal{L}}$. Exactly as before we get:

**Lemma 9** *$\mathbb{P}$ is strongly revisable.*

**Corollary 9.1** *$\mathbb{P}$ is revisable.*

Hence $\mathcal{L}$ is modest, since it has only the core axioms. Exactly as before we get:

**Lemma 10** *$\mathbb{P}$ is subcomplete.*

Lemma 7 does *not* go through, however. In fact $2^\beta$ acquires cardinality $\omega_1$. This follows from the very general theorem:

**Lemma 11** *Let $W$ be an inner model of ZFC and CH. Let $H_{\omega_1} = H^W_{\omega_1}$. Let $\beta > \omega_1$ s.t. $2^{\underset{\sim}{\beta}} = \beta$ in $W$. Suppose that* $\mathrm{cf}(\beta) = \omega$ *and $\overline{\overline{\beta}} = \omega_1$ in* $\mathbf{V}$. *Then* $\mathrm{card}((2^\beta)^W) = \omega_1$ *in* $\mathbf{V}$.

*Proof*. Let $M = L^A_\beta$ where $L_\beta[A] = H_\beta$ in $W$. Let $f$ map $\omega_1$ onto $M$ in $\mathbf{V}$. Let $\langle \beta_i \mid i < \omega \rangle \in \mathbf{V}$ be cofinal in $\beta$. Set: $X_\alpha = f''\alpha$ for $\alpha < \omega_1$. Set:

$$C = \{\alpha < \omega_1 \mid \alpha = \omega_1 \cap X_\alpha \wedge X_\alpha \prec M \wedge \{\beta_i \mid i < \omega\} \subset X_\alpha\}.$$

For $\alpha \in C$ set $\pi_\alpha : M_\alpha \overset{\sim}{\leftrightarrow} X_\alpha$, where $M_\alpha$ is transitive. Then $M_\alpha \in H_{\omega_1}$. For any $B \subset \beta$, $B \in W$, there is $\alpha \in C$, s.t. $B \cap \beta_i \in X_\alpha$ for $i < \omega$. Set:

$$\overline{B} = \bigcup \{\pi_\alpha^{-1}(B \cap \beta_i) \mid i < \omega\}.$$

Then $\langle \alpha, \overline{B} \rangle \in H_{\omega_1}$ and $B$ is recoverable from $\langle \alpha, \overline{B} \rangle$ by:

$$\tilde{\pi}(\alpha, \overline{B}) = \bigcup_{u \in M_\alpha} \pi_\alpha(u \cap \overline{B}).$$

Thus $\tilde{\pi}$ maps a subset of $H_{\omega_1}$ onto $(\mathfrak{P}(\beta))^W$. QED(Lemma 11)

### 6.4. *The extended Namba problem*

Shelah was the first to show that Namba forcing can be iterated without adding reals. If we iterate it out to a strongly inaccessible $\kappa$, then $\kappa$ becomes the new $\omega_2$ and arbitrarily large regular cardinals below $\kappa$ become $\omega$-cofinal. However, many regular cardinals become $\omega_1$-cofinal. The "extended Namba problem" asks whether, without adding reals, one can make $\kappa$ become $\omega_2$ while giving *all* of the regular cardinals in the interval $(\omega_1, \kappa)$ cofinality $\omega$. This problem seemed so difficult that at one point we conjectured a provably negative answer in ZFC for all $\kappa$. Moti Gitik then disproved this conjecture by constructing a ZFC model in which the extended Namba problem had a positive solution for some $\kappa$. His model was a generic extension of a universe containing a supercompact cardinal. Following Gitik's breakthrough we then obtained a positive solution in ZFC for all $\kappa$. It is impossible to give the full proof of that result in these notes, but we shall endeaver to give some account of the methods used. We may assume w.l.o.g. that GCH holds below $\kappa$, since we may achieve this by a prior forcing in which all collapsed regular cardinals acquire a cofinality $\geq \omega_2$. If we then give the surviving regular cardinals in $(\omega_1, \kappa)$ the cofinality $\omega$, the collapsed ones will also become $\omega$-cofinal.

It is natural to try to solve this problem by an iteration $\langle \mathbb{B}_i \mid i \leq \kappa \rangle$. We ask now what the initial steps of this iteration should look like. We follow the convention that $\mathbb{B}_0 = 2$. Thus $\mathbb{B}_1$ is the first stage which "does something". We certainly expect it to give $\omega_2$ the cofinality $\omega$ without adding reals. By Lemma 11 it follows that $\omega_3$ will be collapsed, so $\omega_3$ must acquire cofinality $\omega$. But then $\omega_4$ is collapsed etc. Thus every $\omega_n$ must be collapsed with cofinality $\omega$. By Lemma 11, it then follows that $\omega_{\omega+1}$ is collapsed etc. This chain of implications does not break down until we reach $\omega_{\omega_1}$. There, however, it does break down, since we can use the $\mathbb{P}$ of Example 2 with $\beta = \omega_{\omega_1}$. All regular cardinals in $(\omega_1, \omega_{\omega_1})$ acquire cofinality $\omega$ and $\omega_{\omega_1+1}$ is not collapsed, thus becoming the new $\omega_2$. We take $\mathbb{B}_1 \simeq \mathrm{BA}(\mathbb{P})$. We can then repeat the process, getting $\mathbb{B}_2 \supseteq \mathbb{B}_1$ which collapses $\omega_{\omega_1 \cdot 2}$ to $\omega_1$ etc. This gives us the first $\omega$ stages $\langle \mathbb{B}_i \mid i < \omega \rangle$. Our job now is to find an appropriate limit $\mathbb{B}_\omega$. Since each $\mathbb{B}_i$ is subcomplete, the inverse limit $\mathbb{B}^*$ is also subcomplete. However, a bit of reflection shows that $\mathbb{B}^*$ is too small to do the job: At the limit stage $\omega_{\omega_1 \cdot \omega}$ will be collapsed to $\omega_1$. Hence by Lemma 11 $\omega_{(\omega_1 \cdot \omega)+1}$ will be collapsed and hence must acquire cofinality $\omega$ etc.

Proceeding in this fashion we see that $\omega_{\omega_1 \cdot (\omega+1)}$ must be collapsed. Thus our limit algebra must be large, not containing any dense set of size less than $\omega_{\omega_1(\omega+1)}$. At the same time it should have a dense subset of size $\omega_{\omega_1(\omega+1)}$ in order that the successor is preserved. It turns out that a limit with the requisite properties can be obtained by a construction rather like that of Example 2. We shall now sketch that construction, but a full verification of its properties is beyond the purview of these notes.

Let $M^0 = L^A_\gamma$ where $\gamma = \omega_{\omega_1\omega}$, $L_\gamma[A] = H_\gamma$, and $A$ canonically codes $\langle \mathbb{B}_i \mid i < \omega \rangle$. We define $\Gamma_0$, $\Pi_0$ as follows:

$\Gamma_0 =$ the set of $u = \langle M_u, B_u \rangle$ s.t.

- $M = L^{A_u}_{\gamma_u}$ where $M_u$ models Zermelo set theory and $A_u$ canonically codes a sequence $\langle \mathbb{B}^u_i \mid i < \omega \rangle$ of complete Boolean algebras in the sense of $M$ with $\mathbb{B}^u_i \subseteq \mathbb{B}^u_j$ $(i \leq j < \omega)$.
- $B_u \subset \bigcup\limits_i \mathbb{B}^u_i$ s.t. $B_u \cap \mathbb{B}_i$ is $\mathbb{B}_i$-generic over $M$ for $i < \omega$.
- $\sup\{\delta(\mathbb{B}_i) \mid i < \omega\} = \beta$ and $\mathbb{B}_i$ collapses $\delta(\mathbb{B}_i)$ to $\omega_1^M$ for $i < \omega$.

$\Pi_0 =$ the set of $\langle \pi, u, v \rangle$ s.t. $u, v \in \Gamma_0$, $\pi : M_u \prec M_v$ and $\pi'' B_u \subset B_v$.

We write $\pi : u \lhd_0 v$ for $\langle \pi, u, v \rangle \in \Pi_0$. Setting $B^i_u = B_u \cap \mathbb{B}^u_i$, we see that $\pi$ has a unique extension $\pi^i$ s.t. $\pi^i : M_u[B^i_u] \prec M_v[B^i_v]$ and $\pi^i(B^i_u) = B^i_v$. Set: $M^*_u = \bigcup\limits_i M_u[B^i_u]$ and $\pi^* = \bigcup\limits_i \pi^i$. Then $\pi^* : M^*_u \to M^*_v$ cofinally.

Letting $f^i_u$ be the canonical map of $\omega_1$ onto $L^{A_u}_{\delta(\mathbb{B}^i_u)}$, we see that $\pi$ is uniquely characterized by: $\pi \circ f^i_u = f^i_v$ for $i < \omega$. It follows easily that $\pi = \pi_{uv}$ is the unique $\pi : u \lhd_0 v$ and that $\Gamma^0$, $\Pi^0$ is an approximation system. Now let $M = L^A_\beta$ where $\beta = \omega_{\omega_1(\omega+1)}$, $L_\beta[A] = H_\beta$ and $M^0 = L^A_\gamma$ $(\gamma = \omega_{\omega_1 \cdot \omega})$. Set:

$\Gamma =$ the set of $u = \langle M_u, B_u \rangle$ s.t. $M_u$ is smooth and there is $\gamma = \gamma_u \in M_u$ s.t. $u^0 = \langle M^0_u, B_u \rangle \in \Gamma_0$, where $M_u = L^{A_u}_\beta$ and $M^0_u = L^{A_u}_\gamma$.

We then set: $\Pi =$ the set of $\langle \pi, u, v \rangle$ s.t. $u, v \in \Gamma$ and:

- $\pi^0 : u^0 \lhd v^0$ where $\pi^0 = \pi \upharpoonright M^0$.
- $\pi : M_u \prec M_v$.
- Let $\pi : M_u \to M_{u,v}$ cofinally. Then $\langle M_{u,v}, \pi \rangle$ is the liftup of $\langle M_u, \pi^0 \rangle$.

We again set:

$$\pi : u \lhd v \quad \text{iff} \quad \langle \pi, u, v \rangle \in \Pi.$$

Thus $\langle \Gamma, \Pi \rangle$ is an approximation system which is related to $\langle \Gamma^0, \Pi^0 \rangle$ exactly as in Section 6.2.

Again, letting $M = L^A_\beta$ be as above, and $N = \langle H_{\beta^+}, <, M, \ldots \rangle$, we form the language $\mathcal{L}$ on $N$ containing only the core axioms.

**Lemma 12** *$\mathcal{L}$ is consistent.*

*Proof*. Let $\mathbb{B}^* =$ the inverse limit of $\langle \mathbb{B}_i \mid i < \omega \rangle$. Then $\mathbb{B}^*$ is subcomplete. Let $B^*$ be $\mathbb{B}^*$-generic. We prove the consistency of $\mathcal{L}$ in $\mathbf{V}[B^*]$. Let $B_i =$

$B^* \cap \mathbb{B}_i$, $B = \bigcup_{i<\omega} B_i$. Let $H = H_{(2^\beta)^+}$ in $\mathbf{V}$. Let $\pi : \overline{H} \prec H$ in $\mathbf{V}[B^*]$ s.t. $\overline{H}$ is countable and transitive. Let:

$$\pi(\overline{N}, \overline{M}, \overline{M}^0, \langle \mathbb{B}_i \mid i < \omega \rangle) = N, M, M^0, \langle \mathbb{B}_i \mid i < \omega \rangle.$$

Set $\overline{B}_i = \pi^{-1} {}'' B_i$ for $i < \omega$. Since we are working in $\mathbf{V}[B^*]$ we may assume that $\overline{B}_i$ is $\mathbb{B}_i$-generic over $\overline{M}$ for $i < \omega$. Clearly $\pi$ takes $\overline{M}^0$ to $M^0$ cofinally. Moreover:

$$\pi \restriction \overline{M}^0 : \langle \overline{M}^0, \overline{B} \rangle \lhd_0 \langle M^0, B \rangle.$$

Now let $\langle \tilde{H}, \tilde{\pi} \rangle$ be the liftup of $\langle \overline{H}, \pi \restriction \overline{M}^0 \rangle$. Let: $\tilde{\pi}(\overline{M}, \overline{N}, \overline{\mathcal{L}}) = \tilde{M}, \tilde{N}, \tilde{\mathcal{L}}$, where $\pi(\overline{\mathcal{L}}) = \mathcal{L}$. Since there is $k : \tilde{H} \prec H$ with $k(\tilde{\mathcal{L}}) = \mathcal{L}$, it suffices to prove that $\tilde{\mathcal{L}}$ is consistent. We claim:

**Claim** $\langle H_\kappa, B \rangle$ models $\tilde{\mathcal{L}}$, where $\kappa > 2^\beta$ is regular in $\mathbf{V}$.

*Proof*. The only problematical case is: Let $X \subset \tilde{M}$ be countable. There is $u \in \Gamma \cap H_{\omega_1}$ s.t. $u \lhd \langle \tilde{M}, B \rangle$ and $X \subset \text{rng}(\pi_{u, \langle \tilde{M}, B \rangle})$. Let $Y \prec \tilde{H}$ be countable s.t. $\text{rng}(\tilde{\kappa}) \cup X \subset Y$ and whenever $\Delta \in Y$ is dense in $\mathbb{B}_i$ $(i < \omega)$, then $\Delta \cap B \neq \emptyset$. Let:

$$\pi' : H' \tilde{\leftrightarrow} Y, \quad \pi'(M^{0'}, M', \langle \mathbb{B}'_i \mid i < \omega \rangle) = M^0, \tilde{M}, \langle \mathbb{B}_i \mid i < \omega \rangle.$$

Set: $B' = \pi'^{-1} {}'' B_i$, $\pi'' = \pi' \restriction M'$.

**Claim** $\pi'' : \langle M', B' \rangle \lhd \langle \tilde{M}, B \rangle$.

Clearly: $\pi'' \restriction M^{0\prime} : \langle M^{0\prime}, B' \rangle \lhd_0 \langle M^0, B \rangle$. Since $\pi'' : M' \prec \tilde{M}$, it suffices to show that: If $\pi'' : M' \to M^*$ cofinally, then $\langle M^*, \pi'' \rangle$ is the liftup of $\langle M', \pi'' \restriction M^{0\prime} \rangle$ – i.e. that $\pi''$ takes $M'$ $\gamma'$-cofinally to $M^*$ wheres $\gamma' = (\omega_1 \cdot \omega)^{M'}$. This follows by the usual argument. QED(Lemma 12)

The strong revisability lemma reads:

**Lemma 13** *For sufficiently large $\theta > 2^\beta$ we have: Let $N^* = \langle H_\theta, M, \mathbb{P}, <, \ldots \rangle$. Let $p$ conform to $N^*$ and set: $\overline{N}^* = \overline{N}^*(N^*, p) = \langle \overline{H}, \overline{M}, \overline{\mathbb{P}}, <, \ldots \rangle$. Let $\overline{B} \subset \bigcup_{i<\omega} \mathbb{B}_i^{\overline{M}}$ s.t. $\overline{B} \cap \mathbb{B}_i^{\overline{M}}$ is $\mathbb{B}_i^{\overline{M}}$-generic over $\overline{M}$ for $i < \omega$. Then $q = \langle \langle \overline{M}, \overline{B} \rangle, F^p \rangle \in \mathbb{P}$.*

We must forego the proof of Lemma 13, since it is very long and involves properties of the algebras $\mathbb{B}_i$ which we have not developed here.

An immediate corollary is:

**Corollary 13.1** $\mathbb{P}$ *is revisable, since revisability says that the above holds when $\overline{B} = B^{\overline{G}}$ for a $\overline{G}$ which is $\overline{\mathbb{P}}$-generic over $\overline{N}^*$.*

Since $\mathcal{L}$ has only the core axioms, it is then modest. But then we get:

**Lemma 14** *$\mathbb{P}$ is subcomplete.*

We sketch briefly the proof of Lemma 14, which is largely the same as before. Let $\theta$ be big enough to verify the subcompleteness of $\mathbb{B}_i$ for $i < \omega$. Let $W = L_\tau^A$ be a ZFC$^-$ model with $H_\theta \subset W$ and $\theta < \tau$. Let $\pi : \overline{W} \prec W$ where $\overline{W}$ is countable and full. Let $\pi(\overline{\theta}, \overline{\mathbb{P}}, \overline{s}) = \theta, \mathbb{P}, s$.

**Claim** There is $q \in \mathbb{P}$ s.t. if $G \ni q$ is $\mathbb{P}$-generic, there is $\sigma \in \mathbf{V}[G]$ with:

(a) $\sigma : \overline{W} \prec W$
(b) $\sigma(\overline{\theta}, \overline{\mathbb{P}}, \overline{s}) = \theta, \mathbb{P}, s$
(c) $C_\gamma^W(\text{rng}(\sigma)) = C_\gamma^W(\text{rng}(\pi))$, where $\gamma = On \cap M^0 = \sup\limits_{i<\omega} \delta(\mathbb{B}_i)$.
(d) $\sigma''\overline{G} \subset G$.

(**Note** $\gamma \leq \delta(\mathbb{P})$, since otherwise $\gamma$ would not be collapsed.)
Let $\Omega > \theta$ be big enough to verify the strong revisability of $\mathbb{P}$. Set:

$$N^* = \langle H_\Omega, <, M, N, \mathbb{P}, W, \pi, \ldots \rangle.$$

Let $p$ conform to $N^*$. Set: $\overline{N}^* = \overline{N}^*(N^*, p) = \langle H', M', N', \mathbb{P}', W', \pi', \ldots \rangle$. Set: $\mathbb{B}_i' = \mathbb{B}_i^{M'}$ $(i < \omega)$. Set $\theta', \mathbb{P}', s' = \pi'(\overline{\theta}, \overline{\mathbb{P}}, \overline{s})$. Set $\gamma' = \pi'(\overline{\gamma})$, where $\pi(\overline{\gamma}) = \gamma = On \cap M^0$. Noting that $W'$ is countable and imitating the proof of Section 4, Theorem 2 we get:

**Sublemma 14.1** *There are $\sigma'$ and $B' \subset \bigcup\limits_{i<\omega} \mathbb{B}_i'$ s.t. $B_i' = B' \cap \mathbb{B}_i'$ is $\mathbb{B}_i'$-generic over $W'$ for $i < \omega$ and:*

*(a)* $\sigma' : \overline{W} \prec W'$
*(b)* $\sigma'(\overline{\theta}, \overline{\mathbb{P}}, \overline{s}) = \theta', \mathbb{P}', s'$
*(c)* $C_{\gamma'}^{W'}(\text{rng}(\sigma')) = C_{\gamma'}^{W'}(\text{rng}(\pi'))$
*(d)* $\sigma'{}''\overline{B} \subset B'$, *where* $\overline{B} = B^{\overline{G}}$.

To get this we successively define $\mathring{\sigma}_i, b_i \in \mathbb{B}_i'$ s.t. whenever $B_i' \ni b_i$ is $\mathbb{P}'$-generic over $W'$ and $\sigma_i' = \mathring{\sigma}_i{}^{B_i'}$, then $\sigma_i'$ satisfies (a)–(c) and: $\sigma_i'{}''\overline{B}_i \subset B_i'$ (where $\overline{B}_i = \overline{B} \cap \overline{\mathbb{B}}_i$).
We ensure $h_i(b_{i+1}) = b_i$ for $i < \omega$. We then successively choose $B_i' \ni b_i$ with: $B_i'$ is $\mathbb{B}_i'$-generic over $W'$ and $B_i' \supset B_\ell'$ for $\ell < i$. We set: $B' = \bigcup\limits_i B_i'$ and let $\sigma'$ be the 'limit' of $\sigma_i' = \mathring{\sigma}_i{}^{B_i'}$ $(i < \omega)$ exactly as in the proof of Section 4, Theorem 2.

QED(Sublemma 14.1)

By the strong revisability lemma we have: $q = \langle\langle M', B'\rangle, F^p\rangle \in \mathbb{P}$. Let $G \ni q$ be $\mathbb{P}$-generic. Then $\pi_q^G \cup F^q$ extends uniquely to: $\sigma^* : \overline{N}^* \prec N^*$. Set $\sigma = \sigma^* \cdot \sigma'$. It follows by a virtual repetition of previous proofs that $\sigma$ has the desired properties.

QED(Lemma 14)

Now let $\mathbb{B}' = \mathrm{BA}(\mathbb{P})$. We define a map $\mu : \bigcup_{i<\omega} \mathbb{B}_i \to \mathbb{B}'$ by:

$$\mu(b) = [[\check{b} \in \overset{\circ}{\mathbb{B}}]] \quad \text{where} \quad \overset{\circ}{\mathbb{B}}^G = B^G \quad \text{for all generic } G.$$

Then:

(1) $\mu$ is injective.

*Proof*. It suffices to show: $\mu(b) = 0 \to b = 0$. Let $b \neq 0$. Then $\mathcal{L} + \underline{b} \in \overset{\circ}{\mathbb{B}}$ is consistent by the proof that $\mathcal{L}$ is consistent. Hence there is $p \in \mathbb{P}$, $\overline{b} \in B_p$ s.t. $\pi^p(\overline{b}) = b$. Hence $p \models \check{b} \in \overset{\circ}{\mathbb{B}}$ – i.e. $p \in \mu(b)$. QED(1)

(2) $\mu \upharpoonright \mathbb{B}_i$ is a complete embedding.

*Proof*. $\mu\left(\bigcap_{i\in X} b_i\right) = \left[\left[\check{\bigcap_{i\in X}} b_i \in \overset{\circ}{\mathbb{B}}\right]\right]$. QED(2)

Hence we can take $\mathbb{B} \supset \bigcup_{i<\omega} \mathbb{B}_i$ s.t. for some $k$, $k : \mathbb{B}' \overset{\sim}{\leftrightarrow} \mathbb{B}$ and $k\mu = \mathrm{id}$. $\mathbb{B}$ is then a limit of $\langle \mathbb{B}_i \mid i < \omega\rangle$ which collapses $\varrho = \omega_{\omega_1(\omega_1+1)}$ to $\omega_1$ while making all regular $\tau \in (\omega_1, \varrho)$ become $\omega$-cofinal. A proof like that of Lemma 7 shows that $\varrho^+$ is not collapsed, becoming the new $\omega_2$. Hence we apply Example 2 at the next stage to collapse $\varrho^{(\omega_1)}$ = the $\omega_1$-th successor of $\varrho$ to $\omega_1$. We continue in this fashion. We define an iteration $\langle \mathbb{B}_i \mid i \leq \kappa\rangle$ and a sequence $\langle \varrho_i \mid i \leq \kappa\rangle$ as follows: $\varrho_0 = \omega_1$, $\mathbb{B}_0 = 2$.

$\varrho_{i+1} = \varrho_i^{(\omega_1)}$ and $\mathbb{B}_{i+1}$ is constructed using Example 2 so as to collapse all regular $\tau \in (\omega_1, \varrho_{i+1})$ without collapsing $\varrho_{i+1}^+$. For limit $\lambda$ we proceed as follows:

**Case 1** $\lambda$ has cofinality $\omega$ or has acquired cofinality $\omega$ at an earlier stage (i.e. $\mathrm{cf}(\lambda) < \lambda \wedge \mathrm{cf}(\lambda) \neq \omega_1$ in $\mathbf{V}$).

By essentially the above construction we form a limit $\mathbb{B}_\lambda$ which collapses $\varrho_\lambda = \left(\sup_{i<\lambda} \varrho_i\right)^{(\omega_1)}$ without collapsing $\varrho^+$.

**Case 2** Case 1 fails.

We set $\varrho_\lambda = \sup_{i<\lambda} \varrho_i$ and let $\mathbb{B}_\lambda$ be the direct limit of $\langle \mathbb{B}_i \mid i < \lambda \rangle$. If $\mathrm{cf}(\lambda) = \omega_1$ in $\mathbf{V}$, then $\varrho_\lambda^+$ becomes the new $\omega_2$. Otherwise $\lambda = \varrho_\lambda$ is inaccessible. Using the fact that we took the direct limit stationarily often below $\lambda$ it follows that $\mathbb{B}_\lambda$ satisfies the $\lambda$-chain condition. Hence $\lambda$ is the new $\omega_2$.

⋆ ⋆ ⋆ ⋆ ⋆

By induction on $i$ we verify that $\mathbb{B}_i$ is subcomplete for $i \leq \kappa$, using Section 4, Theorem 4 for Case 2 above. We stress, however, that in order to carry out the induction we must also verify many other properties of the $\mathbb{B}_i$ which have not been dealt with here. These include some strong symmetry properties.

Given that GCH holds below $\kappa$, we can modify the above construction by making selective regular $\tau \in (\omega_1, \kappa)$ $\omega_1$-cofinal. The set of such $\tau$ can be chosen arbitrarily in advance. Hence:

**Theorem** *Let $\kappa$ be inaccessible. Let GCH hold below $\kappa$. Let $A \subset \kappa$. There is a set of conditions $\mathbb{P} \subset \mathbf{V}_\kappa$ s.t. whenever $G$ is $\mathbb{P}$-generic, then in $\mathbf{V}[G]$ we have:*

- *$\kappa$ is $\omega_2$.*
- *If $\tau \in (\omega_1, \kappa)$ is regular in $\mathbf{V}$, then* $\mathrm{cf}(\tau) = \begin{cases} \omega_1 & \textit{if } \tau \in A, \\ \omega & \textit{if not.} \end{cases}$

## References

[ASS] J. Barwise. *Admissible Sets and Structures*, Perspectives in Math. Logic Vol. 7, Springer Verlag, 1976.

[NA] H. Friedman, R. Jensen. *A Note on Admissible Ordinals*, in: The Syntax and Semantics of Infinitary Languages. Springer Lecture Notes in Math. Vol. 72, 1968.

[PR] R. Jensen, C. Karp. *Primitive Recursive Set Functions*, in Axiomatic Set Theory, AMS Proceedings of Symposia in Pure Math. Vol. XIII, Part 1, 1971.

[PF] S. Shelah. *Proper and Improper Forcing*, Perspectives in Math. Logic, Springer Verlag, 1998.

[AS] R. Jensen. *Admissible Sets**.

[LF] R. Jensen. *$\mathcal{L}$-Forcing**

[SPSC] R. Jensen. *Subproper and Subcomplete Forcing**

[ENP] R. Jensen. *The Extended Namba Problem**

[ITSC] R. Jensen. *Iteration Theorems for Subcomplete and Related Forcings**

[DSP] R. Jensen. *Dee-Subproper Forcing**

[FA] R. Jensen. *Forcing Axioms Compatible with CH**

* These handwritten notes can be downloaded from
`http://www.mathematik.hu-berlin/de/~raesch/org/jensen.html`
(or enter 'Ronald B. Jensen' in Google).

# $E$-RECURSION 2012*

Gerald E. Sacks
*Department of Mathematics, Harvard University,*
*Cambridge, MA 02138, USA*
*sacks@math.harvard.edu*
*math.harvard.edu/~sacks*

*In Memory of S. C. Kleene*
*Premier Recursion Theorist*

$E$-recursive versus $\Sigma_1$ admissible, two competing intuitions of recursion theory. Partial $E$-recursive functions, $E$-recursive enumerability and computation trees. Divergence witnesses and reflecting ordinals. Priority arguments and forcing constructions for $E$-closed structures. Logic on $E$-closed structures.

## Contents



*This paper is a recreation of a course given by the author at the Asian Initiative for Infinity (AII) Graduate Summer School held at the National University of Singapore between 20 June 2012 and 17 July 2012. The author thanks Professors C. T. Chong, T. Slaman and H. Woodin for their kind invitation.

## 1. Introduction

### 1.1. *New intuitions for old*

The phrase heard most often in classical recursion theory is $\{e\}(n)$, where $e$ is a code number for a finite instruction derived from schemes and $n$ is a nonnegative integer. The corresponding phrase in $E$-recursion theory is $\{e\}(x)$, where $e$ is again a code number for a finite instruction, but $x$ is an arbitrary set. To put it bluntly, $x \in V$. The transition from classical recursion to $E$-recursion is more than a long jump from $HF$, the set of hereditarily finite sets, to $V$, the class of all sets. The move leaves behind some comfortable intuitive notions of computability. For example there exists a function that fails to be partial $E$-recursive despite the $E$-recursiveness of its graph.

The theorems of $E$-recursion foster the development of new intuitions. By contrast $\Sigma_1$ admissible recursion theory, in particular $\alpha$-recursion theory, greatly extends classical notions of computability without losing sight of them.

### 1.2. *Failure of the least number operator*

In classical recursion theory the proof that a function is partial recursive, if its graph is recursively enumerable, is an application of the least number operator. Suppose $f(n)$ is defined. To evaluate $f(n)$ enumerate the graph of $f$ until an ordered pair of the form $< n, y >$ is found. In $E$-recursion theory the idea of enumerating all computations still makes sense but unbounded search is no longer an effective procedure.

### 1.3. *History*

Objects of type 0 are non-negative integers. An object of type $n + 1$ is a set of objects of type $n$. Kleene [9], [10] defined $\{e\}(x)$, the $e$-th partial recursive function for $x$ an object of finite type. He defined ${}^{n+1}E$, the set of all non-empty objects of type $n$, and proved:

$$\forall x \subseteq \omega[x \text{ is recursive in } {}^2E \longleftrightarrow x \in HYP].$$

The type $n+1$ object, ${}^{n+1}E$, is equivalent to the equality predicate for objects of type $n$. In Kleene's theory equality is not recursive when $n > 0$. But he did investigate what was recursive in ${}^2E$ and ${}^3E$.

$E$-recursion extended his ideas from objects of finite type to arbitrary sets with one major change: equality is $E$-recursive. The initial ground breaking results were due to Gandy, Moschovakis [12] and Normann. The schemes defining $E$-recursion were introduced by Normann, but discovered independently by Moschovakis who drew attention to the concept of divergence witness, whose great importance in the study of computation higher up was not anticipated in classical recursion theory or in $\Sigma_1$ admissible recursion theory.

Further results were obtained by Fenstad, Griffor, Grilliot, Harrington, Kechris, MacQueen, Moldstad, Shoenfield and Slaman. Back in the 70's there was a golden age of $E$-recursion with centers in Cambridge Mass, Oslo and UCLA.

#### 1.3.1. *Sources*

The most complete source for $E$-recursion is Sacks [21]. The most polished sources are Moschovakis [12] and Slaman [22], [23].

### 1.4. *Intuitions behind E-recursion*

Let $x$ and $y$ be arbitrary sets.

The predicate, $x = y$, is $E$-recursive. Hence infinitely long computations are necessary.

Let $\{e\}(x)$ be the $e$-th partial $E$-recursive function; $e$ is a code number, a non-negative integer that encodes one of finitely many $E$-recursion schemes; the result (if there is one) of applying that scheme to $x$ is the value of $\{e\}(x)$.

The function

$$f(e, x) = \{e\}(x)$$

is partial $E$-recursive and its existence is required by the *only* non-trivial $E$-recursion scheme.

## 1.5. *The Normann schemes*

Let $\overrightarrow{x}$ denote $x_1, \ldots, x_n$. $< 1, n, i >$ is a positive integer that effectively encodes $1, n, i$.

There are seven Normann schemes. The first four are needed to show fundamental set theoretic notions are $E$-recursive.

(1) projection $\{e\}(\overrightarrow{x}) = x_i$ if $e =< 1, n, i >$.
(2) difference $\{e\}(\overrightarrow{x}) = x_i - x_j$ if $e =< 2, n, i, j >$.
(3) pairing $\{e\}(\overrightarrow{x}) = \{x_i, x_j\}$ if $e =< 3, n, i, j >$.
(4) union $\{e\}(\overrightarrow{x}) = \cup\{y \mid y \in x_1\}$ if $e =< 4, n, i, j >$.

Scheme (5) is the source of infinitely long computations. If $x_1$ is infinite, then the computation of $\{e\}(\overrightarrow{x})$ leads immediately to infinitely many sub-computations.

(5) $E$-recursive bounding
$\{e\}(\overrightarrow{x}) = \{\{c\}(y, x_2, \ldots, x_n) \mid y \in x_1\}$ if $e =< 5, n, c >$.
(6) composition
$\{e\}(\overrightarrow{x}) = \{c\}(\{d_1\}(\overrightarrow{x}), \ldots, \{d_m)(\overrightarrow{x}))$ if $e =< 6, n, m, c, d_1, \ldots, d_m >$.
(7) enumeration
$\{e\}(c, \overrightarrow{x}, \overrightarrow{y}) = \{c\}(\overrightarrow{x})$ if $e =< 7, n, m >$.

The enumeration scheme, it will be seen, does all the work.

## 1.6. *Classical recursion theory*

A set $y$ is **transitive** iff $(\forall x \in y)(x \subseteq y)$. Let $tc(z)$ denote the transitive closure of $z$, the least transitive $w \supseteq z$. Intuitively, $tc(z)$ is $z \cup (\cup z) \cup (\cup(\cup z)) \ldots$; $tc$ is an $E$-recursive function by Proposition 9.

$HF$ is the set of hereditarily finite sets.

$$x \in HF \longleftrightarrow tc(\{x\}) \text{ is finite.}$$

**Proposition 1.** *Let $f$ be a function whose domain and range are subsets of $\omega$. Then (i) $\longleftrightarrow$ (ii) $\longleftrightarrow$ (iii).*

*(i) $f$ is partial recursive.*
*(ii) $f$ is partial $E$-recursive.*
*(iii) $f$ is $\Sigma_1^{ZF}$ definable over $< HF, \in >$.*

The two major extensions of classical recursion theory, $\Sigma_1$ admissibility and $E$-recursion, differ in spirit and on many initial segments of Gödel's $L$, but agree on $\omega$. The fact that two intuitively different formulations of recursion coincide on $\omega$ adds to the evidence for Church's Thesis.

The classical partial recursive functions can be derived from trivial finitary schemes and one non-trivial scheme, enumeration. The derivation begins with the enumeration scheme, proceeds to the fixed point theorem, then to definition by recursion, and finally to the least number operator.

## 2. $\Sigma_1$ Admissibility versus $E$-Recursion

### 2.1. $\Sigma_1^{ZF}$ *definable versus E-recursive*

A formula in the language of set theory is $\Delta_0^{ZF}$ iff all of its quantifiers are bounded; a formula is $\Sigma_1^{ZF}$ iff it is of the form $\exists x \mathcal{G}$, where $\mathcal{G}$ is $\Delta_0^{ZF}$.

Let $A$ be a transitive set. $A$ is **$\Sigma_1$ admissible** iff

$A$ is closed under the operations of pairing and union;

$A$ satisfies $\Delta_0^{ZF}$ **comprehension**, for every $\Delta_0^{ZF}$ $\mathcal{F}(w)$,

$\forall x \exists z \forall w[w \in z \longleftrightarrow w \in x \wedge \mathcal{F}(w)]$ holds in $A$;

$A$ satisfies $\Delta_0^{ZF}$ **bounding**, for every $\Delta_0^{ZF}$ $\mathcal{G}(x, y)$,

$[\forall x \in z \exists y \mathcal{G}(x, y) \longrightarrow \exists v \forall x \in z \exists y \in v \mathcal{G}(x, y)]$ holds in $A$.

The formulas $\mathcal{F}(w)$ and $\mathcal{G}(x, y)$ are **boldface;** that is, they may contain parameters from $A$. From now on all formulas of ZF are boldface.

Note that a $\Sigma_1$ admissible $A$ satisfies $\Delta_1^{ZF}$ comprehension and $\Sigma_1^{ZF}$ bounding. A set $x$ is: ***A-finite*** iff $x \in A$; ***A-recursive*** iff $x$ is a $\Delta_1^{ZF}$ definable subset of $A$; ***A-recursively enumerable*** iff $x$ is a $\Sigma_1^{ZF}$ definable subset of $A$. A function is ***partial A-recursive*** iff its graph is a $\Sigma_1^{ZF}$ definable subset of $A$.

The range of an $A$-recursive function defined on and restricted to an $A$-finite set is $A$-finite. The intersection of an $A$-recursive set and an $A$-finite set is $A$-finite, but the intersection of an $A$-re set and an $A$-finite set need not be $A$-finite. A set is ***A-recursive*** iff it and its complement relative to $A$ are $A$-re.

According to Lemma 1 every partial $E$-rec function is $\Sigma_1^{ZF}$. The converse is false by Proposition 7.

### 2.2. *Equality is E-recursive*

The **representing function** $f(x)$ of a predicate $P(x)$ is defined by:

$$\forall x[f(x) = 1 \longleftrightarrow P(x)]$$
$$\forall x[f(x) = 1 \vee f(x) = 0].$$

A predicate is ***E*-recursive** iff its representing function is $E$-recursive.

**Proposition 2.** *Equality is E-recursive.*

*Proof.* Let 0 denote $x - x$, and 1 denote $\{0\}$.

$$\begin{aligned} t(y) &= 0 \\ f(z) &= \{t(y) \mid y \in z\} \\ z = 0 &\longleftrightarrow f(z) = 0 \\ z \neq 0 &\longleftrightarrow f(z) = 1 \end{aligned}$$

$g(z) = 1 - f(z)$ represents $z = 0$.
$g((x - y) \cup (y - x))$ represents $x = y$. □

### 2.3. $\Delta_0^{ZF}$ *predicates are E-recursive*

**Proposition 3.** *If $\mathcal{G}(x, y)$ is E-recursive, then $(\exists y \in x)\mathcal{G}(x, y)$ is E-recursive.*

*Proof.* Let $t(x, y) = 0$ if $\mathcal{G}(x, y)$ and $= 1$ otherwise. Define

$$f(x) = \{t(x, y) \mid y \in x \wedge \mathcal{G}(x, y)].$$

$f$ represents $(\exists y \in x)\mathcal{G}(x, y)$. □

**Corollary 1.** *If $P(x)$ is $\Delta_0^{ZF}$, then $P(x)$ is E-recursive.*

*Proof.* By induction on the logical complexity of $P(x)$. □

It follows from Corollary 1 that a $\Sigma_1^{ZF}$ formula $Q(x)$ can be put in the form $\exists y R(x, y)$, where $R$ is $E$-recursive. It follows from Proposition 6 there exists an $E$-recursive $R(x, y)$ such that $\exists y R(x, y)$ is not $E$-re.

### 2.4. *E-recursive evaluations*

$E$ is the class of $E$-recursive evaluations. A tuple $< e, \overrightarrow{x}, y >$ is put in $E$ iff $y$ is a value for $\{e\}(\overrightarrow{x})$ determined by the Normann schemes. $E$ is the range of a $\Sigma_1$ transfinite recursion on the ordinals.

$\{e\}(\overrightarrow{x}) \downarrow y$ **(converges to** $y$**)** iff

$$< e, \overrightarrow{x}, y > \in E.$$

$f$ is a **partial** $E$**-recursive function** iff

$$\exists e \forall \overrightarrow{x} \; f(\overrightarrow{x}) = \{e\}(\overrightarrow{x}).$$

**2.5.** ***The natural enumeration of E***

($\overrightarrow{x}$ is $x_1, \ldots, x_n$)

Stage $\sigma = 0$:

$< e, < \overrightarrow{x}, x_i >>$ is put in $E_0$ if $e =< 1, n, i >$.

Schemes (2), (3) and (4) are treated similarly.

Stage $\sigma > 0$:

$E_{<\sigma} = \cup\{E_\gamma \mid \gamma < \sigma\}$.

$< e, \overrightarrow{x}, z >$ is put in $E_\sigma$ if

$e =< 5, n, c >$,

$\forall y \in x_1 \exists w[< c, < y, x_2, \ldots, x_n >, w >\in E_{<\sigma}]$,

$z = \{w \mid \exists y \in x_1[< c, < y, x_2, \ldots, x_n >, w >\in E_{<\sigma}]\}$.

Schemes (5) and (6) are treated similarly.

$E = \cup\{E_\sigma \mid \sigma \in ORD\}$.

The enumeration of $E$ lend itself to a straightforward induction that shows $\{e\}(x)$ converges to at most one $y$: $\exists_{\leq 1} y < e, < x >, y >\in E$. Thus for any $e$, $\lambda x \mid \{e\}(x)$ is a partial $E$-recursive *function.*

The enumeration of $E$ is a $\Sigma_1^{ZF}$ transfinite recursion on the ordinals, so the graph of any partial $E$-recursive function is $\Sigma_1^{ZF}$. The converse is false according to Proposition 7.

2.5.1. *Transfinite E-recursion*

**Proposition 4.** *(fixed point theorem) Let $f$ be a partial E-recursive function defined for all $n \in \omega$. Then there exists a $c \in \omega$ such that $\{f(c)\} = \{c\}$.*

*Proof.* The enumeration scheme leads to a partial $E$-recursive function such that

$$\forall e\ t(e) \downarrow \wedge \{t(e)\} = \{\{e\}(e)\}.$$

There is a $d$ such that $\{d\} = f \circ t$. Then

$$\{t(d)\} = \{\{d\}(d)\} = \{f(t(d))\}.$$

Let $c$ be $t(d)$. Then $\{f(c)\} = \{c\}$. □

**Proposition 5.** *Let $I$ be a total E-recursive function. Then there exists a partial E-recursive function $f : ORD \longrightarrow V$ such that*

$$\forall\gamma \ f(\gamma)\downarrow \ and \ f(\gamma) = I(f \restriction \gamma).$$

*[$(f \restriction \gamma)$ is the graph of $f$ restricted to $\gamma$.]*

*Proof.* There exists a recursive $g$ such that

$$\forall e \ [g(e)\downarrow \wedge \forall\gamma \ \{g(e)\}(\gamma) = I(\{e\} \restriction \gamma)].$$

By Proposition 5 $\exists c \ g(c) = \{c\}$. By induction on $\gamma$,

$$\{c\}(\gamma)\downarrow \text{ and } \{c\}(\gamma) = I(\{c\} \restriction \gamma).$$

Then $f$ is $\{c\}$. □

Gödel's $L$, the class of constructible sets is defined by a $\Sigma_1^{ZF}$ transfinite recursion, and consequently is a $\Sigma_1^{ZF}$ definable class.

$L(0) = \emptyset.$

$L(\gamma + 1) = Fod((L(\gamma)).$

($Fod(x)$ is the set of first order definable subsets of $x$.)

$L(\lambda) = \cup(L(\gamma) \mid \gamma < \lambda)$. ($\lambda$ is a limit.)

$L = \cup(L(\gamma) \mid \gamma \in ORD).$

It follows from Corollary 1 and Proposition 5 that $L(\gamma)$ is an $E$-recursive function of $\gamma$. Nonetheless according to Proposition 6, $L$ is not $E$-recursively enumerable if $V \neq L$.

### 2.6. *E-recursive enumerability*

**Definition 1.** $A$ is ***E*-recursively enumerable in *y*** ($E$-re in $y$) iff $\exists e \ A = \{x \mid \{e\}\{x, y\}\downarrow\}$.

**Definition 2.** $A$ is ***E*-recursively enumerable** ($E$-re) iff $\exists e \ A = \{x \mid \{e\}\{x\}\downarrow\}$.

According to Gandy selection (Section 4), if $A$ and its complement are $E$-re, then $A$ is $E$-recursive. The proof from classical recursion does not work in $E$-recursion because it depends on the least number operator.

**Proposition 6.** *If $V \neq L$, then $L$ is not $E$-re.*

*Proof.* Suppose $L = \{x \mid \{e\}(x)\downarrow\}$. Let $b \in V - L$. Then $\{e\}(b)\uparrow$. The predicate, $\{e\}(x)\uparrow$, is lightface $\Sigma_1^{ZF}$ according to Lemma 2. By Levy absoluteness, $\exists x \in L$ such that $\{e\}(x)\uparrow$. □

(Levy absoluteness: assume $\mathcal{F}$ is a sentence in the language of ZF whose parameters belong to $L(\omega_1^L)$; if $V \models \mathcal{F}$, then $L \models \mathcal{F}$.)

A stronger result is available.

**Theorem 1.** *If $V \neq L$, then $L$ is not $E$-re in any $c \in L$.*

The proof is a combination of forcing and iterated $\Sigma_n$ hull formation.

### 2.7. *Gödel's O is not E-recursive*

The function $O : L \longrightarrow Ord$ is defined by: $O(x) = (\text{least}\,\gamma)\ x \in L(\gamma)$.

**Proposition 7.** *The function $O$ is not the restriction of a partial $E$-recursive function to $L$.*

*Proof.* Suppose $\forall x\ x \in L \longrightarrow [\{e\}(x)\downarrow \wedge \{e\}(x) = O(x)]$. Let $\alpha \neq \omega$ be countable in $L$ and $\Sigma_1$ admissible. There exists a $b \subseteq \omega$ such that $b$ is generic over $L(\alpha)$ and $b$ is a member of $L$. Thus $L(\alpha, b)$ is $\Sigma_1$ admissible and $b \notin L(\alpha)$. But $\{e\}(b)\downarrow$, so $\{e\}(b) \in L(\alpha)$ by Proposition 10 and Remark 1, hence $b \in L(\alpha)$. □

The above proof is easily stretched to show $O$ is not the restriction to $L$ of a function partial $E$-re in $c$ for any $c \in L$.

Note that the graph of $O$ is lightface $\Sigma_1^{ZF}$ definable.

## 3. Computations

### 3.1. *Computation instructions*

A **computation instruction** is an $(n+1)$-tuple $< e, \vec{x} >$. ($\vec{x}$ is $x_1, \ldots, x_n$.)

$< e, \vec{x} >$ is the top node of the **computation tree** $\boldsymbol{T}_{<e,\vec{x}>}$. Every node of $T_{<e,\vec{x}>}$ other than the top node is an **immediate subcomputation instruction (isi)** of the node above it.

(i) If $e =< 1, n, i >$ then $< e, \vec{x} >$ has no immediate subcomputations. $< 2, n, i, j >$, $< 3, n, i, j >$ and $< 4, n, c >$ are treated similarly.

(ii) If $e =< 5, n, c >$, then $< c, y, x_2, \ldots, x_n >$ is an isi of $< e, \vec{x} >$ for every $y \in x_1$. $< 6, n, m, c, d_1, \ldots, d_m >$ and $< 6, n, m >$ are treated similarly.

(iii) If $e$ is not an index of a Normann scheme or $n$ is not the correct number of arguments, then $< e, \vec{x} >$ is an isi of $< e, \vec{x} >$.

### 3.2. *Convergent equals wellfounded*

Define $b >_i a$ by: $a$ is an isi of $b$. Let $>_U$ be the transitive closure of $>_i$: $b >_U a$ iff $\exists n$, $\exists c_1, \ldots, c_n$ such that

$$b = c_1 >_i \ldots >_i c_n = a.$$

The relation $b >_i a$ is $E$-re, but not $E$-recursive. The relation $b >_U a$ is $\Sigma_1^{ZF}$, but not $E$-re.

$T_{<e,\vec{x}>}$ is an upside down tree consisting of $< e, \vec{x} >$ and the nodes of $>_U$ below $< e, \vec{x} >$. To say $T_{<e,\vec{x}>}$ is wellfounded is to say it has no infinite descending branch.

**Lemma 1.** $\{e\}(\vec{x}) \downarrow \longleftrightarrow T_{<e,\vec{x}>}$ *is wellfounded.*

*Proof.* Suppose $T_{<e,\vec{x}>}$ is wellfounded. Then clause (iii) of the definition of isi in Subsection 3.1 implies $e$ is an index. A transfinite induction on $T_{<e,\vec{x}>}$ shows: if $< c, \vec{z} >$ is a node, then $\{c\}(\vec{z}) \downarrow$.

Now suppose $\{e\}(\vec{x}) \downarrow$. Then $< e, x, \{e\}(x) >$ was put in $E$ at stage $\sigma$ of the natural enumeration of $E$. Proceed by induction on $E$. If $\{c\}(\vec{z})$ is an immed. subcomp. instruc. of $\{e\}(\vec{x})$, then $< c, \vec{z}, \{c\}(\vec{z}) >$ was put in $E$ prior to stage $\sigma$, and so $T_{<c,\vec{z}>}$ is wellfounded. Hence $T_{<e,\vec{x}>}$ is wellfounded. □

### 3.3. *Length of computations*

**Proposition 8.** *There exists a partial E-recursive function g such that for all e and* $\vec{x}$*:*

$$\{e\}(\vec{x}) \downarrow \longleftrightarrow g(e, \vec{x}) \downarrow .$$
$$\{e\}(\vec{x}) \downarrow \longleftrightarrow g(e, \vec{x}) = T_{<e,\vec{x}>}.$$

Suppose $T_{<e,\vec{x}>}$ is wellfounded. Then each node $b$ of $T_{<e,\vec{x}>}$ has an ordinal rank $r(b)$ defined by recursion. If $b$ is terminal, then $r(b) = 0$. Otherwise $r(b)$ is the least ordinal greater than $r(c)$ for every node $c$ immediately below $b$. Let $\mid T_{<e,\vec{x}>} \mid$ be $r(\{e\}(\vec{x}))$.

**Definition 3.** $\mid \{e\}(\vec{x}) \mid = \mid T_{<e,\vec{x}>} \mid$ if $T_{<e,\vec{x}>}$ is wellfounded, and $= \infty$ otherwise.

### 3.4. *Divergence witnesses*

Suppose $\{e\}(\vec{x}) \uparrow$. Then $T_{<e,\vec{x}>}$ is not wellfounded. A **witness $w$ to the divergence of $\{e\}(\vec{x})$** is an infinite descending branch of $T_{<e,\vec{x}>}$.

$w$ is a function with domain $\omega$.
$w(0) = < e, \overrightarrow{x} >$.
$w(n+1)$ is an immed. subcomp. instruc. of $w(n)$.

**Lemma 2.** *(Moschovakis) The predicate,* $\{e\}(\overrightarrow{x})\uparrow$, *is* $\Sigma_1^{ZF}$.

**Lemma 3.** *(Moschovakis) There exists a partial E-recursive function* $g$ *such that for all* $e, \overrightarrow{x}, d, \overrightarrow{y}$*:*

$$[\{e\}(\overrightarrow{x})\downarrow \vee \{d\}(\overrightarrow{y})\downarrow] \longleftrightarrow g(e,\overrightarrow{x},d,\overrightarrow{y})\downarrow.$$

$$g(e,\overrightarrow{x},d,\overrightarrow{y})\downarrow \longrightarrow g(e,\overrightarrow{x},d,\overrightarrow{y}) = \min(|\{e\}(\overrightarrow{x})|, |\{d\}(\overrightarrow{x})|).$$

*Proof.* (A sketch.) $g(e,\overrightarrow{x},d,\overrightarrow{y})$ is defined by transfinite $E$-recursion on

$$\min(|\{e\}(\overrightarrow{x})|, |\{d\}(\overrightarrow{y})|).$$

The essence of the recursion is expressed by:

$$\min(|u|,|v|) = \max\{\min(|a|,|b|) \mid a <_U u \wedge b <_U v\}.$$

($<_U u$ was defined in Subsection 3.2.) But there is a difficulty. If $u\uparrow$, then $\{a \mid a <_U u\}$ may not be $E$-recursive in $u$. The difficulty can be overcome with the help of $E$-recursive predicates

$$|\{e\}(\overrightarrow{x})| < \gamma, \quad |\{e\}(\overrightarrow{x})| = \gamma$$

defined by transfinite $E$-recursion. □

## 4. Gandy Selection

**Theorem 2.** *(Gandy, Moschovakis) There exists a partial E-rec function* $g(e,\overrightarrow{x})$ *such that for* $\forall e \forall \overrightarrow{x}$*:*

$$(\exists n \in \omega)\{e\}(n,\overrightarrow{x})\downarrow \longleftrightarrow g(e,\overrightarrow{x})\downarrow.$$

$$g(e,\overrightarrow{x})\downarrow \longrightarrow g(e,\overrightarrow{x}) \in \omega \wedge \{e\}(g(e,\overrightarrow{x}),\overrightarrow{x}).$$

*Proof.* (as in Moldstad [11]) For notational simplicity, hide the $\overrightarrow{x}$ in $\{e\}(k,\overrightarrow{x})$. The exists a partial recursive function $f$ such that

$$f(c,e,k) = \{c\}(e,k+1)+1$$
$$\text{if } \{c\}(e,k+1)\downarrow \wedge |\{c\}(e,k+1)| \leq |\{e\}\}(k)|;$$
$$f(c,e,k) = 0$$
$$\text{if } |\{e\}\}(k)| < |\{c\}(e,k+1)|.$$

The definition of $f$ makes use of Lemma 3 with respect to the partial $E$-recursiveness of the predicate

$$\min(\mid \{c\}(e, k+1) \mid, \mid \{e\}\}(k) \mid).$$

By the fixed point theorem $\exists d \; f(d, e, k) = \{d\}(e, k)$. Let $h$ be $\{d\}$.

$$h(e, k\} = h(e, k+1) + 1$$
$$\text{if } h(e, k+1) \downarrow \wedge \mid h(e, k+1) \mid \leq \mid \{e\}\}(k) \mid;$$
$$h(e, k\} = 0$$
$$\text{if } \mid \{e\}\}(k) \mid < \mid h(e, k+1) \mid .$$

$$\forall k \; [h(e, k+1) \downarrow \longrightarrow h(e, k) \downarrow] \text{ and } \forall k \; [\{e\}(k) \downarrow \longrightarrow h(e, k) \downarrow].$$

Hence (1) $\exists k \{e\}(k) \downarrow \longrightarrow h(e, 0) \downarrow$ and
(2) $h(e, 0) \downarrow \longrightarrow \exists k \mid \{e\}\}(k) \mid < \mid h(e, k+1) \mid$.
Fix $k$ and suppose $\{e\}(k) \downarrow$; then (1), (2) imply
(3) $h(e, k) = 0$ and $\mid \{e\}\}(k) \mid < \mid h(e, k+1) \mid$.
Let $k_0$ be the least $k$ that satisfies (3); Then

$$h(e, k_0) = 0 \text{ and } \forall i < k_0 \; h(e, i) = h(e, i+1) + 1.$$

So $h(e, 0) = k_0$.
Note that $k_0$ is the least $k$ such that

$$\mid \{e\}\}(k) \mid = \min\{\mid \{e\}(n) \mid \mid n \in \omega\}.$$

Define $g(e, \overrightarrow{x}) = h(e, \overrightarrow{x}, 0) \; (= h(e, 0))$. □

### 4.1. *Existential number quantifiers*

Let $e$ be a variable restricted to $\omega$.

**Lemma 4.** *If $P(e, x)$ is a $E$-re predicate, then so is $\exists e P(e, x)$.*

*Proof.* By Gandy selection there exists a partial $E$-recursive function $g(x)$ such that

$$\exists e P(e, x) \longleftrightarrow g(x) \downarrow;$$
$$g(x) \downarrow \longrightarrow g(x) \in \omega \vee P(g(x), x).$$
□

### 4.2. *E-recursive Skolem functions*

Say $w$ **is $E$-recursive in $z$** ($w \leq_E z$) iff $\exists e[\{e\}(z) \downarrow \wedge w = \{e\}(z)]$.
The predicate $w \leq_E z$ is $E$-re.

**Lemma 5.** *Suppose the predicate $P(x,y)$ is $E$-re and*

$$(\forall x \in z)(\exists y \leq_E x)P(x,y).$$

*Then there exists a partial $E$-recursive function $f$ such that*

$$(\forall x \in z)[f(x)\downarrow \;\wedge\; P(x, f(x))].$$

*Proof.* By Gandy selection there exists a partial $E$-recursive function $g(x)$ such that

$$(\forall x \in z)[g(x)\downarrow \wedge\, g(x) \in \omega \\ \wedge\, \{g(x)\}(x)\downarrow \wedge P(x, \{g(x)\}(x))]. \qquad \square$$

## 5. $E$-Recursion in $L$

### 5.1. *E-closed sets*

**Definition 4.** Assume $b$ is a transitive set; $b$ is **$E$-closed** iff $\{e\}(\overrightarrow{x}) \in b$ whenever $\overrightarrow{x} \in b$ and $\{e\}(\overrightarrow{x})\downarrow$.

**Definition 5.** $\kappa^x = \sup\{\gamma \mid \exists e \exists \overrightarrow{a} \in tc(x)\ \gamma = \{e\}(x, \overrightarrow{a})\}$.

($tc(x)$ is the transitive closure of $x$. $\overrightarrow{a}$ is $a_1, \ldots, a_n$.)

**Definition 6.** $\boldsymbol{E(x)}$ is the least transitive $E$-closed set $b$ such that $x \in b$.

**Proposition 9.** *$tc$ is an $E$-recursive function.*

**Problem 1.** *Proof.* By transfinite $E$-recursion on rank. $\square$

**Proposition 10.** $E(x) = L(\kappa^x, tc(\{x\}))$.

*Proof.* Suppose $\delta < \kappa^x$. By induction on $\delta$, $L(\delta, tc(\{x\})) \leq_E x$, $L(\delta, tc(\{x\})) \in E(x)$, and $L(\delta, tc(\{x\})) \subseteq E(x)$. Hence $L(\kappa^x, tc(\{x\})) \subseteq E(x)$.

Suppose $\overrightarrow{a} \in E(x)$ and $\{e\}(x, \overrightarrow{a})\downarrow$. Then $|\ \{e\}(x, \overrightarrow{a})\ | \leq_E x, \overrightarrow{a}$, so $|\ \{e\}(x, \overrightarrow{a})\ | < \kappa^x$. By induction $T_{<e,x,\overrightarrow{a}>}$ is first order definable over $L(|\ \{e\}(x, \overrightarrow{a})\ |, tc(\{x\}))$. Hence $E(x) \subseteq L(\kappa^x, tc(\{x\}))$. $\square$

Let $Ad_1(x)$ be the least $\Sigma_1$ admissible set with $x$ as a member; $Ad_1(x)$ will be of the form $L(\alpha, tc(\{x\})$.

**Remark 1.** $E(x) \subseteq Ad_1(x)$.

**Remark 2.** $E(\omega) = L(\omega_1^{CK})$, hence $\Sigma_1$ admissible. ($\omega_1^{CK}$ is the least infinite ordinal not the order type of *a recursive wellordering of* $\omega$.)

**Remark 3.** $E(\omega_1)$ is not $\Sigma_1$ admissible, but $E(\omega_\omega)$ is $\Sigma_1$ admissible (Moschovakis).

### 5.2. *Reflection*

**Definition 7.** $\kappa_0^x = \sup\{\gamma \mid \gamma \leq_E x\}$.

**Definition 8.** $\delta$ is **$x$-reflecting** iff

$$L(\delta, tc(\{x\})) \models \mathcal{F} \text{ implies } L(\kappa_0^x, tc(\{x\})) \models \mathcal{F}$$
$$\text{for every } \Sigma_1^{ZF} \text{ sentence } \mathcal{F} \text{ whose only parameter is } x.$$

**Proposition 11.** *If $a \in tc(x)$ and $\delta$ is $< x, a >$-reflecting, then $\delta \leq \kappa^x$.*

*Proof.* Suppose $\delta > \kappa^x$. Then $E(< x, a >) = E(x) \in L(\delta, tc(\{x\}))$.

So $E(< x, a >) \in L(\kappa_0^{x,a}, tc(\{x\})$, an impossibility since $L(\kappa_0^{x,a}, tc(\{x\}) \subseteq E(< x, a >)$. □

**Definition 9.** $\kappa_r^x$ is the greatest $x$-reflecting ordinal.

**Corollary 2.** *If $a \in tc(x)$, then $\kappa_r^{x,a} \leq \kappa^x$.*

### 5.3. *Kechris's basis theorem*

Kechris's result is needed to establish a Harringtonesque connection between divergence witnesses and reflection in Subsection 5.5.2.

**Proposition 12.** *The predicate, $\delta$ is not $x$-reflecting, with $\delta$ as its only free variable and $x$ as its sole parameter, is E-rec in $x$.*

*Proof.* Let $\mathcal{F}$ be a $\Sigma_1^{ZF}$ sentence with sole parameter $x$ such that

$$L(\kappa_r^x + 1, tc(\{x\})) \models \mathcal{F} \text{ and } L(\kappa_r^x, tc(\{x\})) \models \neg\mathcal{F}.$$

(Note that the choice of $\mathcal{F}$ depends on $x$. Slaman has shown there is no way to choose $\mathcal{F}$ uniformly in $x$.)

Then $\delta$ is not $x$-reflecting $\longleftrightarrow L(\delta, tc(\{x\})) \models \mathcal{F}$. □

**Theorem 3.** *Assume $y \leq_E x$, $A$ is E-re in $x$, and $(y - A) \neq \emptyset$.*

$$\text{Then } \exists b\; b \in (y - A) \wedge \kappa_r^{x,b} \leq \kappa_r^x.$$

*Proof.* Let $\mathcal{F}$ be $\Sigma_1^{ZF}$ with sole parameter $x$. Then $\mathcal{F}$ reflects down from $\kappa_r^{x,b}$ to $\kappa_0^{x,b}$. Thus

$$\kappa_0^{x,b} \leq \kappa_r^x \longrightarrow \kappa_r^{x,b} \leq \kappa_r^x.$$

So it suffices to find a $b \in (y - A)$ such that $\kappa_0^{x,b} \leq \kappa_r^x$. Suppose there is no such $b$. Then

$$y \subseteq A \cup \{b \mid \kappa_r^x < \kappa_0^{x,b}\}.$$

By Proposition 12 the predicate, $\kappa_r^x < \kappa_0^{x,b}$, is $E$-re in $x$ because it is equivalent to

$$\exists\delta \;\; \delta \leq_E x, b \;\wedge\; \delta \text{ is not } x\text{-reflecting}.$$

It follows that $\forall b \in y \;\; \exists \delta_b \leq_E x, b$ such that (i) or (ii) holds.

(i) $\delta_b$ is the length of a computation that puts $b$ in $A$.
(ii) $\delta_b$ is not $x$-reflecting.

By Gandy selection $\delta_b$ is a partial $E$-rec function of $b$ (with $x$ as a parameter) convergent for all $b \in y$, hence bounded on $y$ by some $\delta^x \leq_E x$.

If some $b \in y$ satisfies (ii) then it satisfies $\delta_b > \kappa_r^x > \delta^x$. Hence (ii) is false for all $b \in y$ and so $(y - A) = \emptyset$. □

## 5.4. *From Gandy to Kechris*

**Definition 10.** For $z \subseteq \omega$: $\delta$ is recursive in $z$ iff $\delta$ is finite or $\delta$ is the order type of a recursive wellordering of $\omega$.

**Kechris**: Suppose $y \leq_E x$, $A$ is $E$-re in $x$, and $(y - A) \neq \emptyset$.

Then $\exists b \; b \in (y - A) \wedge \kappa_r^{x,b} \leq \kappa_r^x$.

**Gandy:** Suppose $b, x \subseteq \omega$, $\emptyset \neq y \subseteq 2^\omega$, and $y$ is $\Sigma_1^1$ with parameter $x$.

Then $\exists b \; b \in y \wedge \omega_1^{x,b} = \omega_1^x$.

Assume $z \subseteq \omega$: $\omega_1^z$ is the least ordinal not recursive in $z$; $\omega_1^z = \kappa_0^z = \kappa_r^z$; for all $y \subseteq 2^\omega$: $y$ is $\Sigma_1^1$ with parameter $z$ iff $(2^\omega - y)$ is $E$-re in $z$.

Thus Kechris's basis theorem can be viewed as a generalization of Gandy's.

## 5.5. *Divergence witness definability sketch*

Suppose $\{e\}(x) \uparrow$. A divergence witness for $\{e\}(x)$ is an infinitely long branch $w$ of $T_{<e.x>}$.

$w = \{< e_t, x_t >| \ t \in \omega\}$, $e_0 = e$, and $x_0 = x$. $< e_{t+1}, x_{t+1} >$ is an immediate subcomputation instruction (*isi*) of $< e_t, x_t >$.

5.5.1. *Scheme T*

It is safe to pretend in inductive arguments and recursive definitions involving computations that there is only one computation scheme, namely scheme $T$, a derived scheme that incorporates the principal features of the Normann schemes. The code number of scheme $T$ is $< 2^m \cdot 3^n >$. The isi's of $< 2^m \cdot 3^n, x >$ are:

$$< m, x >;$$
$$< n, y > \text{ iff } \{m\}(x) \downarrow \text{ and } y \in \{m\}(x).$$

5.5.2. *Definition of* $\boldsymbol{w}$ *by recursion on* $\boldsymbol{t}$

Assume $e_t =< 2^{m_t} \cdot 3^{n_t} >$ and $\{e_t\}(x_t) \uparrow$. If $\{m_t\}(x_t) \uparrow$, then $< e_{t+1}, x_{t+1} >=< m_t, x_t >$. If $\{m_t\}(x_t) \downarrow$ then $e_{t+1} = n_t$ and $x_{t+1} \in \{m_t\}(x_t)$. ($x_0 = x$.) The choice of $x_{t+1}$ is guided by Kechris's basis theorem to insure that

$$\kappa_r^{x_0,\ldots,x_{t+1}} \leq \kappa_r^x \quad (x_0 = x)$$

and that consequently $w$ will be first order definable over $L(\kappa_r^x, tc(\{x\}))$. It turns out that $x_{t+1}$ is the "least" member of $\{m_t\}(x_t)$ and $w$ is the "left-most" infinite branch of $T_{<e,x>}$.

Assume there is a wellordering of $x$ $E$-recursive in $x$.

## 5.6. *Divergence witness details*

Assume $x_t \in L(\kappa_r^x, tc(\{x\}))$, $e_t =< 2^{m_t} \cdot 3^{n_t} >$, $\{e_t\}(x_t) \uparrow$, and $\kappa_r^{x_0,\ldots,x_t} \leq \kappa_r^x$.

Then $\kappa_0^{x_t} \leq \kappa_r^x$. If $\{m_t\}(x_t) \downarrow$, then $T_{<m_t,x_t>} \in L(\kappa_0^{x_t}, tc(\{x\}))$. Hence examination of $L(\kappa_r^x, tc(\{x\}))$ reveals whether or not $\{m_t\}(x_t)$ converges; if it diverges, then $< e_{t+1}, x_{t+1} >=< m_t, x_t >$.

Assume $\{m_t\}(x_t) \downarrow$ Define $e_{t+1} = n_t$. The assumed wellordering $x$ $E$-recursive in $x$ induces a wellordering $v$ of $\{m_t\}(x_t)$ $E$-recursive in $x$. Define

$$x_{t+1} = v\text{-least } u[u \in \{m_t\}(x_t) \wedge \mid \{n_t\}(u) \geq \kappa_r^x \mid].$$

It must be seen that $\{n_t\}(x_{t+1}) \uparrow$ and $\kappa_r^{x_0,\ldots,x_t+1} \leq \kappa_r^{x_0,\ldots,x_t}$. By Kechris's basis theorem there is a $z$ such that

$$z \in \{m\}(x_t) \wedge \{n_t\}(z) \uparrow \wedge \textbf{(0)} \ \kappa_r^{x_0,\ldots,x_t,z} \leq \kappa_r^{x_0,\ldots,x_t}.$$

Let $z_0$ be the $v$-least such $z$. Only $z_0 = x_{t+1}$ remains to be shown: $\{n_t\}(z_0)\uparrow$, so $x_{t+1} \leq_v z_0$.

$$\forall z <_v z_0\ [\{n_t\}(z)\downarrow \vee \textbf{(1)}\ \kappa_r^{x_0,\ldots,x_t,z} > \kappa_r^{x_0,\ldots,x_t}.$$

As in the proof of Kechris's basis theorem, (1) is equivalent to **(2)** $\kappa_0^{x_0,\ldots,x_t,z} > \kappa_r^{x_0,\ldots,x_t}$, and (2) is equivalent to

$$\exists e\ \{e\}(x_0,\ldots,x_t,z)\downarrow \wedge \{e\}(x_0,\ldots,x_t,z) > \kappa_r^{x_0,\ldots,x_t}.$$

By Gandy selection there is a partial $E$-recursive function $f(x_0,\ldots,x_t,z)$ defined for all $z <_v z_0$ such that

$$f(x_0,\ldots,x_t,z) =\mid \{n_t\}(z) \mid \vee\ f(x_0,\ldots,x_t,z) > \kappa_r^{x_0,\ldots,x_t}.$$

Let $\gamma = \sup\{f(x_0,\ldots,x_t,z) \mid z <_v z_0\}$. Then $\gamma \leq_E x_0,\ldots,x_t,z_0$, and (0) implies $\gamma < \kappa_r^{x_0,\ldots,x_t}$. So $(\forall z <_v z_0)\{n_t\}(z)\downarrow$.

If $x_t + 1 <_v z_0$, then $\mid \{n_t\}(x_t+1) \mid < \kappa_r^{x_0,\ldots,x_t} \leq \kappa_r^x \leq \mid \{n_t\}(x_t+1) \mid$. Hence $x_{t+1} \geq_v z_0$.

## 5.7. $\Sigma_1$ *admissibility and divergence*

**Lemma 6.** *Let $x$ be a set of ordinals. Suppose $L(\kappa, tc(\{x\}))$ is $E$-closed but not $\Sigma_1$ admissible. Then $\kappa_r^{x,y} < \kappa$ for all $y \in L(\kappa, tc(\{x\}))$.*

*Proof.* Suppose $\exists y\ \kappa_r^{x,y} \geq \kappa$. Assume: $L(\kappa, tc(\{x\})) \models \forall u \in d\ \exists v \mathcal{F}(u,v,p)$; $d, p \in L(\kappa, tc(\{x\})$; and $\mathcal{F}$ is $\Delta_0^{ZF}$. There is a wellordering of $tc(\{x,y\})$ in $L(\kappa, tc(\{x\})$, hence by reflection a wellordering $E$-recursive in $x, y$. The discussion of divergence witness construction in Subsection 5.6 can be extended to show $\kappa_r^{x,y,p,b} \geq \kappa_r^{x,y}$ for all $b \in d$.

By reflection $\forall b \in d$, $\exists c\ [c \in L(\kappa_0^{x,y,p,b}, tc(\{x,y,p,b\})) \wedge \mathcal{F}(b,c,p)]$. Hence $\forall b \in d$, $\exists c \leq_E x,y,p,b\ \ \mathcal{F}(b,c,p)]$.

Gandy selection implies $c$ is a partial $E$-recursive function of $b$, with parameters $x, y, p$, and defined for all $b \in d$. Let $r$ be the range of $c$ restricted to $d$. Then $r \in L(\kappa, tc(\{x\}))$ and $L(\kappa, tc(\{x\})) \models \forall u \in d\ (\exists v \in r)\mathcal{F}(u,v,p)$. □

## 5.8. *The divergence-admissibility split*

Assume $L(\kappa)$ is $E$-closed.

**Definition 11.** $L(\kappa)$ **admits divergence witnesses** iff $\forall e \forall x \in L(\kappa)$; if $\{c\}(x)\uparrow$, then $\exists w \in L(\kappa)$ that witnesses the divergence of $\{c\}(x)$.

**Theorem 4.** $(a) \longleftrightarrow (b)$.

*(a) $L(\kappa)$ does not admit divergence witnesses.*
*(b) $L(\kappa)$ is $\Sigma_1$ admissible and for all $A \subseteq L(\kappa)$, $A$ is $\Sigma_1^{L(\kappa)}$ iff $A$ is E-re on $L(\kappa)$.*

If $(a)$ is false then $L(\kappa)$, as will be seen, is a suitable platform for $E$-recursion theory. The presence of divergence witnesses is just what is needed to develop unfamiliar proofs of familiar statements about degrees of $E$-re sets. If (a) is true, then the fundamental concepts of $\Sigma_1$ recursion (i.e. alpha recursion) and $E$-recursion coincide.

The proof of **Theorem 4** is in the same spirit as that of **Lemma 6**.

### 5.9. *Relativization and reducibility*

**Relativizing $E$-recursion to $B$** means adding the scheme

$$\{e\}^B(\overrightarrow{x}) = B \cap x_i \qquad (e =< 8, n, i >)$$

to the seven Normann schemes.

Assume for the remainder of this Subsection that $L(\kappa)$ is $E$-closed.

**Definition 12.** Assume $A, B \subseteq L(\kappa)$. **$A$ is $E$-reducible to $B$** ($A \leq_\kappa B$) iff

$$\exists e \exists p \in L(\kappa)\ \forall x \in L(\kappa) : \{e\}^B(x,p) \downarrow,$$
$$T_{<e,x,p;B>} \in L(\kappa), \text{ and } [x \in A \longleftrightarrow \{e\}^B(,p) = 1].$$

**Definition 13.** **$A$ is $E$-recursively enumerable on $L(\kappa)$** iff $\exists e \exists p \in L(\kappa)$ $A = \{x \mid x \in L(\kappa) \wedge \{e\}(x,p) \downarrow\}$.

$E$-reducibility is inspired by Turing reducibility. It is transitive for sets $E$-re on an $E$-closed $L(\kappa)$, but fails to be transitive in general.

### 5.10. *Transitivity of E-reducibility*

Assume $L(\kappa)$ is $E$-closed throughout this subsection.

**Definition 14.** $D \subseteq L(\kappa)$ is **subgeneric** iff $\forall e\ \forall x \in L(\kappa)$

$$\{e\}^D(x) \downarrow \longrightarrow \{e\}^D(x) \in L(\kappa).$$

If $C$ is subgeneric and $B \leq_\kappa C$, then $B$ is subgeneric.

**Definition 15.** $D \subseteq L(\kappa)$ is **regular** iff $\forall z \in L(\kappa)\ (D \cap z) \in L(\kappa)$.

Note that subgenericity implies regularity.

**Lemma 7.** *Suppose $A, B, C \subseteq L(\kappa)$ are E-re on $L(\kappa)$. If $A \leq_\kappa B$ and $B \leq_\kappa C$, then $A \leq_\kappa C$.*

*Proof.* If $C$ is subgeneric, then $A \leq_\kappa C$ by composition. Assume $C$ is not subgeneric. Suppose

$$(1)\ L(\kappa) = E(x) \text{ for some set } x \text{ of ordinals.}$$

If $E(x)$ is $\Sigma_1$ admissible, then by Theorem 6 $\leq_\kappa$ defined for $E$-recursion agrees with its $\Sigma_1$ admissible recursion counterpart and hence is transitive. If $E(x)$ is not $\Sigma_1$ admissible, or if (1) is false, then $E(x)$ admits divergence witnesses.

Thus it is safe to assume $L(\kappa)$ admits divergence witnesses. Assume $C$ is regular. Then every computation relative to $C$ of height less than $\kappa$ belongs to $L(\kappa)$. Since $C$ is not subgeneric there is a computation relative to $C$ of height $\kappa$. Then $\exists e \exists y \in L(\kappa) \cap 2^\kappa$:

$$(\forall x \in y)\ T_{<e,x;C>} \in L(\kappa), \text{ and } \kappa = \sup\{|\ \{e\}(x)\ |\ |\ x \in y\};$$

suppose $\exists p \in L(\kappa)\ \forall z \in L(\kappa)\ [z \in A \longleftrightarrow \{d\}(z,p)\downarrow]$; then $z \notin A$ iff $\exists x \exists w$

$$(2)\ x \in y \wedge w \in L(|\ \{e\}^C(x)\ |) \wedge w \text{ witnesses } \{d\}(z,p)\uparrow].$$

It follows from (2) and the regularity of $C$ that $A \leq_\kappa C$. The least $x$ satisfying (2) is computable from $C$ by effective transfinite recursion.

Assume $C$ is not regular. Suppose $\exists q \in L(\kappa)\ \forall x \in L(\kappa)\ x \in C \longleftrightarrow \{f\}(x,q)\downarrow$. Choose $y \in L(\kappa)$ so that $(y \cap C) \notin L(\kappa)$. Then

$$\kappa = \sup\{|\ \{f\}(x,q)\ |\ |\ x \in (y \cap C)\}.$$

Now proceed as before (when $C$ was regular) to show $A \leq_\kappa C$. □

## 5.11. *Regularity of E-re degrees*

Assume $L(\kappa)$ is $E$-closed.

**Definition 16.** Suppose $A, B \subseteq L(\kappa)$ are $E$-re on $L(\kappa)$. $A$ and $B$ are of the same **degree** ($A \equiv_{L(\kappa)} B$) iff each is $E$-reducible to the other.

By Lemma 7, $\equiv_{L(\kappa)}$ is an equivalence relation.

**Definition 17.** Suppose $B$ is $E$-re on $L(\kappa)$. $B$ is **complete** iff every set $E$-re on $L(\kappa)$ is $E$-reducible to $B$.

**Theorem 5.** *Suppose $C \subseteq L(\kappa)$ is E-re on $L(\kappa)$. Then there exists a regular $B \subseteq L(\kappa)$, E-re on $L(\kappa)$, such that $B \equiv_{L(\kappa)} C$.*

*Proof.* (sketch) If $L(\kappa)$ is not of the form $E(x)$ for some $x \in L(\kappa)$, then $C$ is regular. Assume $L(\kappa) = E(x)$. If $E(x)$ is $\Sigma_1$ admissible, then the "regular sets theorem" of $\alpha$-recursion theory yields the desired $B$. Assume that $E(x)$ is $\Sigma_1$ admissible and $C$ is not regular. As in the proof of Lemma 7, $C$ is complete. Thus there is a computation relative to $C$ of height $\kappa$; that computation is used to define $B$. □

## 5.12. *Projecta*

Assume $L(\kappa)$ is $E$-closed.

**Definition 18.** Suppose domain$(f) \subseteq \kappa$. **$f$ is partial $E$-recursive on $L(\kappa)$** iff $\exists e\ p \in L(\kappa)\ \forall b < \kappa\ f(b) = c \longleftrightarrow \{e\}(b,p) = c$.

**Definition 19.** Suppose $\gamma \leq \kappa$, domain $f \subseteq \gamma$, and range $f = L(\kappa)$. $f$ is a **partial $E$-recursive-on-$L(\kappa)$ map of $\gamma$** onto **$L(\kappa)$** iff $f$ is partial $E$-recursive on $L(\kappa)$.

**Definition 20.** $\rho =$ least $\gamma \leq \kappa\ \exists f\ f$ is a partial $E$-rec-on-$L(\kappa)$ map of $\gamma$ onto $L(\kappa)$.

**Definition 21.** $\eta =$ least $\gamma \leq \kappa\ \exists A\ \ A \in (2^\gamma - \kappa)$ and $A$ is $E$-re on $L(\kappa)$.

**Lemma 8.** $\eta = \rho$.

*Proof.* Let $f$ be a be a partial $E$-recursive-on-$L(\kappa)$ map of $\rho$ onto $L(\kappa)$. If $\rho < \eta$, then $\kappa \in L(\kappa)$.

To show $\eta \geq \kappa$, fix $\gamma < \rho$ and let $A \subseteq \gamma$ be $E$-re in $\tau \in \kappa$ via $e$. Let $g$ be a universal partial $E$-recursive function.

$$g[x] = \{\{c\}(z_1,\ldots,z_n) \mid \{c\}(z_1,\ldots,z_n)\downarrow; c,n \in \omega; z_i \in x\}.$$

Let $H = g[\gamma \cup \{\tau\}]$. By induction on complexity, $H \preceq_0 L(\kappa)$ : Every $\Delta_0^{ZF}$ sentence with parameters in $H$ is true in $L(\kappa)$ if true in $H$. $H$ is isomorphic to a transitive set $H_0$ via a collapsing map $t$. $(\forall z \in H)[O(z) \in H]$. ($O(z) =$ least $\delta[z \in L(\delta+1) - L(\delta)]$.) Hence $H_0 = L(\beta)$ for some $\beta < \kappa$, and $L(\beta) = g[\gamma \cup \{t(\tau)\}]$.

A slight alteration of $g$ maps a bounded initial segment of $\rho$ onto $L(\beta)$. $\rho$ is a cardinal in the sense of $L(\kappa)$, so $\beta < \rho$. $A = \{x \mid x < \gamma \wedge \{e\}(x, t(\tau\}\}$, so $A \in L(\beta+1) \subseteq L(\kappa)$. □

Suppose $B \subseteq L(\kappa)$.

Let $\rho^B$ be the least $\gamma \leq \kappa$ such that for some $p \in L(\kappa; B)$, there exists a partial map $\{e\}(x,p)$ of $\gamma$ onto $L(\kappa; B)$ via computations in $L(\kappa; B)$; "partial" means $\{e\}(x,p)$ might not converge for all $x \in \gamma$.

Let $\eta^B$ be the least $\gamma \leq \kappa$ such that for some $R \in 2^\gamma - L(\kappa; B)$, $R$ is $E$-re on $L(\kappa; B)$ relative to $B$. This means $\exists e \exists q \in L(\kappa; B)$ such that $R \cap \gamma$ is $\{x \mid \{e\}^B(x, q) \downarrow$ via computations in $L(\kappa; B)\}$.

**Lemma 9.** *(Slaman) Assume $L(\kappa)$ is E-closed. If $B \subseteq \kappa$ is regular and E-re-on-$L(\kappa)$, then $\eta^B = \rho^B$.*

**Proposition 13.** *Assume $L(\kappa)$ is E-closed and admits divergence witnesses, If $A \subseteq \kappa$ is E-re-on-$L(\kappa)$ and not complete, then A is subgeneric.*

The next theorem is needed for priority constructions in which requirements are indexed by ordinals less than $\rho$.

**Theorem 6.** *Assume $L(\kappa)$ is E-closed and admits Moschovakis witnesses. If $p \in \kappa$ and $\gamma < \rho$, then* $\sup\{\kappa_r^{p,\delta} \mid \delta < \gamma\} < \kappa$.

### 5.13. *Post's problem*

Assume $L(\kappa)$ is $E$-closed and $A, B, C \subseteq L(\kappa)$.

$A$ is $E$-reducible to $B$ iff $\exists e \exists p \in L(\kappa)\ \forall x \in L(\kappa)$:

$$\{e\}^B(x, p\} \downarrow,\ T_{<e,x,p;B>} \in L(\kappa),\ \text{and } x \in A \longleftrightarrow \{e\}^B(x, p\} = 1.$$

$C$ is $E$-re on $L(\kappa)$ iff $\exists e \exists p \in L(\kappa)\ \forall x \in L(\kappa)$: $x \in C \longleftrightarrow \{e\}(x, p\} \downarrow$.

**Theorem 7.** *There exist two subsets of $L(\kappa)$, both E-re on $L(\kappa)$, such that neither is E-reducible to the other.*

*Proof.* (A sketch) If $L(\kappa)$ lacks divergence witnesses, then $L(\kappa)$ is $\Sigma_1$ admissible and the solution to Post's problem supplied by $\alpha$-recursion theory is also solution in the sense of $E$-recursion.

Suppose $L(\kappa)$ admits divergence witnesses. An inequality requirement is handled by waiting for a computation $c$ to converge. If it does, then an inequality is created and preserved forever. The creation may add some $x$ to $B$, one of the two sets, but $x$ cannot wait forever. It must be added prior to stage $\kappa_0^{x,p}$ of the construction. ($p$ is a parameter needed to enumerate $B$.) If $c$ does not converge, then witness to the divergence of $c$ eventually appears and is preserved forever.

Requirements are indexed by ordinals less than $\rho$. If $v < \rho$ and $p \in \kappa$, then $\sup\{\kappa_r^{p,y} \mid y < v\}$ is less than $\kappa$. It follows that a set of requirements indexed by ordinals less than $v$ can be satisfied by some stage less than $\kappa$.

Define *recf*$(\lambda)$, the **E-re-on-$L(\kappa)$ cofinality of $\lambda$**, to be (least $\gamma$) $\exists B \subseteq \kappa$ [$\sup B = \lambda =$ order type of $B$ and $B$ is $E$-re on $L(\kappa)$]. Requirements are met by a Shore blocking argument based on the fact that $recf(\rho) = recf(\kappa)$. No requirements are injured. $\square$

### 5.14. *Splitting and density*

A typical requirement in the following result is injured $\kappa$-finitely often. As yet there is no priority argument in $E$-recursion theory in which a requirement is injured unboundedly often.

**Theorem 8.** *(Slaman) Suppose $C \subseteq L(\kappa)$ is regular and $E$-re, but not $E$-recursive, on $L(\kappa)$. Then there exist $A$ and $B$, each $E$-re on $L(\kappa)$, such that $A \cup B = C$ and $A \cap B = \emptyset$.* ***(Splitting)***
*Suppose $A, B \subseteq L(\kappa)$ are regular and $E$-re on $L(\kappa)$, and $A <_{L(\kappa)} B$. Then there exists $C \subseteq L(\kappa)$ such that $C$ is $E$-re on $L(\kappa)$ and $A <_{L(\kappa)} C <_{L(\kappa)} B$.* ***(Density)***

## 6. Forcing Computations to Converge

### 6.1. *Genericity versus E-closure*

Assume $L(\kappa)$ is $E$-closed, $\mathcal{P} \in L(\kappa)$ is a set forcing relation, and $G$ is $\mathcal{P}$-generic. If $L(\kappa)$ is $\Sigma_1$ admissible, then $L(\kappa, G)$ is $\Sigma_1$ admissible. If $L(\kappa)$ is not $\Sigma_1$ admissible, then $L(\kappa, G)$ need not be $E$-closed.

**Example 1.** $E(\omega_1)$ is $E$-closed but not $\Sigma_1$ admissible. Let $E(\omega_1) = L(\kappa)$. Suppose $G$ is a Levy forcing collapse of $\omega_1$ to $\omega$. Then $L(\kappa, G) = E(a)$ for some $a \subseteq \omega$, and so $L(\kappa, G)$, hence $L(\kappa)$, is $\Sigma_1$ admissible.

If $\mathcal{P}$ is c.c.c., or is countably closed, and $G$ is $\mathcal{P}$-generic, then $L(\kappa, G)$ is $E$-closed.

### 6.2. *The forcing language*

Assume $L(\kappa)$ is $E$-closed but not $\Sigma_1$ admissible. Then $L(\kappa) \models$ [$\exists$ greatest cardinal]. Let $gc(\kappa)$ denote the greatest cardinal in the sense of $L(\kappa)$. From now on $G$ denotes a subset of $gc(\kappa)$.

Every member of $L(\kappa, G)$ is of the form $\{e\}(a, G)$: $a \in G$ and $| \{e\}(a, G) |$. The terms of the forcing language $\mathcal{L}(\kappa, \mathcal{G})$ name the elements of $L(\kappa, G)$. The primitive terms are $\mathcal{G}$ and $\underline{a}$ for each $a \in L(\kappa)$. A general term is of the form $\{e\}(t_1, \ldots, t_n)$, where $t_i$ $(1 \leq i \leq n)$ is a term. Every term is equivalent to one of the form $\{e\}(\underline{a}, \mathcal{G})$.

Typical formulas of $\mathcal{L}(\kappa, \mathcal{G})$ are:

(1) $| \{e\}(\underline{a}, \mathcal{G}) | = \sigma$, for $\sigma < \kappa$;

(2) $\exists\delta[|\{e\}(\underline{a},\mathcal{G})|=\delta]$; and (3) $x\in\{e\}(\underline{a},\mathcal{G})$.

Formula (1) is ranked. (2) is unranked; "$\exists\delta$" means $\exists\delta<\kappa$. The meaning of (3) is conditional. If $\{e\}(a,G)\downarrow$ via a computation of height less than $\kappa$, then $\{e\}(\underline{a},\mathcal{G})$ names $\{e\}(a,G)$.

### 6.3. *Forcing relations*

Let $\mathcal{P}\in L(\kappa)$ be a notion of set forcing designed to force sentences of $\mathcal{L}(\kappa,\mathcal{G})$ and to create generic $G$'s that are subsets of $gc(\kappa)$. $\mathcal{P}$ is $<P,\geq>$. The elements of $P$, denoted by $p,q,r,\ldots$, are called forcing conditions; each one says something about $G$. $\geq$ is a binary relation on $P$. If $p\geq q$ ($p$ is extended by $q$), then $q$ says as much as, or more than, $p$ does about $G$.

$G\in p$ means: what $p$ says about $G$ is true. $\mathcal{P}$ determines a forcing relation $\Vdash$ defined by recursion on $\sigma<\kappa$. Three formal entities are defined simultaneously:

(i) $p\Vdash|\{e\}(\underline{a},\mathcal{G})|=\sigma$, (ii) $\mathcal{T}(p,e,a,\mathcal{G})$, and (iii) $q\Vdash s\in\{e\}(\underline{a},\mathcal{G})$.

If $G$ is generic, (i) holds and $G\in p$, then (ii) is a set of terms that name the elements of $\{e\}(\underline{a},\mathcal{G})$. In (iii) $q$ extends the $p$ of (i) and $s\in$ (ii). Assume that $\mathcal{P}$ specifies all ground zero forcing facts of the form $p\Vdash\delta\in\mathcal{G}$ and $q\Vdash\delta\notin\mathcal{G}$ and that these facts satisfy standard consistency and completeness assumptions about forcing relations.

$\boldsymbol{\sigma=0}$. Take $\{e\}(\underline{a},\mathcal{G})$ to be $\mathcal{G}$. Then $\forall p\forall q_{p\geq q}$

$$p\Vdash|\{e\}(\underline{a},\mathcal{G})|=\underline{0},\ \mathcal{T}(p,e,a,\mathcal{G})=\{\delta\mid\delta<gc(\kappa)\},\ \text{and}$$
$$q\Vdash\delta\in\{e\}(\underline{a},\mathcal{G})\ \text{iff "}q\Vdash\delta\in\mathcal{G}\text{" is a ground zero forcing fact.}$$

$\boldsymbol{\sigma>0}$. Let $e$ be $2^m\cdot3^n$ (Scheme $T$ of 5.5.1). Then $\forall p\forall q_{p\geq q}$
$p\Vdash|\{e\}(\underline{a},\mathcal{G})|=\sigma$ iff

$$\exists\gamma<\sigma\ p\Vdash|\{m\}(\underline{a},\mathcal{G})|=\gamma,$$
$$p\Vdash\forall x\exists\tau<\sigma[x\in\{m\}(\underline{a},\mathcal{G})\longrightarrow|\{n\}(x)|=\tau],$$
$$p\Vdash\forall\tau<\sigma[|\{m\}(\underline{a},\mathcal{G})|=\tau\vee$$
$$\exists x[x\in\{m\}(\underline{a},\mathcal{G})\wedge|\{n\}(x)|\geq\tau].$$

Keep in mind:

$$p\Vdash\exists x\mathcal{F}(x)\ \text{iff}\ p\Vdash\mathcal{F}(t)\ \text{(for a suitable term } t).$$

$$p\Vdash(\mathcal{F}\vee\mathcal{H})\ \text{iff}\ p\Vdash\mathcal{F}\ \text{or}\ p\Vdash\mathcal{H}.$$

$$p\Vdash\neg\mathcal{F}\ \text{iff}\ \forall q[p\geq q\longrightarrow q\nVdash\mathcal{F}].$$

$$p\Vdash\forall x\mathcal{F}(x)\ \text{iff}\ \forall t\forall q_{p\geq q}\exists r_{q\geq r}[r\Vdash\mathcal{F}(t))].$$

### 6.4. *Definability of forcing*

From now on replace $\underline{a}, \mathcal{G}$ by $t$, an arbitrary term of $\mathcal{L}(\kappa, \mathcal{G})$.

**Proposition 14.** *The relations* $p \Vdash\mid \{e\}(t) \models \underline{\sigma}$, $s \in \mathcal{T}(p, e, t, \sigma)$, *and* $q \models s \in \{e\}(t)$ *are E-recursive in* $\mathcal{P}, p, e, t, \sigma$.

*Proof.* By transfinite $E$-recursion. □

## 7. The Tree of Possibilities

The binary relation, $>_V$, is the forcing counterpart of $>_U$. A node on $>_V$ is of the form $< p, e, t >$. Define $< p, e, t >>_V < q, n, s >$ by

$$p \geq q \text{ and } q \Vdash^* [< e, t\dot{>} >_U < n, s >].$$

(Recall that $q \Vdash^* \mathcal{F}$ means $\forall r[q \geq r \longrightarrow r \not\models \mathcal{F}]$.) Suppose $e$ is an instance of scheme $T$ as in Subsection 5.5.1. Let $e$ be $2^m \cdot 3^n$. Define $p \Vdash [< e, t >>_i < n, s >]$ by

$$p \Vdash\mid \{m\}(t) \models \underline{\gamma},\ s \in \mathcal{T}(p, m, t, \gamma), \text{ and } p \Vdash s \in \{m\}(t).$$

Define $p \Vdash a >_U b$ by $\exists a_0, \ldots, a_k$ such that $a_0 = a$, $a_k = b$, and $\forall j < k$ $p \Vdash a_j > a_{j+1}$.

### 7.1. *Effective bounding*

Define $p \Vdash^* \mathcal{F}$ by $p \Vdash \neg\neg\mathcal{F}$.

Suppose $\mathcal{P} \in L(\kappa)$. $\mathcal{P}$ satisfies **effective bounding** iff

$$\forall p, e, t \text{ if } p \Vdash^* \exists\sigma \mid \{e\}(t) \models \sigma,$$

then $p \Vdash^*\mid \{e\}(t) \models \gamma$ for some $\gamma \leq_E p, t, P$.

**Lemma 10.** *If* $>_V$ *is wellfounded below* $< p, e, t >$ *whenever* $p \Vdash^*$ $\exists\sigma \mid \{e\}(t) \models \sigma$, *then* $\mathcal{P}$ *satisfies effective bounding.*

*Proof.* $\gamma$ is computed by transfinite recursion on $>_V$ below $< p, e, t >$. $\gamma$ is (approximately) the height of the wellfounded subtree below $< p, e, t >$.

Suppose $e$ is $2^m \cdot 3^n$. By recursion $p \Vdash^*\mid \{m\}(t) \mid\leq \delta$ for some $\delta \leq_E$ $p, t, \mathcal{P}$. Define $< p', \sigma > \in K$ by $p \geq p' \wedge \sigma \leq \delta \wedge p' \Vdash\mid \{m\}(t) = \sigma$. Note $K \leq_E p, t, \mathcal{P}$.

Fix $< p', \sigma > \in K$ and $s \in \mathcal{T}\ (p', m, t, \sigma)$. Then

$$p' \Vdash \exists\beta[s \in \{m\}(t) \longrightarrow\mid \{n\}(s) \models \beta].$$

Define $q \in J(p', \sigma, s)$ by $p' \geq q \wedge q \Vdash s \in \{m\}(t)$. By recursion $\forall q \in J(p', \sigma, s)$ $\exists \rho \leq_E q, s, \mathcal{P}$ such that $q \models^* | \{n\}(s) | \leq \rho$. $\rho$ is a partial $E$-recursive function of $q, s, \mathcal{P}$. On $J(p', \sigma, s)$, $\rho$ is bounded by $\rho_{p',m,t,\sigma} \leq_E p', \sigma, t, \mathcal{P}$. On $K$, $\rho_{p',m,t,\sigma}$ is bounded by $\rho_K \leq_E p, t, \mathcal{P}$.

The desired bound $\gamma$ is the strict sup of $\delta$ and $\rho_K$. □

## 7.2. *Genericity*

Assume $L(\kappa)$ is $E$-closed and $\mathcal{P} \in L(\kappa)$.

**Definition 22.** $G$ is **$\mathcal{P}$-generic** iff for every sentence $\mathcal{F}$ of $\mathcal{L}(\kappa, \mathcal{G})$, $\exists p \in G$ such that $p \Vdash \mathcal{F}$ or $p \Vdash \neg \mathcal{F}$.

If $G$ is $\mathcal{P}$-generic, then $L(\kappa, G) \models \mathcal{F}$ iff $(\exists p \in G)\ p \Vdash \mathcal{F}$ (by induction on the rank of $\mathcal{F}$).

**Lemma 11.** *If $\mathcal{P}$ satisfies effective bounding and $G$ is $\mathcal{P}$-generic, then $L(\kappa, G)$ is $E$-closed.*

*Proof.* Suppose not. Then $\exists e \exists a \in L(\kappa)$ such that $| \{e\}(\underline{a}, \mathcal{G}) | = \kappa$. Assume $e$ is $2^m \cdot 3^n$. $\exists p \in G$ such that $p \Vdash | \{e\}(\underline{a}, \mathcal{G}) | = \kappa$. By effective bounding, $p \Vdash^* | \{m\}(\underline{a}, \mathcal{G}) | \leq \gamma$ for some $\gamma \leq_E p, a, \mathcal{P}$. Also

$$p \models^* [\forall x \in \{m\}(\underline{a}, \mathcal{G}) \exists \beta \mid \{n\}(x) | = \beta];$$

$x$ ranges over $\vee \{\mathcal{T}(p, m, a, G, \delta) \mid \delta \leq \gamma\}$, a set $E$-recursive in $p, a, \mathcal{P}$. Hence $p$ weakly forces a bound on $| \{n\}(x) |$ $E$-recursive in $p, a, \mathcal{P}$. □

## 7.3. *Enumerable forcing relations*

**Theorem 9.** *(Slaman) Assume $L(\kappa)$ is $E$-closed but not $\Sigma_1$ admissible. Let $\mathcal{P} \in L(\kappa)$ be a set forcing relation, Then (i) if and only if (ii).*

*(i) $\mathcal{P}$ satisfies effective bounding.*
*(ii) The relation, $p \Vdash^* \exists \sigma \mid \{e\}(t) | = \sigma$, is $E$-recursively enumerable on $L(\kappa)$.*

*Proof.* Suppose (ii) holds. Suppress $\mathcal{P}$. Then an obscure fixed point argument shows

$$\forall p \in \mathcal{P}\ p \text{ does not weakly force } | \{e\}(t) | \geq \kappa_r^{t,p}.$$

By reflection, if $q \Vdash^* | \{e\}(t) | < \kappa_r^{t,q}$, then $q \Vdash^* | \{e\}(t) | \leq \gamma$ for some $\gamma \leq_E t, q$. Now (i) follows from Kechris's Basis Theorem. □

### 7.3.1. *An Annoying Assumption (AAF) needed for forcing*

The study of forcing makes repeated use of a key reflection fact: if some wellordering of $x$ is $E$-recursive in $x$, then $\kappa_r^x \geq \kappa_r^{x,u}$ for all $u$. The latter is a consequence of the definability of divergence witnesses proved in Subsections 5.5.2 and 5.6.

**Assumption AAF.** $\mathcal{P}$, all $p \in P$, and all $\mathcal{P}$-generic $G$ are effectively wellorderable.

The annoyance is mild because $G$ is usually a set of ordinals. From now on AAF will always be in force.

## 8. Countably Closed Forcing

**Definition 23.** $\mathcal{P}$ is **countably closed** iff for every sequence

$$\{p_n \mid n \in \omega\},\ \forall n[p_n \geq p_{n+1}] \longrightarrow \exists q \forall n[p_n \geq q].$$

**Example 2.** Suppose $L(\kappa)$ thinks: there is a greatest cardinal denoted by $gc(\kappa)$ and its cofinality is greater than $\omega$. In short $L(\kappa) \models gc(\kappa) > \omega$. Define $P$ by $p \in P \longleftrightarrow \exists \delta < gc(\kappa)\ p : \delta \longrightarrow 2$. Define $p \geq q$ by domain $p \subseteq$ domain $q$ and $\forall x \in \text{domain}\, p\ \ p(x) = q(x)$.

The above example satisfies Assumption AAF.

**Theorem 10.** *Suppose $L(\kappa)$ is $E$-closed but not $\Sigma_1$ admissible, and $\mathcal{P} \in L(\kappa)$ is a countably closed forcing relation. Then every $\mathcal{P}$-generic extension of $L(\kappa)$ is $E$-closed.*

*Proof.* Assume (**1**) $p \Vdash^* \exists \sigma \mid \{e\}(t) \models \sigma$. It suffices to show (**2**) $>_V$ is wellfounded below $< p, e, t >$, by Lemmas 10 and 11. (**3**) Suppose not. An infinite descending path below $< p, e, t >$ will be converted to a $q \leq p$ that weakly forces a witness to the divergence of $\{e\}(t)$. There is a $g$ such that if (1) and (2) hold, then $\{g\}(p, e, t) \downarrow$ and $p \Vdash^* \mid \{e\}(t) \mid \leq \{g\}(p, e, t)$. By (3) $\{g\}(p, e, t) \uparrow$; let $z \in L(\kappa)$ witness $\{g\}(p, e, t) \uparrow$: $z_0 =< g, < p, e, t >>$ and $\forall n\ z_n > z_{n+1}$. Now an infinite descending sequence $w$ below $< p, e, t >$ in $>_V$ is extracted from $z$.

$$w_0 =< p, e, t >$$

$$w_r >_U w_{r+1}$$

$$w_r =< p_r, e_r, t_r >$$

$$\{g\}(p_r, e_r, t_r) \uparrow \text{ and } \exists n\ z_n =< g, < p_r, e_r, t_r >>$$

$p_r \geq p_{r+1}$. $\exists q\ \forall r\ p_r \geq q$; $q$ weakly forces $w$ to witness $\{e\}(t) \uparrow$. □

(The assumption "$L(\kappa)$ is not $\Sigma_1$ admissible" is not needed in the above proof.)

### 8.0.1. *Enumerability*

Kleene proved the set of hyperarithmetic reals is $\Pi^1_1$. His result, restated, says $2^\omega \cap E(\omega)$ is $E$-re in $\omega$. The next two theorems clarify the role of $\Sigma_1$ admissibility in Kleene's result.

**Theorem 11.** *If $z \subseteq Ord$ and $E(z)$ is $\Sigma_1$ admissible, then $E(z)$ is $E$-re in some element of $L(\kappa)$.*

*Proof.* $\exists \kappa\ E(z) = L(\kappa, z)$. The $\Sigma_1$ admissibility of $E(z)$ mplies $\kappa \leq \kappa_r^{z,y}$ for some $y \in E(z)$. $\exists v \in L(\kappa)$ such that $v \subseteq L(\kappa)$ and $z, y \leq_E v$. Then $\kappa \leq \kappa_r^{v,x}$ for **all** $x$. Let $O_z(x)$ be the least $\delta < \kappa$ such that $x \in L(\delta+1, z) - L(\delta, z)$. By reflection, $O_z(x) \leq_E v, x$. So $\exists e\ O_z(x) = \{e\}(x, v)$.

Enumerate $x$ if $\{e\}(x,v)\downarrow$ and $E(z) \in L(\{e\}(x,v)+1, z)$. □

**Theorem 12.** *Assume $L(\kappa)$ is $E$-closed but not $\Sigma_1$ admissible and $L(\kappa) \models [gc(\kappa)$ is regular$]$. Then $2^{gc(\kappa)} \cap L(\kappa)$ is not $E$-re in any $b \in L(\kappa)$.*

*Proof.* Suppose $2^{gc(\kappa)} \cap L(\kappa) = \{x \mid \{e\}(b,x)\uparrow\}$ for some $b \in L(\kappa)$. Define $P$ by $p \in P$ iff $\exists \delta < gc(\kappa)\ p : \delta > 2$. First assume $gc(\kappa) > \omega$. Then $\mathcal{P}$ is countably closed, and Theorem 10 implies that $L(\kappa, G)$ is $E$-closed. Then $L(\kappa, G)$ admits divergence witnesses since it is not $\Sigma_1$ admissible.

Fix $e$ and $b \in L(\kappa)$. Then (1) or (2) holds.

**(1)** $\emptyset \Vdash^* \exists\sigma \mid \{e\}(\underline{b}, \mathcal{G}) \models \sigma$.
**(2)** $\exists q, w, \delta < \kappa\ q \Vdash w \in L(\delta, \mathcal{G}) \wedge w$ witnesses $\{e\}(\underline{b}, \mathcal{G})\uparrow$.
**(a)** (1) implies $\exists G \subseteq gc(\kappa)\ G \notin L(\kappa) \wedge \{e\}(b, \mathcal{G})\downarrow$.
**(b)** (2) implies $\exists G \subseteq gc(\kappa)\ G \in L(\kappa) \wedge \{e\}(b, \mathcal{G})\uparrow$.

In both cases if $\kappa$ is uncountable, then $G$ is generic over $L(\lambda)$ for some $\lambda < \kappa$. The proof of **(a)** uses an $S \in 2^{gs(\kappa)} - L(\kappa)$ such that $\forall\delta < \kappa\ (S \cap \delta) \in L(\kappa)$. $S$ is coded into $G$ to insure that $G \notin L(\kappa)$. □

## 8.1. *A general forcing fact*

**Proposition 15.** *Assume $L(\kappa)$ is $E$-closed but not $\Sigma_1$ admissible and $\mathcal{P} \in L(\kappa)$. If $p \Vdash^* \mid \{e\}(\underline{a}, \mathcal{G}) \mid \leq \kappa_r^{p,a,\mathcal{P}}$, then $p \Vdash^* \mid \{e\}(\underline{a}, \mathcal{G}) \mid \leq \gamma$ for some $\gamma \leq_E p, a, \mathcal{P}$.*

*Proof.* Assume $e = 2^m \cdot 3^n$ (Scheme $T$). Then $\forall q(p \geq q)\ \exists r(q \geq r)$ $\exists \delta\ r \Vdash \mid \{m\}(\underline{a}, \mathcal{G}) \mid = \delta$. Also $\delta < \kappa_r^{p,a,\mathcal{P}} \leq \kappa_r^{p,a,q,\mathcal{P}}$. $\exists r(q \geq r)\ \exists \delta$ $r \Vdash \mid \{m\}(\underline{a}, \mathcal{G}) \mid = \delta$ and $\delta < \kappa_r^{p,a,\mathcal{P}} \leq \kappa_r^{p,a,q,\mathcal{P}}$. By reflection $r$ and $\delta$ are $E$-recursive in $p, a, q, \mathcal{P}$. By Gandy selection $r$ and $\delta$ are partial $E$-recursive functions of $p, a, q, \mathcal{P}$. For all $q \leq p$, $\delta$ is bounded above by some $\delta_0 \leq_E p, a, \mathcal{P}$. Thus $p \Vdash^* \mid \{m\}(\underline{a}, \mathcal{G}) \mid \leq \delta_0$. The above mode of argument yields the desired bound $\gamma$ on $\mid \{e\}(\underline{a}, \mathcal{G}) \mid$. □

## 9. Countable Chain Condition Forcing

### 9.1. *Less-than-$\omega_1$ selection*

Suppose $L(\kappa) = E(\omega_1)$. By Gandy selection, there is an $f$ partial $E$-recursive in $\omega_1$ such that $\forall e\ \forall \beta < \omega_1\ \forall a \in L(\kappa)$: if $(\exists \gamma < \beta)\ \{e\}\{a, \gamma) \downarrow$, then $f(e, \beta, a) \downarrow \wedge f(e, \beta, a) \in \omega_1 \wedge \{e\}(a, f(e, \beta, a)) \downarrow$.

### 9.2. *Effectiveness of c.c.c. forcing*

$\mathcal{P} \in L(\kappa)$ is said to be **c.c.c.** iff for every $Q \in L(\kappa) \cap 2^P$, if the conditions in $Q$ are pairwise incompatible, then $Q$ is countable in $L(\kappa)$. (Two conditions are incompatible iff no condition extends them both.)

**Definition 24.** $\min(p, e, t) = \min\limits_{\delta} \exists r\ p \geq r\ \ r \Vdash \mid \{e\}(t) \mid = \delta$.

**Lemma 12.** *Assume (for simplicity) $L(\kappa) = E(\omega_1)$. Then*

$$\min(p, e, t) \leq_E p, t, \omega_1.$$

*Proof.* Assume: $e = 2^m \cdot 3^n$ and $\min(p, e, t)$ is defined by the so-called main recursion on the length of computations. Thus $\min(p, m, t) = \delta_0 \leq_E p, t, \omega_1$. Let $\mathcal{T}(p, m, t, \gamma_0)$ be $\{t_i \mid i < \theta\}$. Note that $\theta \leq \omega_1$ and $\theta \leq_E p, t, \omega_1$. Let $Q$ denote a set of forcing conditions. The following will be shown: $\forall i < \infty\ \exists Q_i \subseteq P\ \forall q \in Q_i\ p \geq q$ and

$$(\mathbf{1})\ q \Vdash t_i \in \{m\}(t_i) \text{ or } (\mathbf{2})\ q \Vdash t_i \notin \{m\}(t_i)$$

and if (1) holds, then $q \Vdash \exists \delta \mid \{n\}(t_i) \mid = \delta$.

A recursion of length $\omega_1$ will add conditions to the $Q_i$'s. Initially each $Q_i = \varnothing$. But finally $\cap_{i<\theta} Q_i \neq \varnothing$. This last means $\exists r\ \forall r_1(r \geq r_1)\ \forall i < \theta$ $\exists q \in Q_i$ such that $q$ is compatible with $r$. Suppose at the end of stage $\beta < \omega_1$, $\cap_{i<\theta} Q_i = \varnothing$. A subrecursion of length $\omega_1$ adds conditions to $Q_i$ $(i < \beta)$. The conditions are computed via less-than-$\omega_1$ recursion and the main recursion on min. The subrecursion succeeds because $\mathcal{P}$ is c.c.c. The main recursion succeeds for the same reason. Thus $\bigcap_{i<\theta} \neq \varnothing$. In other words $\exists r\ \forall i < \theta\ Q_i \geq r$ and $r \Vdash \{e\}(t) = \delta$. ($Q_i \geq r$ means: $\forall r_1 \leq r$ $\exists q \in Q_i$ such that $r_1$ and $q$ are compatible.) □

## 10. Selection

Suppose $A$ is $E$-re in $b$ and $A \cap x \neq \varnothing$. Is there a way to compute an element of $A \cap x$ from $b, x$? (The answer as usual is yes and no, but more no than yes.) Or a nonempty subset of $A \cap x$ from $b, x$? Or an ordinal $\theta$ from $b, x$ such that computations of length $\leq \theta$ enumerate elements of $A \cap x$?

**Selection($x$)** says: $\exists$ partial $E$-recursive function $f$ such that $\forall e \forall b$

$$\exists z \in x\ \{e\}(z,b) \downarrow \longleftrightarrow \exists z \in x \mid \{e\}(z,b) \mid \leq f(e,b) < \infty.$$

Sometimes an additional parameter $p$ is present in $f$.

### 10.1. *Grilliot selection*

Let $f, g : x \to \omega \times (2^x \times \{2^x\})$. Assume $x$ is transitive, closed under pairing, and $\omega \subseteq x$. Define $g < f$ by $\forall z \in x\ \ g(z)$ is an immediate subcomputation instruction of $f(z)$. Define $\min(f) = \min\{\mid f(z) \mid\mid z \in x\}$. Assume $\min(f) < \infty$. Grilliot's recursion equation is

$$\min(f) = \max\{\min(g) \mid g < f\}.$$

His intention was to compute $\min(f)$ from $f, 2^x$. An immediate obstacle is the failure of $\{g \mid g < f\}$ to be $E$-recursive in $f, 2^x$. Harrington and MacQueen worked around the obstacle with effective approximations of $\{g \mid g < f\}$. Define $g <^\beta f$ by $\forall z \in x\ g(z)$ is an immediate subcomputation instruction of $f(z)$ via a computation of length $\leq \beta$. Note that $\{g \mid g <^\beta f\} \leq_E \beta, f, 2^x$.

### 10.2. *Harrington-MacQueen selection*

An instance of the axiom of choice, $\mathbf{AC}_x$ plays a curious role in the proof of Theorem 13.

$\mathbf{AC}_x : \forall f$ If $\forall z \in x\ f(z) \neq \varnothing$, then $\exists g\ \forall z \in x\ g(z) \in f(z)$.
(Assume $x = \text{domain}(f)$.)

**Theorem 13.** *Assume* $f : x \to \omega \times (2^x \times \{2^x\})$ *and* $\exists z \in x \mid f(z) \mid < \infty$. *Then* $\min\{\mid f(z) \mid \mid z \in x\} \leq_E f, 2^x$.

*Proof.* Let $\theta = \sup\{\beta \mid \exists f\ f$ maps $x$ onto $\beta\}$. Note $\theta \leq_E 2^x$. $t(\gamma)$ is defined by recursion on $\gamma < \theta$. $\min(f)$ will be $\sup\{t(\gamma) \mid \gamma < \theta\}$. Let

$t^-(\gamma) = \sup(t(\delta) \mid \delta < \gamma\}$.

$$\text{If } t^-(\gamma) \geq \min(f), \text{then } t(\gamma) = \min(f).$$
$$\text{If } t^-(\gamma) < \min(f), \text{then}$$
$$t(\gamma) = \sup^+\{\min(g) \mid g <^{t^-(\gamma)} f \wedge t^-(\gamma) \leq \min(g)\}.$$

$AC_x$ implies $\exists g\ g <^{t^-(\gamma)} f \wedge t^-(\gamma) \leq \min(g)$. To find such a $g$, let $f(z)$ be $< 2^m \cdot 3^n, u >$.

If $t^-(\gamma) \leq |< m, u >|$, then $g(z) =< m, u >$.

Otherwise $\exists y \in \{m\}(u)$ such that $t^-(\gamma) \leq |< n, y >|$. If there is a $\gamma$ such that $t(\gamma) = \min(f)$, then $\sup\{t(\gamma) \mid \gamma < \theta\}$ is $\min(f)$, If there is no such $\gamma$, then there is a map of $x$ onto $\theta$. □

## 10.3. *Selection and admissibility*

For each $< e, b > \in E(x)$, let $K^x_{e,b}$ be $\{z \mid \{e\}(z, x, b) \uparrow\} \cap tc(x)$. ($tc$ is transitive closure.) $f$ is a **Grilliot Selection Function for $E(x)$** iff $\forall < e, b > \in E(x)$, $K^x_{e,b} \neq \varnothing \to f(e, b) \in E(x) \cap 2^{K^x_{e,b}}$.

**Lemma 13.** *If $f$, a Grilliot selection function for $E(x)$, is partial E-recursive in some $b \in E(x)$, then $E(x)$ is $\Sigma_1$ admissible.*

Suppose $g$ is a many-valued $\Sigma_1^{E(X)}$ map with domain $d$. Admissibility requires an $r \in E(x)$ such that $\forall v \in d$, $g(v)$ has a value in $r$ $\exists D \in \Delta_0^{ZF}$ such that $g(v) = w$ iff $E(x) \models \exists y D(v, w, y)$. Let $t$ be a partial $E$-recursive (in $x$) map of $\omega \times tc(x)$ onto $E(x)$. Then for each $v \in d$,

$$\{z \mid t(z) \downarrow\ \wedge\ D(v, (t(z))_0, (t(z))_1)\}$$

is $E$-re in $v, x, a$. The required $r$ is $\{(t(z))_0 \mid \exists v\ z \in f(v) \wedge v \in d\}$.

### 10.3.1. *Moschovakis selection*

**Theorem 14.** *(Moschovakis) Suppose $E(R(\alpha)) \models cofinality(\alpha) = \omega$. Let $k \in R(\alpha)$ be an $\omega$-sequence through $\alpha$. Assume $f : R(\alpha) \to \omega \times R(\alpha) \times \{R(\alpha\}$. If $\exists x \in R(\alpha) \mid f(x) \mid < \infty$, then*

$$\min\{| f(x) || x \in R(\alpha)\} \leq_E f, R(\alpha), k.$$

**Corollary 3.** *If $E(R(\alpha)) \models cofinality(\alpha) = \omega$, then $E(R(\alpha))$ is $\Sigma_1$ admissible.*

**Theorem 15.** *(Sacks & Slaman) Let $x$ be a set of ordinals. Suppose in $E(x)$: $\forall i < \omega$ ($k_i$ is a cardinal and $k_i < k_{i+1}$) and $\sup(x) = \sup\{k_i \mid i < \omega\}$. If $A \subseteq x$ is nonempty and $E$-re in $x$, then*

$$\min A \leq_E x, \{k_i \mid i < \omega\}.$$

**Corollary 4.** *If $x \subseteq Ord$ and $E(x) \models$ cofinality$(gc(\kappa) = \omega)$, then $E(x)$ is $\Sigma_1$ admissible.*

## 11. van de Wiel's Theorem

Let $f$ map $V$ into $V$.

**Definition 25.** $f$ is **uniformly $\Sigma_1$ definable** iff $\exists$ *lightface* $\mathcal{F}(x,y) \in \Sigma_1^{ZF}$ such that $\forall \Sigma_1$ admissible $A$, $f[A] \subseteq A$ and $f \restriction A = \{< a, b > \mid < A, \in > \models \mathcal{F}(\underline{a}, \underline{b})$. (Lightface means nonnegative integer parameters only.)

**Theorem 16.** *Let $f : V \to V$. Then (i)$\longleftrightarrow$(ii)*

*(i) $f$ is $E$-recursive.*
*(ii) $f$ is uniformly $\Sigma_1$ definable.*

*Proof.* (Slaman) Assume (ii). Fix $x$ and define $K_x = L(\kappa_r^x, tc(\{x\}))$. The plan is to show: $K_x \models \exists v \mathcal{F}(\underline{x}, v)$. Then $L(\kappa_0^x, tc(\{x\})) \models \exists v \mathcal{F}(\underline{x}, v)$ by reflection, and $f(x) \leq_E$ by Gandy selection.

Choose $p \in K_x$ so that $\kappa_r^p = \min\{\kappa_r^q \mid q \in K_x\}$. Let $K_{x,p}$ be $L(\kappa_r^{x,p}, tc(\{x,p\}))$. Assume $tc(x)$ is countable. Now a hull $Z \subseteq K_{x,p}$ is constructed whose transitive collapse, $\overline{Z}$, is $\Sigma_1$ admissible; Let $y_0$ denote the collapse of $y \in Z$. Then $\overline{Z} \models \exists v \mathcal{F}(\underline{x_0}, v)$, $Z \models \exists v \mathcal{F}(\underline{x}, v)$, $f(x) \in K_{x,p}$, $f(x) \leq_E x, p$ and $f(x) \in K_x$.

Let $Z$ be $\{z_i \mid i < \omega\}$. $z_0$ is $< x, p >$. Choose $z_i$ ($i > 0$) so that $\kappa_r^{z_0,\ldots,z_i} = \kappa_r^{z_0}$. Also make choices so that $Z$ is $E$-closed. If $\exists w \in K_{x,p}$ such that $\{e\}(z_i, w) \uparrow$, then $\exists w \in Z$ such that $\{e\}(z_i, w) \uparrow$. The choices exist thanks to Kechris's basis theorem.

Now show $\overline{Z}$ is $\Sigma_1$ admissible. Suppose $\overline{Z} \models (\forall u \in \underline{a_0}) \exists v \mathcal{G}(u, v, \underline{c_0})$. $\mathcal{G}$ is $\Delta_0^{ZF}$. First show $K_{x,p} \models (\forall u \in \underline{a}) \exists v \mathcal{G}(u, v, \underline{c})$. If not, then

$$U = \{u \mid u \in a \wedge K_{x,p} \models \forall v \neg \mathcal{G}(\underline{u}, v, \underline{c_0})\} \neq \varnothing.$$

If $\exists v \; K_{x,p} \models \mathcal{G}(u, v, c)$, then by reflection

$$\exists v \; v \leq_E x, p, c, u \; \wedge \; K_{x,p} \models \mathcal{G}(\underline{u}, \underline{v}, \underline{c}).$$

Thus $U$ is co-$E$-re in $x, p, c, a$, so $\exists u \in U \cap Z$, hence

$$Z \models \exists u \in \underline{a} \forall v \neg \mathcal{G}(u, v, \underline{c}) \text{ and } \overline{Z} \models \exists u \in \underline{a_0} \forall v \neg \mathcal{G}(u, v, \underline{c_0}).$$

If $u \in a$ and $\exists v\ K_{x,p} \models \mathcal{G}(\underline{u}, v, \underline{c})$, then by reflection

$$\exists v\ v \leq_E x, p, c, a, u \ \wedge \ K_{x,p} \models \mathcal{G}(\underline{u}, v, \underline{c});$$

the range of $v$ restricted to $a$ is some $b \in Z$. Thus $Z \models \forall u \in \underline{a}\ \exists v \in \underline{b}\mathcal{G}(u, v, \underline{c})$. Then $\overline{Z} \models \forall u \in \underline{a_0}\ \exists v \in \underline{b_0}\mathcal{G}(u, v, \underline{c_0})$.

The assumption of countability of $tc(x)$ has to be lifted. At this point $\exists e \forall x$ if $TV(x)$is countable, then $f(x) = \{e\}(x)$. Suppose $\exists y\ f(y) \neq \{e\}(y)$. The formula '$f(y) \neq \{e\}(y)$' is $\Sigma_1^{ZF}$ and solvable, hence has a solution $y$ such that $tc(y)$ is countable. □

## 12. Logic on *E*-Closed Sets

Assume $L(\kappa)$ is $E$-closed. Let $\mathcal{L} \in L(\kappa)$ be a set of atomic symbols. $\mathcal{L}_{\infty,\omega}$ denotes a minimal extension of first order logic. The class of formulas is built from $\mathcal{L}$ via the finitary formation rules of first order logic and two infinitary rules, arbitrary conjunctions and arbitrary disjunctions. In addition there is a restrictive proviso: a formula must not contain infinitely many free variables. On the other hand a formula may contain arbitrarily many distinct individual constants. Note that each sequence of quantifiers is finite; that is the intended meaning of the '$\omega$' in the subscript '$\infty, \omega$'. The axioms and rules of deduction come from first order logic with only one significant addition; if for each $i \in I$, there is a deduction of $\mathcal{F}_i$, then there is a deduction of the conjunction $\wedge\{\mathcal{F}_i \mid i \in I\}$.

$\mathcal{L}_{\kappa,\omega}$ is the restriction of $\mathcal{L}_{\infty,\omega}$ to $L(\kappa)$. Let $\mathcal{F} \in L(\kappa)$.

Assume $\Delta$ is an $E$-recursive-on-$L(\kappa)$ set of sentences of $\mathcal{L}_{\kappa,\omega}$.

### 12.1. *Deductions*

**Definition 26.** $\boldsymbol{\Delta} \vdash \mathcal{F}$ iff $\mathcal{F}$ is deducible from $\Delta$ via a deduction in $V$ in the sense of $\mathcal{L}_{\infty,\omega}$.

**Definition 27.** $\boldsymbol{\Delta} \vdash_\kappa \mathcal{F}$ iff $\Delta \vdash \mathcal{F}$ via a deduction in $L(\kappa)$.

**Definition 28.** $\boldsymbol{\Delta} \vdash_\kappa^E \mathcal{F}$ iff $\Delta \vdash \mathcal{F}$ via a deduction $E$-recursive in $\mathcal{F}$.

**Definition 29. $\boldsymbol{\Delta}$ admits effectivization of deductions** iff for all $\mathcal{F} \in L(\kappa)$, if $\Delta \vdash_\kappa \mathcal{F}$, then $\Delta \vdash_\kappa^E \mathcal{F}$.

### 12.2. *Completeness*

**Definition 30. $\boldsymbol{\Delta}$ is $\boldsymbol{\kappa}$ -consistent** iff $\Delta \nvdash_\kappa (\mathcal{F} \wedge \neg\mathcal{F})$.

Assume below that $\Delta$ admits effectivization of deductions.

**Lemma 14.** *If $\Delta$, $\vee\{\mathcal{F}_i \mid i \in I\}$ is $\kappa$-consistent, then $\Delta$, $\mathcal{F}_i$ is $\kappa$-consistent for some $i \in I$.*

**Theorem 17.** *If $L(\kappa)$ is countable and $\Delta$ is $\kappa$-consistent, then $\Delta$ has a model.*

**Corollary 5.** *If $\Delta$ is $\kappa$-consistent, then $\Delta$ is consistent in the sense of $\mathcal{L}_{\infty,\omega}$.*

**Question** Is there an interesting application of Theorem 17.

## References


1. Barwise J (1979) Admissible structures, Springer, Berlin, Heidelberg, New York
2. Fenstad JE (1980) General recursion theory, Springer, Berlin, Heidelberg, New York
3. Gandy RO (1967) General recursive functionals of finite type and hierarchies of functionals, Ann. Fac. Sci. Univ. Clermont-Ferrand 35; 215-223
4. Green J (1974) $\Sigma_1$ compactness for next admissible sets, J. Symb. Log. 39; 105-116
5. Griffor ER (1980) *E*-recursively enumerable degrees, Ph.D Thesis, Massachusetts Institute of Technology
6. Harrington L (1973) Contributions to recursion theory in higher types, Ph.D Thesis, Massachusetts Institute of Technology
7. Harrington L, MacQueen D (1976) Selection in abstract recursion theory, J. Symb. Log. 41: 153-158
8. Houle T (1982) Abstract extended 2-sections, Ph.D Thesis, Oxford University (Note: date is approximate)
9. Kleene, SC (1959) Recursive functionals and quantifiers of finite types I, Trans Amer. Math. Soc. 91; 1-52
10. Kleene, SC (1963) Recursive functionals and quantifiers of finite types II, Trans Amer. Math. Soc. 108; 106-142
11. Moldstad J (1972) Computations in Higher Types, Lecture Notes in Math. 574, Springer. Berlin, Heidelberg, New York
12. Moschovakis YN (1967) Hyperanalytic predicates, Trans. Amer. Math. Soc. 129; 249-282
13. Moschovakis YN (1980) Descriptive set theory, North-Holland, Amsterdam
14. Normann D (1975) Degrees of functionals, Preprint Series in Math. 22, Oslo
15. Normann D (1978a) Set recursion, In: Generalized recursion theory II, North-Holland, Amsterdam, 303-320
16. Normann D (1978b) Recursion in $^3E$ and a splitting theorem, In: Essays on mathematical and philosophical logic, D. Reidel, Dordrecht, 275-285

17. Sacks GE (1974) The 1-section of a type $n$-object, In: Generalized recursion theory, North-Holland, Amsterdam, 81-96
18. Sacks GE (1977) The $k$-section of a type $n$-object, Amer. Jour. Math. 99; 901-917
19. Sacks GE (1985) Post's problem in $E$-recursion, In: Proc. of symposia in pure mathematics, Amer. Math. Soc. 42; 177-193
20. Sacks GE (1986) On the limits of $E$-recursive enumerability, Ann. Pure and Applied Log. 31; 87-120
21. Sacks GE (1990) Higher recursion theory, Springer, Berlin, Heidelberg, New York
22. Slaman TA (1985a) Reflection and forcing in $E$-recursion, Ann. Pure and Applied Log. 29; 79-106
23. Slaman TA (1985b) The $E$-recursively enumerable degrees are dense, In: Proc. of symposia in pure mathematics, Amer. Math. Soc. 42; 195-213
24. van de Wiele J (1982) Recursive dilators and generalized recursion, In: Proc. Herbrand symposium, North-Holland, Amsterdam, 325-332

Printed in the United States
By Bookmasters